Bacterial Control of Mosquitoes & Black Flies

Bacterial Control of Mosquitoes & Black Flies

Biochemistry, Genetics & Applications of *Bacillus thuringiensis israelensis* and *Bacillus sphaericus*

Huguette de Barjac

Donald J. Sutherland

EDITORS

UNWIN HYMAN

London

Softcover reprint of the hardcover 1st edition 1990

ISBN-13: 978-94-011-5969-2 e-ISBN-13: 978-94-011-5967-8
DOI: 10.1007/ 978-94-011-5967-8

British Library Cataloguing in Publication Data

Information available

Library of Congress Cataloging-in-Publication Data

Bacterial control of mosquitoes and black flies: biochemistry genetics, and applications of Bacillus thuringiensis israelensis and Bacillus sphaericus / H. de Barjac, D. J. Sutherland, editors.
p. cm.
Includes bibliographic references

1. Bacillus thuringiensis. 2. Bacillus sphaericus.
3. Mosquitoes—biological control. 4. Simuliidae—Biological control. I. Barjac, Huguette de. II. Sutherland, Donald J.
QR82.B3B34 1990
614.4'32-dc20

89-70095
CIP

Contents

PART 2 *Bacillus sphaericus*

Contributors

Huguette de Barjac, Institut Pasteur, 28, Rue Du Dr Roux, 75724 Paris, France

Amaret Bhumiratana, Department of Biotechnology, Faculty of Science, Mahidol University, Rama VI Road, Bangkok 10400, Thailand

Thomas M. Boyle, Department of Biochemistry, Ohio State University, 484 West 12th Avenue, Columbus, Ohio 43210

Lee A. Bulla, Jr., Department of Molecular Biology, College of Agriculture, University of Wyoming, Laramie, Wyoming 82071–3944

William F. Burke, Jr., Department of Microbiology, Arizona State University, Tempe, Arizona 85287–2701

Chris N. Chilcott, DSIR, Entomology Division, Mt. Albert Research Centre, Auckland 3, New Zealand

Jose A. Correa, U.S. Department of Agriculture, ARS-SPA SR, Weslaco, Texas 78529

Elizabeth W. Davidson, Department of Zoology, Arizona State University, Tempe, Arizona 85287–1501

Donald H. Dean, Department of Biochemistry, Ohio State University, 484 West 12th Avenue, Columbus, Ohio 43210

Francis A. Drobniewski, St. Bartholemew's Hospital Medical College, West Smithfield, London, EC1, UK

Howard T. Dulmage, H. D. Associates, P. O. Box 4113, Brownsville, Texas 78520

David J. Ellar, Department of Biochemistry, University of Cambridge, Tennis Court Road, Cambridge CB2 1QW, UK, England

Robert M. Faust, Plant Protection Institute, U.S. Department of Agriculture, Beltsville, Maryland 20705

Brian A. Federici, Department of Entomology, University of California, Riverside, California 92521

Gabriel Gallegos-Morales, Facultad de Ciencias Biologicas, Universidad Autonoma de Nuevo Leon, APDO. Postal 414, San Nicolas de Los Garza, N.L., C.P. 66400

Pierre Guillet, O.M.S.-ONCHO, O.C.P., Boite Postale 2279, Bamako, Mali

Jean-Marc Hougard, Centre Pasteur Du Cameroun, BP 1274, Yaounde, Cameroun

Jorge E. Ibarra, Centro de Investigacion y de Estudios Avanzados del Instituto Politecnico Nacional, Unidad Irapuato, Irapuato, Mexico

and conclusions. Several authors present imaginative and provocative new hypotheses and suggest experimental approaches that might be applied in the future. Others survey the status of research and determine areas in which research is limited or lacking. The contributors bring into sharp focus current directions of research, and they collate the large body of information and the vast array of data that continue to accumulate in the field of microbial control of mosquitoes and black flies.

The primary purpose of this treatise is to provide a stimulating forum for discussion of new ideas about controlling mosquitoes and black flies. Since no comparable book is available, the volume will be virtually indispensable for all those interested in the control of mosquitoes and black flies, especially those specifically interested in microbial control. It will be of special interest to medical entomologists, protozoologists, physicians, veterinarians, ecologists, research workers, science teachers, and graduate students.

Ecological problems created by chemical insect control methods and their relevance to human health are receiving increased attention everywhere. The combined efforts of eminent contributors to this volume will undoubtedly benefit all interested in biological control and in ways to improve the environment.

Karl Maramorosch
Robert L. Starkey Professor of Microbiology, Rutgers University

Preface

The genesis of this volume was a symposium entitled the "Future of *Bacillus thuringiensis israelensis* and *Bacillus sphaericus* in Vector Control," held at the Fifty-first Annual Meeting of the American Mosquito Control Association, 1985, in Atlantic City, New Jersey. At that time, Professor de Barjac, the symposium organizer, was a Rutgers University Visiting Professor in the Mosquito Research and Control Laboratories of Dr. Sutherland, the program chairman for the meeting. The symposium was an important forum providing for an exchange of information between microbiologists and vector control scientists. From the symposium it was evident that the subject warranted a more complete review, analysis, and publication to encourage continued interaction among such scientists and thereby further stimulate the development of these and other bacteria as vector control agents or resources. To this end, symposium participants and other leaders in the field were invited to contribute chapters to this volume, covering aspects such as biochemistry, mode of action, genetics, and practical aspects associated with the use of these bacteria.

The preparation of this volume has been a global effort. Contributors from many countries and regions have provided their analyses of and perspectives on subjects in which advancements are being rapidly made, and they have incorporated such advancements as much as possible in the final presentation. The editors gratefully acknowledge the authors for their outstanding contributions. Excellent assistance in the preparation of final text has been provided by Mrs. A. Hajek and Mrs. P. Horan at Rutgers University and Mrs. M. F. Blanc at Pasteur Institute, and we gratefully acknowledge the subvention support of the Rutgers University Research Council.

Invariably, advancements are accompanied by the coinage of new terms such as *mosquiticidal*. It joins a long series of terms including insecticide, nematicide, fungicide, larvicide, adulticide, entomicide, and even homicide, which may be convenient in indicating toxicity to or death of a target. However, if the goal is to find more specific agents to control specific pests, terms such as *culicicide* (for Diptera:Culicidae; mosquitoes) and *simuliicide* (for Diptera:Simuliidae; black flies) may come into usage for the two bacilli, the subjects of this volume. The value and appropriateness of such terms remain to be seen. Certainly, in recognition of the current perception of the specificity of these agents, and their value, they should *not* be termed *biocides*.

Bacterial Control of Mosquitoes & Black Flies

PART 1

Bacillus thuringiensis subsp. *israelensis* (*B.t.i.*)

1

Discovery of *Bacillus thuringiensis israelensis*

JOEL MARGALIT

1.1 INTRODUCTION

For the past four decades humans have been almost completely dependent upon synthetic organic insecticides. However, the very properties that made these chemicals useful—long residual action and toxicity for a wide spectrum of organisms—have brought about serious environmental problems. The emergence and spread of insecticide resistance in many species of vectors, the concern with environmental pollution, and the high cost of the new chemical insecticides make it apparent that vector control can no longer be safely dependent upon the use of chemicals.

Thus, increasing attention has been directed toward natural enemies such as predators, parasites, and pathogens. Unfortunately, none of the predators or parasites can be mass-produced and stored for long periods of time. They all must be reared in vivo. It became evident that there was an urgent need for a biological agent that possessed the desirable properties of a chemical pesticide; that is, it must be highly toxic to the target organism, able to be mass-produced on an industrial scale, have a long shelf life, and be transportable.

In the mid seventies, the World Health Organization (WHO) and other international institutions initiated studies and development of existing and new biological control agents. During the years 1975 and 1976 an extensive survey of mosquito breeding sites was launched to determine the occurrence of natural pathogens and parasites of mosquitoes in Israel. In the course of this survey, 310 breeding sites were sampled. Over 120,000 larvae of 27 mosquito species (out of 42 indigenous species) were collected, identified, and examined for presence of pathogens and parasites. As a result of this effort a new mosquito pathogen was detected and isolated in the Negev Desert (Goldberg and Margalit 1977). This pathogen appeared to be a new variety of *Bacillus thuringiensis* demonstrating highest larvicidal activity. Isolates of the new strain were delivered through WHO to Dr. Huguette de Barjac at the reference laboratory of the Pasteur Institute in Paris. Later the new strain was identified and designated by Dr. de Barjac as *Bacillus thuringiensis* subsp. *israelensis* (*B.t.i.*) serotype H14 (de Barjac 1978).

1.2 GEOGRAPHY, CLIMATE, AND ENVIRONMENTAL CONDITIONS

B.t.i. was first isolated from a stagnant pond (fig. 1.1) located in the Nahal Besor Desert river basin near Kibbutz Zeelim in the northwestern Negev Desert of Israel (fig. 1.2). The Negev occupies approximately 12,000 km^2 (about 60% of the state). On the map it forms a triangle, with its base in the north and its apex in the south at Elat. In the northwest the Negev reaches the Mediterranean Sea.

The northwestern Negev is a winter rainfall desert. Most of the rainfall occurs erratically between November and April. It varies from 150 to 350 mm per annum, with an average of about 200 mm. The mean monthly evaporation from an open water surface is 40–50 mm in July (Rosenan 1970). Minimum daily temperature for the coldest month (February) in the Zeelim region is 7°C and for the hottest month (August) 20°C. Maximum daily temperature for the coldest month (January) is 17°C and for the hottest month (July) 35°C (fig. 1.3.). Hot, dry desert winds may raise the temperature to 40°C during late spring. The fluctuation of relative humidity at Zeelim is shown in figure 1.3.

There is only one major stream in the northwestern region, Nahal Besor, which flows down from the Negev hills and sometimes carries strong, violent floods, estimated at 20 to 30 million M^3 annually (Orni and Efrat 1971). Most of the area is covered with loess soil, which is transported by the winds or carried by winter streams descending from the Negev hills (Yaalon 1966). The yellowish brown loess is fine grained. In the first winter rain the surface grains swell and coalesce into a hard crust that is impenetrable to seepage of additional rainwater into the subsoil. Rainwater then collects on the surface, and wild flash floods tear open deep gullies, leaving behind temporary ponds along the riverbeds, which may last until late spring or even through the early summer months. The loess soil at the bottom of the ponds is potentially fertile, containing substances such as silica, calcium, alumina, and iron and giving rise to the development of a productive ecosystem in which mosquitoes constitute an important component (Dimentman and Margalit 1981). Eventually, the temporary pools dry out and the loess surface soil crumbles in summer so that the fine particles are transported in dust storms, carrying with them and scattering different organisms, including microorganisms and their spores. This phenomenon may explain the prevalence of spore-forming *Bacilli,* including *Bacillus sphaericus* and *Bacillus thuringiensis* types, in the Negev environ. In a recent survey for mosquito larval pathogens conducted in 1985–1986, out of 130 samples obtained either from edges of dry-

FIGURE 1.1 Drying stagnant pond located at the site where *B. t. i.* was first recovered. Photograph taken in 1984.

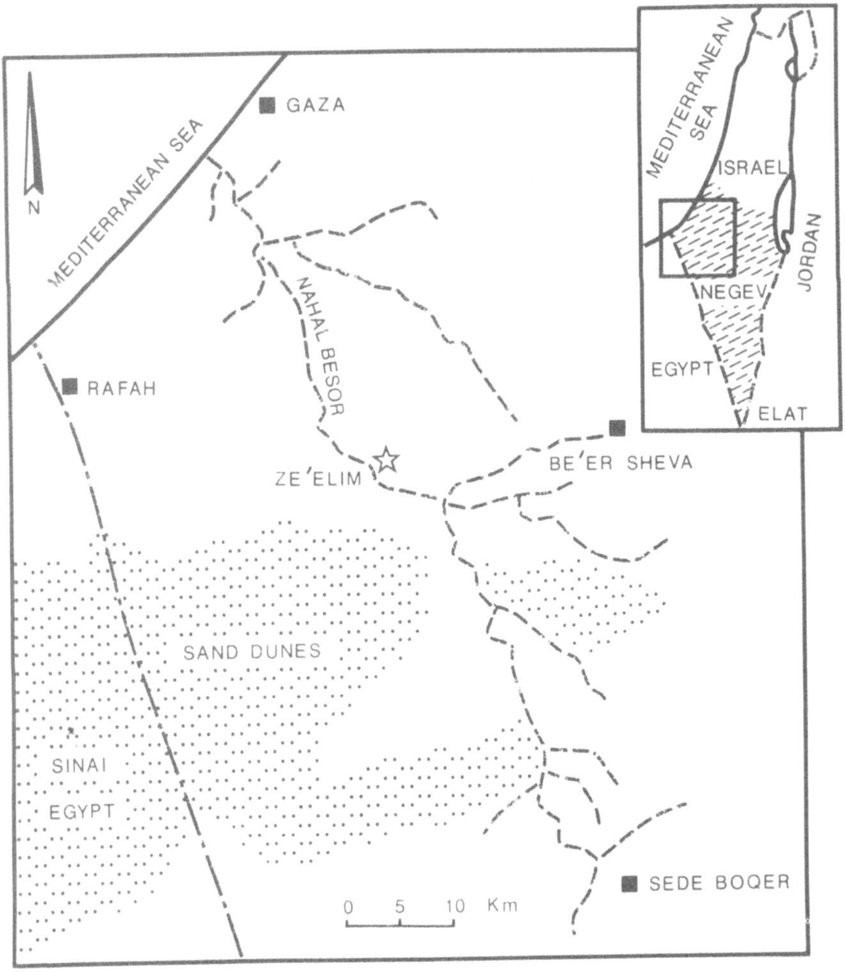

FIGURE 1.2 Map of Ze'elim and Nahal Besor River Basin

ing rainpools with mosquito larvae, or later from dry soil samples at the bottom of dried larval habitats, several hundred spore-forming bacteria were isolated and assayed. Of them, *B. thuringiensis* types were toxic to both *Culex* and *Aedes* larvae, and dozens of *B. sphaericus* were toxic to *Culex* larvae only (Brownbridge and Margalit 1986, 1987).

1.3 DETECTION AND ISOLATION

The new strain of *B. thuringiensis* was isolated in Israel in the summer of 1976 from a breeding site of *Culex pipiens* complex mosquitoes.

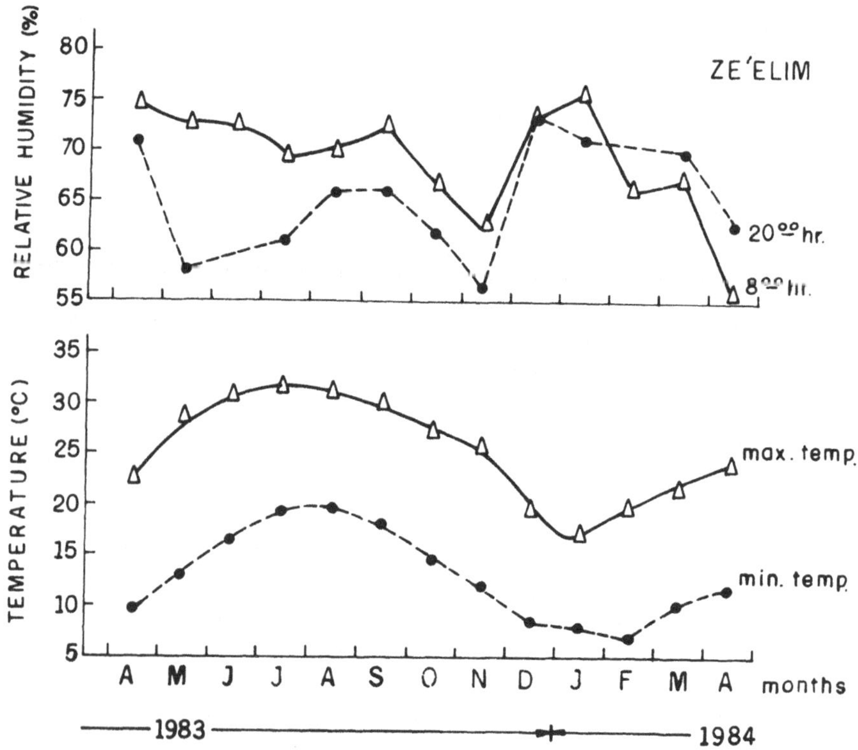

FIGURE 1.3 Temperature and relative humidity at Ze'elim in 1983–1984. Data obtained from the Israel Metereological Service.

(Goldberg and Margalit 1977). This strain was found in an isolated stagnant pond of approximately 15 by 60 m, with a maximum depth of 30 cm, located in a dry desert riverbed of Nahal Besor, adjacent to Kibbutz Zeelim. The larval site contained brackish water with approximate salinity of 900 mg Cl/l and a heavy load of organic material such as excreta of the nearby Bedouin tribes' sheep and camels and decomposing algae and other detritus. A very high concentration of dead and dying *Cx. pipiens* larvae in an epizootic situation was found as a thick grayish white carpet on the surface of the water body. Green water of unicellular algae contained live arthropods such as cyclopid and ostrocod crustacea, as well as live aquatic insects of ephemerid, libellulid, corixid, chironomid, and hydrophilid families. From the edge of the pond a sample containing dead and decomposing larvae, water, and silty mud was taken to the laboratory and refrigerated before processing for isolation and assay of larvicidal activity.

Subsamples taken from the homogenate of the parent sample were cultured on standard media and then processed for larvicidal activity. Processing in the laboratory was mostly done by Leonard Goldberg, who was on

sabbatical leave in Israel from the Office of Naval Research, Berkeley, California. The samples were heat treated to select for spore formers and diluted to a predetermined volume to give 30–50 colonies when plated. Ten to 15 surface agar plates were prepared, containing several hundred single-colony isolates. Each was used in a mosquito larval bioassay. Only about one in a hundred such bacterial preparations had larvicidal activity, and only one was extremely toxic to mosquito larvae. This was designated ONR-60A.

The larvicidal activity of the very first strain was tested in 1976 and found to be extremely effective against 5 species of mosquitoes: *Cx. pipiens* ($LC_{50} = 6 \times 10^3$ spores/ml), *Culex univitattus* (2×10^4), *Aedes aegypti* (1×10^4), *Uranotaenia unguiculata* (3×10^4), and *Anopheles sergentii* (5×10^5), thus demonstrating much higher activity than any other previously known bacterial pathogen. Since its detection, *B.t.i.* has been tested by scientists all over the world and found to be toxic against practically all filter-feeding mosquitoes and black fly larvae tested; namely, 72 species of mosquitoes (of 11 genera) and 14 species of black flies (7 genera) (Margalit and Dean 1985). From the ONR-60A single colony, 12 surface plates were made, and each was pooled and lyophilized in 12 small glass vials. Several of the vials were delivered to WHO in Geneva; Pasteur Institutes Reference Laboratory in Paris; and WHO Collaborating Center for the Biological Control of Vectors of Human Diseases at Ohio State University, Columbus, Ohio, headed by Dr. John Briggs. The strain was given two separate accession numbers (WHO 1884 and WHO 1897). Other glass vials were sent to Dr. D. Dean at the *Bacillus* Genetic Stock Center and to Dr. S. Singer of Western Illinois University, Macomb, Illinois.

Thus, from a single colony were derived all the known cultures now in use of what we know as *Bacillus thuringiensis* subsp. *israelensis* (serotype H14) (de Barjac 1978), and the first report of the extensive mosquito activity was published by Goldberg and Margalit (1977).

References

Barjac, H. de. 1978. A new subspecies of *Bacillus thuringiensis* very toxic for mosquitoes: *Bacillus thuringiensis* var. *israelensis* serotype 14 (in French). *C. R. Acad. Sci.* (Paris) 286D: 797–800.

Brownbridge, M., and Margalit, J. 1986. New *Bacillus thuringiensis* strains isolated in Israel are highly toxic to mosquito larvae. *J. Invertebr. Pathol.* 48: 216–222.

______. 1987. Mosquito active strains of *Bacillus sphaericus* isolated from soil and mud samples collected in Israel. *J. Invertebr. Pathol.* 50: 106–122.

Dimentman, C., and Margalit, J. 1981. Rainpools as breeding and dispersal sites of mosquitoes and other aquatic insects in the Central Negev Desert. *J. Arid Environment* 4: 123–129.

Goldberg, L. J., and Margalit, J. 1977. A bacterial spore demonstrating rapid larvicidal activity against *Anopheles sergentii, Uranotaenia unguiculata, Culex univitattus, Aedes aegypti,* and *Culex pipiens. Mosq. News* 37: 355–358.

Margalit, J., and Dean, D. 1985. The story of *Bacillus thuringiensis* var. *israelensis* (*B.t.i.*) *J. Amer. Mosq. Control Assoc.* 1: 1–7.

Orni, E., and Efrat, E. 1971. *Geography of Israel.* Jerusalem: Israel University Press.
Rosenan, N. 1970. Evaporation from open water surfaces. In *Climate IV/3: Atlas of Israel,* ed. H. K. Amiran. Jerusalem and Amsterdam: Survey of Israel and Elsevier.
Yaalon, D. H. 1966. Factors and processes in the formation of soils in Israel. In *Encyclopedia of Agriculture,* ed. H. Halperin, 1: 178–186 (in Hebrew). Tel Aviv: Publication of the Encyclopedia of Agriculture.

2

Characterization and Prospective View of *Bacillus thuringiensis israelensis*

HUGUETTE DE BARJAC

When the bacterial sample later identified as *Bacillus thuringiensis* serovar H14 *israelensis* (*B. t. i.*) was collected from the Negev Desert, it was a little like the fairy tale "Sleeping Beauty." Interest in *Bacillus thuringiensis* (*B. t.*), which had been more or less in dormancy for years, was quickly revived and attracted the attention of many workers in universities and industry. A flood of studies resulted, thereby advancing the field of biological control of insects in general.

What was this spectacular awakening due to? As stated at that time (de Barjac 1978c, 1978d), *B. t. i.*, with its toxicity to mosquito and black fly larvae, represented an opportunity to expand biological control of insects into the public health area for the prevention of tropical diseases. Until then, the battlefield of *B. thuringiensis* was limited to agriculture and forestry, with the known strains killing the larvae of insects feeding on crops or trees. With *B. t. i.* the value of entomopathogenic bacteria was greatly enhanced.

In 1977, from dead mosquito larvae in a larval habitat located in the Negev Desert, Goldberg and Margalit isolated a sporulated bacterium that demonstrated rapid and high larvicidal power. Bioassays performed by these researchers indicated that different species in four genera (*Anopheles, Uranotaenia, Culex,* and *Aedes*) were sensitive. This bacterium was identified to be in the genus *Bacillus,* and its pathogenicity was attributed to its spores.

Having received a sample of this *Bacillus* from the World Health Organization (WHO) at the end of 1977 under the reference ONR-60A, I subsequently identified it as a new serotype of *B. thuringiensis* H14, named it after its origin, serovar *israelensis,* and proposed it for the control of mosquito larvae (de Barjac 1978c). At the same time, its larvicidal action on various mosquito species was confirmed, described at the cellular level (de Barjac 1978a) and related not to the spore but to the crystal toxic proteins (de Barjac 1978b, 1978d).

Another isolate from the same origin, received a little later from the World Health Organization (reference 1884), proved to have the same characteristics as the foregoing reference strain ONR-60A, but apparently contained larger crystals and caused somewhat higher and more stable mortality. This explains our selection of strain 1884 for preparing our second and third international standards, IPS 80 and IPS 82, after having made the first standard IPS 78 (Institut Pasteur standard 1978) with the type-strain ONR-60A.

Since then, many other strains with similar pathogenic power have been isolated from 15 different countries and from insects, soils, or waters, all but two belonging to serotype H14. In 1989, this serotype contains 228 strains; in several cases, these so-called strains are mere reisolates from the same sample, the real history of each isolate being often very difficult to obtain from its author. According to their larvicidal power and main phenotypic characters as used in current taxonomy, all of the known strains grouped into the serotype H14 appeared to be the same until recently, when two new isolates from Japan did not show toxicity to mosquito larvae (Ohba, Yu, and Aizawa 1988).

On the other hand, two exceptions are known to serotype H14 extremely mosquiticidal strains. The first one is a strain with similar toxicity first referred to as PG-17-03, later as PG-14, isolated from soil in the Philippines (Padua et al. 1982; Padua, Ohba, and Aizawa 1984), which belongs to serotype H8a,8b. Roughly speaking, when working with 48 hour final whole cultures (FWC) in current usual medium (UG medium, de Barjac and Lecadet 1976), LC_{50}s of *B.t.i.* and PG-14 are on the same dilution level (around 10^{-6}) with *Aedes aegypti* or *Culex pipiens* larvae and around 10^{-5} with *Anopheles stephensi* larvae. Based on a mean of about 10^8 spores ml^{-1} of FWC, these dilutions correspond respectively with 10^2 and 10^3 spores ml^{-1}. Secondly, a strain, Dak-Fe, with the same properties as PG-14 has been isolated from soil in Dakar (Jelusic, unpub. data).

In addition, several other *B.t.* strains exhibit to mosquito larvae a significant but lower toxicity than *B.t.i.*, PG-14, or Dak-Fe. One of these is an isolate BA 068 from Reeves (1970) belonging to serotype H1. Later, Ohba, Aizawa, and Furusawa (1979) isolated from silkworm litter two strains, 73-E-10-2 and 73-E-10-16, belonging to serotype H10, active on mosquito and black fly larvae (Padua, Ohba, and Aizawa 1980; Finney and Harding 1982). Another example is HD-1 from serotype H3a,3b, which produces a protein (P2) toxic to mosquito larvae (Yamamoto and McLaughlin 1981). LC_{50}s of all these isolates are usually lower than *B.t.i.* LC_{50}s by 2 to 4 log when expressed in FWC dilutions.

In fact, various other *B.t.* strains belonging to serotypes H1; H3a,3b; H5a,5b; H9; and H10 have a similar toxicity to mosquito larvae. Depending on the strain and the target species, their LC_{50}s vary from 10^{-3} to 10^{-5} in FWC dilutions, corresponding with 10^5 to 10^3 spores ml^{-1}. Moreover, using different mosquito species, or even Lepidoptera species, there are revealed among these strains different patterns of activity (Jelusic and de Barjac, unpub. data).

Thus, in *B.t.* species, in addition to the one serotype (H14) highly

TABLE 2.1.
Classification of *B. thuringiensis* Strains According to the H-serotype

H-serotype	Serovar	Supposed Biovars or Pathovars	Abbreviation	First Mention and First Valid Description
1	thuringiensis		THU	Berliner 1915; Heimpel and Angus 1958
2	finitimus		FIN	Heimpel and Angus 1958
3a	alesti		ALE	Toumanoff and Vago 1951; Heimpel and Angus 1958
3a,3b	kurstaki		KUR	de Barjac and Lemille 1970
4a,4b	sotto		SOT	Ishiwata 1905; Heimpel and Angus 1958
id	id	dendrolimus	DEN	Talalaev 1956; Bonnefoi and de Barjac 1963
4a,4c	kenyae		KEN	Bonnefoi and de Barjac 1963
5a,5b	galleriae		GAL	Shvetsova 1959; de Barjac and Bonnefoi 1962
5a,5c	canadensis		CAN	de Barjac and Bonnefoi 1972
6	entomocidus		ENT	Heimpel and Angus 1958
id	id	subtoxicus	SUB	Heimpel and Angus 1958
7	aizawai		AIZ	Bonnefoi and de Barjac 1963
8a,8b	morrisoni		MOR	Bonnefoi and de Barjac 1963
id	id	tenebrionis	TEN	Krieg, Huger, Langenbruch, and Schnetter 1983
8a,8c	ostriniae		OST	Gaixin, Ketian, Minghua, and Xingmin, 1975
8b, 8d	nigeriensis		NIG	de Barjac et al. unpub. data
9	tolworthi		TOL	Norris 1964; de Barjac and Bonnefoi 1968
10	darmstadiensis		DAR	Krieg, de Barjac, and Bonnefoi 1968
11a,11b	toumanoffi		TOU	Krieg 1969
11a,11c	kyushuensis		KYU	Ohba and Aizawa 1979
12	thompsoni		THO	de Barjac and Thompson 1970
13	pakistani		PAK	de Barjac, Cosmao, Shaik, and Viviani 1977
14	israelensis		ISR	de Barjac 1978
15	dakota		DAK	DeLucca, Simonson, and Larson 1979
16	indiana		IND	DeLucca, Simonson, and Larson 1979
17	tohokuensis		TOH	Ohba, Aizawa, and Shimizu 1981
18	kumamotoensis		KUM	Ohba, Ono, Aizawa, and Iwanami 1981
19	tochigiensis		TOC	Ohba, Ono, Aizawa, and Iwanami 1981
20a,20b	yunnanensis		YUN	Wan-yu, Qi-fang, Xue-ping, and You-wei 1979
20a,20c	pondicheriensis		PON	de Barjac et al. unpub. data
21	colmeri		COL	DeLucca, Palmgren, and de Barjac 1984

TABLE 2.1.
(Continued)

H-serotype	Serovar	Supposed Biovars or Pathovars	Abbreviation	First Mention and First Valid Description
22	shandongiensis		SHA	Ying, Jie, and Xichang 1986
23	japonensis		JAP	Ohba and Aizawa 1986
24	neoleonensis		NEO	Rodriquez-Padilla et al. 1988
25	coreanensis		COR	de Barjac et al. unpub. data
26	siloensis		SIL	de Barjac et al. unpub. data
27	mexicanensis		MEX	Wong and Rodriguez-Padilla lab 1988

NOTE: A nonmotile type is wuhanensis (Hubei Inst. Microbiology 1976), WUH.

larvicidal to mosquitoes and two potent strains (PG-14 and Dak-Fe) of serotype H8a,8b, there exist strains of different serotypes (H1;H3a,3b; H5a,5b; H9; and H10) that exhibit lower but significant and various levels of potency on mosquito larvae. Furthermore, the complexity of *B. t.* is far from being restricted to Lepidoptera or Diptera, as we know strains (*tenebrionis* and *san diego*) pathogenic for Coleoptera (Krieg et al. 1983; Herrnstadt et al. 1986). Also, toxicities different by quality or quantity can coexist in separate strains.

All the *B. t.* strains known are classified according to their H antigens into 27 groups and 7 subgroups, which enable researchers to distinguish the 34 serovars reported in table 2.1. Description of biovars and pathovars has no taxonomic value and appears illusive due to their extreme diversity and variability (de Barjac and Frachon 1990). Currently, the 34 serovars include 1,720 strains present in our laboratory collection and listed in a catalogue, which is available on request.

By comparison, the *B. sphaericus* species represented in 1989 in our collection by 300 strains, grouped into 48 H-serotypes, contains 3 serotypes highly larvicidal to mosquitoes and 3 other ones much less toxic (see chapter 14). Another difference between *B. thuringiensis* and *B. sphaericus* mosquito larvicidal strains lies in their respective spectrum of action, which is wider for *B. t. i.* strains and has black fly larvae as supplementary targets. Larvae of Sciarid flies (Cantwell and Cantelo 1984) and horn flies (Temeyer 1984) as well as larvae of *Phlebotomus* subsp. (de Barjac, Larget, and Killick-Hendrick 1981) have also been reported as susceptible to *B. t. i.*

Since its discovery, *B. t. i.* has been commercialized by various firms, such as Sandoz (Teknar®), Abbott (Vectobac®), and Solvay (Bactimos®), and used on a large scale in many countries. Its success in controlling mosquitoes in Europe and Asia is not a point that allows for discussion. Still more well established is its success in controlling black flies in Africa (see chapter 11).

Nevertheless, the lack of persistance of *B. t. i.* presents a problem in its application. Although such a problem could be partially overcome either by the use of a more persistent *Bacillus* like *B. sphaericus* or by the production of transgenic bacteria expressing cloned larvicidal toxins, the discovery of other wild microorganisms is highly desirable.

With the renewal of interest in bacteriological control of insects caused by the arrival and success of *B.t.i.* on the bioinsecticide market, other interesting pathogenic serotypes of *B. thuringiensis* and *B. sphaericus,* other *Bacillus* species, or even other genera of bacteria should be isolated. These, together with genetic engineering techniques (see chapters 6 and 7), will be the main sources of progress in the biological control of insect vectors and pests in the near future.

References

Barjac, H. de. 1978a. Etude cytologique de l'action de *Bacillus thuringiensis* var. *israelensis* sur larves de moustiques. *C. R. Acad. Sci.* (Paris) 286D: 1629–1632.

______. 1978b. Toxicité de *Bacillus thuringiensis* var. *israelensis* pour les larves d'*Aedes aegypti* et d'*Anopheles stephensi. C. R. Acad. Sci.* (Paris) 286D: 1175–1178.

______. 1978c. Une nouvelle variété de *Bacillus thuringiensis* très toxique pour les moustiques: *B. thuringiensis* var. *israelensis* sérotype 14. *C. R. Acad. Sci.* (Paris) 286D: 797–800.

______. 1978d. Un nouveau candidat à la lutte biologique contre les moustiques: *Bacillus thuringiensis* var. *israelensis. Entomophaga* 23: 309–319.

Barjac, H. de, and Frachon, E. 1990. Classification of *Bacillus thuringiensis* strains. *Entomophaga.* In press.

Barjac, H. de; Larget, I., and Killick-Hendrick, R. 1981. Toxicité de *Bacillus thuringiensis* var. *israelensis* serotype H14 pour les larves de phlébotomes vecteurs de leishmanioses. *Bull. Soc. Pathol. Exot.* 74: 485–489.

Barjac, H. de, and Lecadet, M.-M. 1976. Dosage biochimique de l'exotoxine thermostable de *B. thuringiensis* d'après l'inhibition d'ARN-polymérases bactériennes. *C. R. Acad. Sci.* (Paris) 282D: 2119–2122.

Cantwell, G. C., and Cantelo, W. W. 1984. Effectiveness of *Bacillus thuringiensis* var. *israelensis* in controlling a Sciarid fly, *Lycoriella mali,* in mushroom compost. *J. Econ. Entomol.* 77: 473–475.

Finney, J. R., and Harding, J. B. 1982. The susceptibility of *Simulium verecundun* (Diptera: Simuliidae) to three isolates of *Bacillus thuringiensis* serotype 10 (*Darmstadiensis*). *Mosq. News* 42: 434–435.

Goldberg, L. J., and Margalit, J. 1977. A bacterial spore demonstrating rapid larvicidal activity against *Anopheles sergentii, Uranotaenia unguiculata, Culex univitattus, Aedes aegypti,* and *Culex pipiens. Mosq. News* 37: 355–358.

Herrnstadt, C.; Soares, G. G.; Wilcox, E. R.; and Edwards, D. L. 1986. A new strain of *Bacillus thuringiensis* with activity against Coleopteran insects. *Bio/Technology* 4: 305–308.

Krieg, A.; Huger, A. M.; Lagenbruch, G. A.; and Schnetter, W. 1983. *Bacillus thuringiensis* var. *tenebrionis:* Ein neuer, gegenüber Larven von Coleopteren wirksamer Pathotyp. *Z. Ang. Ent.* 96: 500–508.

Ohba, M.; Aizawa, K.; and Furusawa, T. 1979. Distribution of *Bacillus thuringiensis* serotype in Ehime Prefecture (in Japanese). *Appl. Entomol. Zool.* 14: 340–345.

Ohba, M.; Yu, M.; and Aizawa, K. 1988. Occurrence of non-insecticidal *Bacillus thuringiensis* flagellar serotype 14 in the soil of Japan. *System. Appl. Microbiol.* 11: 85–89.

Padua, L. E.; Gabriel, B. P.; Aizawa, K.; and Ohba, M. 1982. *Bacillus thuringiensis* isolated from the Philippines. *Philipp. Ent.* 5: 199–208.

Padua, L. E.; Ohba, M.; and Aizawa, K. 1980. The isolates of *Bacillus thuringiensis* serotype 10 with a highly preferential toxicity of mosquito larvae. *J. Invertebr. Pathol.* 36: 180–186.

———. 1984. Isolation of a *Bacillus thuringiensis* strain (serotype 8a:8b) highly and selectively toxic against mosquito larvae. *J. Invertebr. Pathol.* 44 (1): 12–17.

Reeves, E. L. 1970. Pathogens of mosquitoes. *Proc. Calif. Mosq. Control Assoc.* 38: 20–22.

Temeyer, K. B. 1984. Larvicidal activity of *Bacillus thuringiensis* subsp. *israelensis* in the Dipteran *Haematobia irritans. Appl. Environ. Microbiol.* 48: 952–955.

Yamamoto, T., and McLaughlin, R. E. 1981. Isolation of a protein from the parasporal crystal of *Bacillus thuringiensis* var. *kurstaki* toxic to the mosquito larvae, *Aedes taeniorhynchus. Biochem. Biophys. Res. Commun.* 103 (2): 414–421.

3

Parasporal Body of *Bacillus thuringiensis israelensis*

STRUCTURE, PROTEIN COMPOSITION, AND TOXICITY

BRIAN A. FEDERICI
PETER LÜTHY
JORGE E. IBARRA

3.1 INTRODUCTION

Since its discovery in Japan and Germany during the early part of this century, more than 25 subspecies of the spore-forming insecticidal bacterium *Bacillus thuringiensis* Berliner have been described (de Barjac 1985). The most distinctive characteristic of this bacterium is a parasporal body produced during sporulation that consists primarily of insecticidal proteins (Angus 1965; Heimpel 1967; Aronson, Beckman, and Dunn 1986; Höfte and Whiteley 1989). In most subspecies, the parasporal body is a bipyramidal crystal containing one or more similar proteins of about 135 kDa that are toxic to lepidopterous larvae. When ingested by a larva, this toxin-containing inclusion dissolves in the alkaline gut juices, and midgut proteases cleave the protoxin, yielding an active peptide toxin of 60–70 kDa, the δ-endotoxin. Although the toxin's precise mode of action is not fully understood, intoxication results in an osmotic imbalance across the midgut epithelial cell membrane, which leads quickly to hypertrophy and lysis of midgut cells. Lysis is followed by disruption of the basement membrane, leakage of digestive juices into the hemocoel, and larval death (Lüthy and Ebersold 1981). The δ-endotoxins of different subspecies of *B. thuringiensis* can vary considerably in toxicity to larvae. These variations are thought to be due to differences in the amino acid sequence of the toxins, and are currently the subject of much interest because of the potential for increasing toxicity and host spectrum through site-directed mutagenesis.

The first isolate of *B. thuringiensis* shown to possess characteristics markedly different from those described above was ONR-60A. This isolate was collected by Goldberg and Margalit (1977) from a mosquito larval habitat in the Negev Desert of Israel and demonstrated to be highly toxic to the larvae of several mosquito species. Shortly after its discovery, de Barjac (1978a, 1978b) identified ONR-60A as a new serotype (H14) of *B. thuringiensis,* which she named *B. thuringiensis* subsp. *israelensis* (*B.t.i.*). She confirmed *B.t.i.*'s toxicity for mosquitoes and showed that, as for other subspecies, ingestion of sporulated cells destroyed the larva's midgut epithelium, resulting in death (de Barjac, 1978a). Other studies showed that *B.t.i.* was also toxic to black fly larvae yet was safe for nontarget organisms (Undeen and Nagel 1978; Mulla, Federici, and Darwazeh 1982). *B.t.i.*'s unusual host range and potential for development as a larvicide attracted substantial interest, and subsequent studies have shown that its toxicity is associated with a unique parasporal body. In contrast to those of most other subspecies, *B.t.i.*'s parasporal body is spherical, enveloped, and contains four major parasporal body proteins (27, 65, 128, 135 kDa) assembled into three different types of inclusions. Moreover, after solubilization in alkali it is cytolytic to a wide range of vertebrate and invertebrate cells, including erythrocytes, and toxic to mice if injected. Thus, *B.t.i.* is characterized as being mosquiticidal, cytolytic, hemolytic, and even neurotoxic; and a considerable controversy has arisen over the past few years regarding which of the parasporal body proteins accounts for its toxicity to mosquitoes.

In this chapter, we will review the literature on the parasporal body of *B.t.i.* with emphasis on its structure, composition, and toxicity. We will also attempt to resolve some of the controversy over the toxicity of different proteins and briefly summarize information available on mosquiticidal parasporal bodies produced by other subspecies of *Bacillus thuringiensis.*

3.2 SYNTHESIS

The formation of the parasporal body within the sporangium of *B.t.i.* follows a time course similar to that of other subspecies of *B. thuringiensis.* Synthesis of the major parasporal body proteins is initiated within three hours of the onset of sporulation and continues for several hours (Lee, Eckblad, and Bulla 1985). Inclusion formation begins during sporulation phase 2 and coincides with the formation of the forespore septum (Charles and de Barjac 1982). Within an hour of initiation, inclusions of different densities are already apparent within the parasporal body (figs. 3.1 and 3.2). Synthesis and assembly of the parasporal body continue until the end of stage 5, which is about the same time spores become refractile (fig. 3.1a, b). The sporangium wall lyses during stage 7, the last stage, releasing the intact parasporal body. Under standard growth conditions in commonly used media such as nutrient broth or peptonized milk, parasporal body formation is complete within 24 hours of culture initiation, although autolysis of the sporangium usually requires another 24–48 hours.

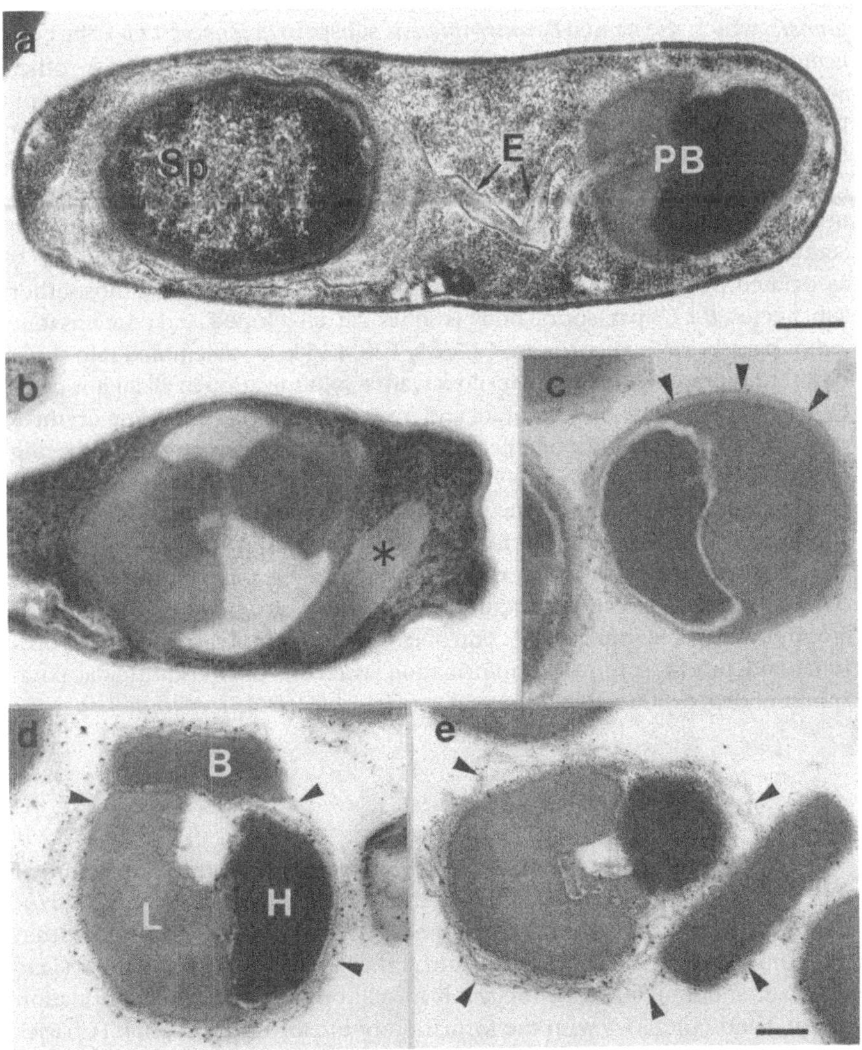

FIGURE 3.1 Electron micrographs of the parasporal body of *B.t.i.*: (*a*), developing spore (*Sp*) and parasporal body (*PB*) during stage 5 of sporulation; *E*, exosporium. Note the inclusions of different densities; (*b*), completely formed parasporal body just prior to lysis of the sporangium. Note the inclusions of different densities and the bar-shaped inclusion (*) adjacent to the main body. The latter inclusion type often occurs at the periphery of the parasporal body and apparently is the least tightly bound of the 3 types; (*c*), parasporal body recently released from a sporangium. The multilayered envelope is still tightly bound around the inclusions. Sections through parasporal bodies typically reveal only 2 different inclusion types, as observed in this parasporal body; (*d*), parasporal body illustrating the 3 different inclusion types; *L*, large inclusion of low electron density thought to contain the 27-kDa protein; *B*, bar-shaped body that contains the 65-kDa protein; *H*, inclusion of high electron density that may

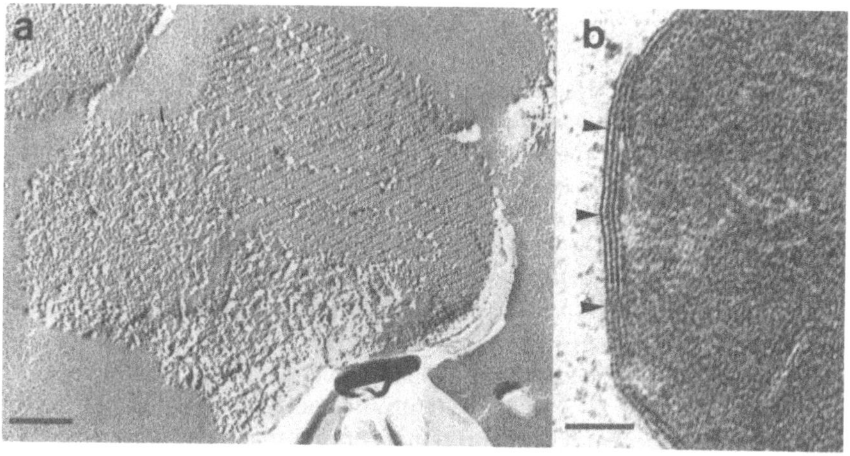

FIGURE 3.2 Electron micrographs illustrating structural characteristics of *B.t.i.* parasporal body inclusions: (*a*), freeze-etched fracture through a parasporal body illustrating crystalline and noncrystalline regions; (*b*), section through the most dense inclusion type illustrating the alternating electron-dense and electron-translucent layers at the periphery. This alternation of layers only occurs in inclusions of this type. Bar in (*a*) = 250 nm; bar in (*b*) = 50 nm.

3.3 STRUCTURE

The parasporal body of *B.t.i.* is basically spherical and averages about 1 μm in diameter, ranging from 0.7 to 1.2 μm (fig. 3.1c–e). As it is currently understood, the parasporal body consists of three different types of protein inclusions bound together by a laminated netlike envelope of undetermined composition (Huber and Lüthy 1981; Tyrell et al. 1981; Charles and de Barjac 1982; Mikkola et al. 1982; Insell and Fitz-James 1985; Lee, Eckblad, and Bulla 1985; Ibarra and Federici 1986a). Each inclusion is also surrounded individually by one or more layers of the material that composes the envelope. This envelope—which is typically not a feature of the bipyramidal crystals of *B. thuringiensis* subspecies active against lepidopterous insects, but may be similar to the membranous formation observed in strain 1715 of *B.t.* subsp. *thuringiensis* (Ribier and Lecadet 1973)—is stable in the presence of detergents and proteases. Its composition and origin are

---

contain the 128- and 135-kDa proteins; (*e*), parasporal body in which the envelope is partially disrupted. Note that the bar-shaped body remains only loosely associated with the parasporal body. Bar in (*a*) = 250 nm; (*b*)–(*e*) are approximately the same magnification, with the bar in (*e*) equal to 250 nm. Arrowheads indicate envelope.

unknown, though it resembles the exosporium and is permeable to polypeptides as large as 135 kDa.

When examined by transmission electron microscopy, the parasporal body varies considerably in regard to the number and shape of the inclusions it appears to contain. Typically, the parasporal body appears round and composed of only one or two inclusions: a large inclusion of low electron density, and a smaller, very electron dense inclusion (fig. 3.1c). However, a third type of inclusion, smaller than the other two, polyhedral, and of moderate electron density, is occasionally observed along with the other two types (fig. 3.1b, d, and e). As a result, it is currently thought that the parasporal body consists of at least three different inclusion types. All three are generally not observed in ultrathin sections through individual parasporal bodies because thin sections are usually less than a tenth the width of the parasporal body.

The three different types of protein inclusions that occur within the parasporal body can be differentiated from one another ultrastructurally using a combination of characteristics, including size, shape, electron density, and, in areas where the inclusions are crystalline, the lattice spacing (fig. 3.2a). The largest inclusion makes up 40–50% of the parasporal body and is characterized as being rounded to polyhedral and the least electron dense of the three types (fig. 3.1d). This inclusion has a lattice spacing of 4.3 nm and is thought to contain the 27-kDa protein, based primarily on the relative high abundance of this protein in comparison to others in parasporal bodies analyzed by sodium dodecylsulfate-polyacrylamide gel electrophoresis (SDS-PAGE) and because it matches the solubility properties of this protein (Insell and Fitz-James 1985; Ibarra and Federici 1986b). The second type of inclusion is often bar-shaped (usually appearing rectangular in transverse section), is of moderate electron density, and constitutes approximately 15–20% of the parasporal body (fig. 3.1d and e). This inclusion, referred to as small or large "dots" by Lee, Eckblad, and Bulla (1985), has a lattice spacing of 7.8 nm and consists almost exclusively of a 65-kDa protein (Insell and Fitz-James 1985; Lee, Eckblad, and Bulla 1985; Ibarra and Federici, 1986b). The third inclusion type is highly electron dense; hemispherical to spherical; and, based on its size as observed in ultrathin sections, makes up somewhere in the range of 20–25% of the parasporal body's protein composition (fig. 3.1d and e). Mikkola et al. (1982) have reported two inclusions of this type within the parasporal body. Because this inclusion type is thought to contain the proteins of 128 and 135 kDa, it may be that each of these proteins is assembled into a separate inclusion. An unusual feature of this inclusion type is that alternating electron-dense and electron-translucent lamellae are frequently found within the inclusion matrix or at its periphery (fig. 3.2b). These lamellae resemble those found within the spore wall. In this regard, Delafield, Somerville, Rittenberg (1968); Lecadet, Chevrier, and Dedonder (1972); and Tyrell et al. (1981) have reported immunological homologies between spore walls and parasporal body protein in two different subspecies of *B. thuringiensis.*

It must be emphasized here that the assignments of specific proteins to different types of inclusions are tentative. These assignments are based primarily on correlations between the relative abundance of major parasporal body proteins, determined by SDS-PAGE, with different inclusion types based on relative size. Furthermore, even if these correlations are accurate, they may not be absolute. For example, it has been demonstrated that the bar-shaped body contains a 65-kDa protein (Ibarra and Federici 1986b). However, it cannot be concluded, at least at present, that all of the 65-kDa protein is located in inclusions of this type.

3.4 PURIFICATION AND SOLUBILIZATION

The parasporal body of *B.t.i.* can be isolated and purified quite easily using buoyant density-gradient centrifugation techniques developed for purification of the parasporal bodies of other subspecies of *B. thuringiensis.* After autolysis is complete, the washed sediment from a culture is layered onto a gradient made of Renografin (Tyrell et al. 1981), sucrose (Thomas and Ellar 1983a; Lee, Eckblad, and Bulla 1985), or NaBr (Pfannenstiel et al. 1984) and centrifuged for one to several hours. Intact parasporal bodies usually form a well-defined band in the gradient, whereas the cellular debris remains near the top of the gradient, and the spores form a pellet at the bottom.

Attempts to isolate the different types of parasporal body inclusions from one another have met with only limited success. Parasporal bodies can be partially disrupted by subjecting them to sonication, high gravitational force, or freeze-thaw cycles. When parasporal bodies are disrupted using such methods and then centrifuged through density gradients, a band containing the bar-shaped inclusion is found above the parasporal bodies (Lee, Eckblad, and Bulla 1985; Ibarra and Federici 1986b). However, neither of the other two types of inclusions have yet been isolated in pure form from the parasporal body. Using sonication and sucrose density-gradient centrifugation, Lee, Eckblad, and Bulla (1985) were able to disrupt and separate parasporal bodies into four major particle sizes, which they referred to as small dots, large dots, small refractile bodies, and large refractile bodies. Analysis of these particles by SDS-PAGE indicated the small and large dots were enveloped inclusions of the 65-kDa protein, whereas the small and large refractile bodies were essentially slightly disrupted parasporal bodies of two different size classes.

Solubilization of the parasporal body of *B.t.i.* requires alkaline conditions. However, in contrast to the bipyramidal crystals of other subspecies active against lepidopterans, reducing agents such as dithiothreitol (DTT) are only required to solubilize certain portions of the parasporal body. Thomas and Ellar (1983a) were able to extract about 40% of the total parasporal body protein with 50 mM $Na_2CO_3 \cdot HCl$ at pH 10.5. The extracted fraction consisted mostly of the 27-kDa protein. Insell and Fitz-James (1985) found that the major low (27-kDa) and high (128- and 135-kDa) molecular weight proteins were solubilized in 1% SDS-50 mM DTT-Tris-HCl at 37°C over a pH range of 8.3–9.2, but that the 65-kDa under the same conditions required a pH of 10.5 for solubilization. The latter protein was contained in electron-dense inclusions, some of which were bar-shaped. Lüthy (unpublished observations) also found that most of the protein of low electron density was solubilized after treatment with carbonate buffer, but that the bar-shaped bodies were only solubilized in the presence of DTT (10 mM) at a pH of 9.5 or above. Chestukhina et al. (1985) have reported similar results. Thus, the inclusions containing the 27-kDa protein dissolve readily under alkaline conditions above pH 8.3 without the aid of a reducing agent. A reducing agent and a pH of 8.3 to 9.2 will solubilize the inclusions containing the 128- and 135-kDa proteins, whereas the bar-shaped inclusion containing the 65-kDa protein requires a reducing agent and a high pH for solubilization.

3.5 PROTEIN COMPOSITION

B. t. i.'s unique properties attracted the interest of many investigators, and as a result, the protein composition of the parasporal body has been

TABLE 3.1.
Major Mosquiticidal Proteins in the Parasporal Body of *Bacillus thuringiensis* subsp. *israelensis* and Nomenclature for the Encoding Genes

Mass (kDa) of Encoded Protein (Protoxin)	Proteolytic Cleavage Product (Toxin)	Gene Nomenclature	Other Nomenclature for the Gene	Reference
134.4	53–67	*cryIVA*	130-kDa endotoxin gene	Ward and Ellar 1988
			125-kDa protein gene	Bourgouin et al. 1988
			ISRH4	Sen et al. 1988
			pCH 130	Ward and Ellar 1988
127.8	53–67	*cryIVB*	130-kDa protein gene	Sekar 1986
			135-kDa protein gene	Bourgouin et al. 1988
			Bt8	Chunjatupornchai et al. 1988
			135-kDa protein gene	Delecluse et al. 1988
			ISRH3	Sen et al. 1988
			pPC 130	Ward and Ellar 1988
			130-kDa endotoxin gene	Yamamoto et al. 1988
72.4[a]	30–38	*cryIVD*	Cry D gene	Donovan et al. 1988
27.4	25	*cytA*	27-kDa toxin gene	Waalwijk et al. 1985

SOURCE: From Höfte and Whiteley (1989) with slight modification.
[a]Referred to commonly in the literature as the 65-kDa protein.

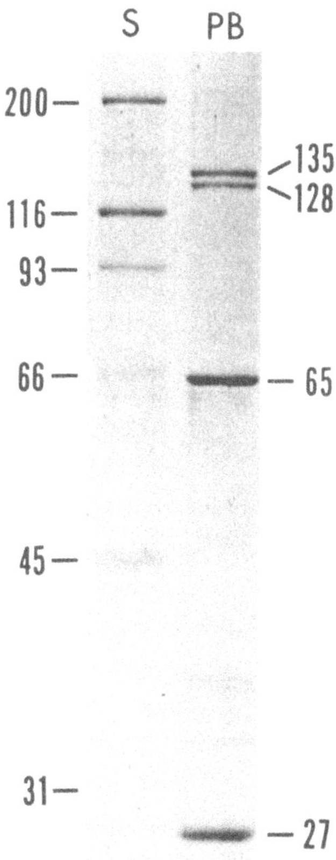

FIGURE 3.3 Protein composition of the intact parasporal body of *Bacillus thuringiensis* subsp. *israelensis* as determined by sodium dodecyl sulfate-polyacrylamide gel electrophoresis: (*S*), Molecular size standards: carbonic anhydrase, 31 kDa; ovalbumin, 45 kDa; bovine serum albumin, 66 kDa; phosphorylase *b*, 92 kDa; β-galactosidase, 116 kDa; and myosin, 200 kDa. (*PB*), *B.t.i.* parasporal body.

analyzed using gel chromatography or SDS-PAGE in numerous studies (Tyrell et al. 1981; Thomas and Ellar 1983a; Yamamoto et al. 1983; Pfannenstiel et al. 1984; Armstrong, Rohrmann, and Beaudreau 1985; Chestukhina et al. 1985; Hurley et al. 1985; Insell and Fitz-James 1985; Sriram, Kamdar, and Jayaraman 1985; Wu and Chang, 1985; Ibarra and Federici 1986b; Visser et al. 1986; Chilcott and Ellar 1988). Although there are some discrepancies in the number and size of proteins detected, two basic properties of the parasporal body are evident from these studies. First, the intact parasporal body contains four major proteins, with masses of, respectively, 27, 65, 128, and 135 kDa based on their migration in SDS-polyacrylamide gels (fig. 3.3). The genes encoding these proteins have been cloned and sequenced, and the masses deduced from the nucleotide sequence of each are as follows: 27.4, 72.4, 127.8, and 134.4 kDa (table 3.1). For consistency with the masses commonly reported in the literature, these will be referred to here as the 27-, 65-, 128-, and 135-kDa proteins.

Using column chromatography, only three major proteins are usually detected, those of 27, 65, and 130 kDa. However, in addition to the discovery of genes encoding two different polypeptides of high mass, the 130-kDa

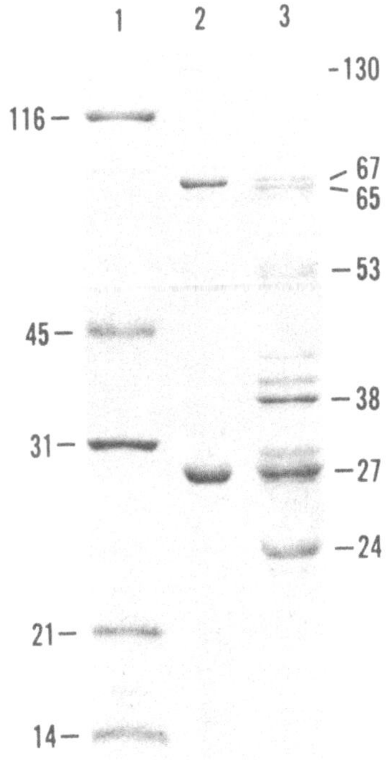

FIGURE 3.4 Comparison of the protein composition of intact and solubilized parasporal bodies: (1), protein standards as in fig. 3; (2), intact parasporal bodies; (3) parasporal bodies solubilized in alkali (50 mM Na_2Co_3·HCl, pH 10.5) for 1 hour at 37°C. Note that solubilization under alkaline conditions yields additional proteins of 67, 53, 36–40, 30, and 25 kDa resulting from proteolysis of the major parasporal body protcins.

protein has been resolved by SDS-PAGE into two proteins in numerous studies. Values reported for the 128-kDa protein range from 93 to 135 kDa, whereas those reported for the 135-kDa protein range from 98 to 145 kDA (Thomas and Ellar 1983a; Pfannenstiel et al. 1984; Insell and Fitz-James 1985; Lee, Eckblad, and Bulla 1985; Ibarra and Federici 1986a; Gill et al. 1987).

In addition to the major proteins, a series of minor proteins in the range of 25, 30–40, 53, and 67 kDa are often observed in polyacrylamide gels, particularly if they are stained with silver. Most of these, as noted below, are proteolytic cleavage products of the major parasporal body proteins. An exception to this is a possible 58-kDa protein. The gene for this protein has been cloned and sequenced and expressed in *Bacillus subtilis* and *Escherichia coli* by Thorne et al. (1986). Though the expressed protein is toxic to mosquitoes, either it is not expressed in *B. t. i.* or its relative abundance in the parasporal body is very low. Its role in the toxicity of *B. t. i.* therefore remains unclear.

The second basic property of the parasporal body that emerges from the above studies is that the number of proteins detected by SDS-PAGE increases substantially if the parasporal bodies are solubilized in alkali, particularly at pH 9.5 or above (fig. 3.4). Most of the additional proteins have molecular weights in the range of the minor proteins detected in intact parasporal bodies. Moreover, solubilization under alkaline conditions results in a significant decrease in the amounts of 27-, 65-, 128-, and 135-kDa proteins. Thus, the increase in minor proteins that occurs concomitantly with the decrease in major proteins is thought to be due to proteolytic degradation brought about by bacterial alkaline proteases associated with the parasporal body (Chilcott, Kalmakoff, and Pillai 1983; Pfannenstiel et al. 1984). Similar proteases likely exist in the alkaline midguts of dipterous larvae and may increase the rate of toxin activation in vivo. In fact, Charles and de Barjac (1983) demonstrated a gradual but complete dissolution of the *B.t.i.* parasporal body in the larval midgut of *Aedes aegypti,* suggesting a major involvement of proteolytic gut enzymes.

3.6 TOXICITY

The toxicity of *B.t.i.*'s parasporal body varies considerably depending on whether it is intact or solubilized and on how it is assayed. When ingested, either intact or solubilized, the parasporal body is toxic to mosquitoes, black flies, and several other nematocerous dipterans, but not to other insects nor to vertebrates against which it has been tested in feeding studies (Lacey and Undeen 1986). In addition to being toxic to mosquitoes, however, the solubilized parasporal body is toxic to mice upon injection and cytolytic to many types of cells in vitro. Thomas and Ellar (1983a) were the first to detect the broad cytolytic activity of the solubilized parasporal body; they attributed this activity to a 25-kDa protein that is cleaved from the 27-kDa protein when the parasporal body is solubilized under alkaline conditions. They, along with several other investigators, also attributed the mosquiticidal toxicity of *B.t.i.* to the 25-kDa protein (Thomas and Ellar 1983a, 1983b; Yamamoto, Iizuka, and Aronson 1983; Davidson and Yamamoto 1984; Armstrong, Rohrmann, and Beaudreau 1985; Insell and Fitz-James 1985; Sriram, Kamdar, and Jayaraman 1985). Alternatively, other investigators have reported that the 25-kDa protein is cytolytic but not mosquiticidal, with the latter activity residing in a mixture of peptides of 31–35 kDa (Cheung and Hammock 1985b), or the 65-kDa protein (Hurley et al. 1985; Lee, Eckblad, and Bulla 1985), or the 130- to 135-kDa proteins (Bourgouin, Klier, and Rapoport 1986; Sekar 1986; Visser et al. 1986). In other studies it has been suggested that the high toxicity of the parasporal body is due not to a single protein, but rather to a synergistic interaction of the 25-kDa protein with one or more of the higher molecular weight proteins (Wu and Chang 1985; Ibarra and Federici 1986b; Chilcott and Ellar 1988).

At present, there is broad agreement that the 25-kDa protein accounts for most if not all of the cytolytic activity the solubilized parasporal body exhibits against a very broad range of cell types in vitro. Some disagreement remains over whether this protein is also mosquiticidal, and if so, to what degree. Additionally, though current evidence indicates that 65-, 128-, and 135-kDa proteins are mosquiticidal, the role they play in toxicity and the extent to which they may be potentiated by the 25-kDa protein have not been clearly determined. The interpretation of toxicological data on the parasporal body and individual proteins is complicated by the variety of methods used to prepare and assay preparations for toxicity, as well as the manner in which the data are reported. The species of mosquito and the number and instar of larvae used in bioassays vary greatly among different studies, as does the method of toxic preparation; that is, conditions for solubilization or form in which putative toxins are assayed (intact, solubilized, bound to latex beads, or precipitated). In some studies data are analyzed statistically, whereas in others only raw data are reported. Moreover, in studies where a gene encoding a toxin has been cloned and expressed in another bacterial species, more often than not, very little quantitative analysis of toxicity is reported, making it virtually impossible to draw conclusions regarding the relative toxicity of different proteins. In most studies, the tendency has been to assign larvicidal toxicity to a single protein while excluding others from having any such role. This is exemplified by phrasing such as the "mosquiticidal protein" or *the* "δ-endotoxin" of *B. t. i.* Nevertheless, despite problems encountered in the interpretation of data, comparison of the results obtained for each parasporal body protein indicates that no single protein by itself is as toxic to mosquitoes as the intact parasporal body. The question thus becomes, what accounts for the high toxicity of the parasporal body? Though a definitive answer to this question is not possible at present, a comparative analysis of the data from a series of studies does suggest the high toxicity is due to the interaction of two or more parasporal body proteins. The evidence in support of this hypothesis is reviewed below through an examination of the pertinent data available on the properties and toxicity of the parasporal body and each of its major proteins.

3.6.1 Intact or Solubilized Parasporal Body

One of the most interesting properties of *B. t. i.* is the very high toxicity of the parasporal body to a wide range of mosquito and black fly species. Although toxicity varies to some extent among different species due to real differences in susceptibility as well as differences in feeding behavior, the data for *Ae. aegypti* shown in table 3.2, selected on the basis of suitability for comparison, reflects the trends observed for most species. Basically, the purified, intact parasporal body has an LC_{50} in the range of <4 to 20 ng/ml for fourth instars. Earlier instars are more susceptible. For example, an LC_{50} of around 1 ng/ml has been reported for first instars (Ibarra and Federici 1986b). Once the parasporal body is solubilized in alkali, however, the toxicity drops dramatically. Typical reductions in toxicity range from 50-fold to

TABLE 3.2.
Toxicity of the Intact and Solubilized Parasporal Body of *Bacillus thuringiensis* subsp. *israelensis* to Larvae of *Aedes aegypti*

LC_{50} (ng/ml)			
Intact	**Solubilized**	**Instar**	**Reference**
0.19	—	2nd	Tyrell et al. 1979
1–5	—	2nd or 3rd	Insell and Fitz-James 1985
3.5	—	4th	Ibarra and Federici 1986a
0.3	1.0	3rd or 4th	Chilcott and Ellar 1988
5–50	600–6,000	3rd	Thomas and Ellar 1983a
—	3,630	3rd	Cheung and Hammock 1985b
17.1	7,600	3rd	Cheung and Hammock 1985a
18	1,280	4th	Klowden, Held, and Bulla 1983

greater than 100-fold (table 3.2). These reductions are attributed, not to a real loss in toxicity, but rather to a very reduced level of toxin ingestion by larvae due to their filter-feeding behavior. The toxicity of test preparations can be increased by attaching solubilized toxin to latex beads (Schnell, Pfannenstiel, and Nickerson 1984), by encapsulation (Cheung and Hammock 1985a), or by precipitating the proteins (Insell and Fitz-James 1985; Chilcott and Ellar 1988). These techniques, however, do not restore toxicity to levels characteristic of the intact parasporal body.

In regard to cells in culture, the intact parasporal body is not toxic. However, after solubilization in alkali, the parasporal body at concentrations in the range of 2,500–25,000 ng/ml causes lysis of a wide range of cells, including erythrocytes, within 15–120 minutes (Thomas and Ellar 1983a, 1983b; Davidson and Yamamoto 1984; Cheung and Hammock 1985b; Chilcott and Ellar 1988). Additionally, Thomas and Ellar (1983a) showed that solubilized parasporal body preparations are lethal to suckling mice when injected at concentrations of 25,000 ng/g of body weight. Using solubilized parasporal bodies, Thomas and Ellar (1983b) provided good evidence that *B.t.i.*'s cytolytic activity was due to a detergentlike action in which the toxic disrupted membranes by binding to specific lipids. They postulated that the 27-kDa protein was the toxin responsible for cytolytic activity, and acted by binding to the fatty acids phosphatidyl choline and sphingomyelin, among others, as long as these contained unsaturated acyl residues.

3.6.2 The 27-kDa Protein

Of the various parasporal body proteins, the 27-kDa protein has received the most study because it was the first implicated in mosquiticidal

TABLE 3.3.
Toxicity of Different Parasporal Body Proteins from *Bacillus thuringiensis* subsp. *israelensis* to Mosquito Larvae

Protein	LC_{50} (ng/ml) Particulate	Solubilized	Instar[a]	Reference
25/27 kDa	125[b]	—	3rd	Ward et al. 1986
	—	12,500	1st	Davidson and Yamamoto 1984
	—	18,500[c]	3rd	Armstrong, Rohrmann, and Beaudreau 1985
	—	>25,000	3rd	Cheung and Hammock 1985b
	>1,000[d]	>10,000	3rd	Visser et al. 1986
30–35 kDa		220	1st	Yamamoto, Iizuka, and Aronson 1983
	—	6,680	3rd	Cheung and Hammock 1985b
	12.0	—	3rd/4th	Chilcott and Ellar 1988
65 kDa	43	—	1st	Ibarra and Federici 1986a
	600[e]	—	3rd	Insell and Fitz-James 1985
	—	180	4th	Hurley et al. 1985
	—	400	1st	Yamamoto, Iizuka, and Aronson 1983
	—	720	3rd	Kim, Ohba, and Aizawa 1984
	4.0[e]	—	3rd/4th	Chilcott and Ellar 1988
128, 135 kDa	40[d]	350	3rd	Visser et al. 1986
	32	—	3rd/4th	Chilcott and Ellar 1988
128	5,000–10,000[b]	—	3rd/4th	Ward and Ellar 1988
135	500–1,000[b]	—	3rd/4th	Ward and Ellar 1988

[a]Various species.
[b]Expressed products of cloned genes.
[c]Estimated from original data.
[d]Bound to latex beads.
[e]Precipitated.

activity and because it differed markedly in size and cytolytic properties from the 135-kDa toxin proteins of other subspecies of *B. thuringiensis*. This protein, which is the most abundant of those in the parasporal body, is the only one known to cause lysis of a wide range of invertebrate and vertebrate cells in vitro. It is therefore considered the protein primarily responsible for the cytolytic activity of the solubilized parasporal body (Thomas and Ellar 1983a; Davidson and Yamamoto 1984; Armstrong, Rohrmann, and Beaudreau 1985; Cheung and Hammock 1985b; Hurley et al. 1985; Visser et al. 1986; Chilcott and Ellar 1988). Due to the unique, broadly cytolytic properties of this protein, Höfte and Whiteley (1989) have termed the gene that encodes it the *cyt A* gene, differentiating it from the *cry* I–IV genes, which encode all other known endotoxins of *B. thuringiensis*.

When parasporal bodies are solubilized in the course of purifying the 27-kDa protein, this protein is cleaved to 25 kDa by proteases associated with the parasporal body. The cleaved product is quite resistant to further proteolysis and retains the cytolytic activity (Armstrong, Rohrmann, and Beaudreau 1985). Against cells in culture, or erythrocytes, full lysis occurs within an hour at concentrations ranging from 1,000 to 4,000 ng/ml (Davidson and Yamamoto 1984; Cheung and Hammock 1985b). Thus, the level of cytolytic activity is in the range of what would be expected for a purified fraction of the parasporal body considered to be responsible for this activity.

The mosquiticidal activity of the purified 27-kDa protein, alternatively, is much lower than what would be expected if it were the primary toxin (table 3.3). For example, Ward et al. (1986) determined an LC_{50} of 125 ng/ml for this protein. This value, which is only 4–5% the toxicity of the intact parasporal body, was obtained for inclusions of the 27-kDa protein isolated from *B. subtilis* cells in which the cloned gene for this protein had been expressed. More recently, Chilcott and Ellar (1988) reported a similar value (LC_{50} = 115 ng/ml) for the 27-kDa protein isolated by column chromatography from the parasporal body and precipitated for assay. The 25-kDa protein derived from the 27 had an LC_{50} of 74 ng/ml.

When bioassayed in a soluble form, the toxicity of the 27- and 25-kDa proteins is 4-fold to 10-fold less toxic than the solubilized parasporal body (table 3.3). These results show that the 27-kDa protein is mosquiticidal, but not nearly toxic enough to account for the toxicity of the parasporal body.

3.6.3 The 65-kDa Protein

Before discussing the toxicity of the 65-kDa protein, it must be noted that there are very likely three proteins of about this size in parasporal body preparations, particularly in alkali-solubilized preparations in which proteases have not been inactivated. The predominant protein observed by SDS-PAGE in this size range is the native 65-kDa protein found in the bar-shaped inclusion described by Ibarra and Federici (1986b). This protein is assembled into the bar-shaped body as a 65-kDa protein; it is not a product of proteolytic cleavage resulting from solubilization as has been suggested in some studies. The gene encoding this protein, termed the *cryIVD* gene using the nomenclature of Höfte and Whiteley (1989), has been cloned and sequenced by Donovan, Dankocsik, and Gilbert (1988); based on the nucleotide sequence, this protein has a mass of 72.4 kDa. The other proteins in this size range are approximately 67 kDa, migrate slightly slower than the 65-kDa protein in SDS-polyacrylamide gels, and are much more prominent in parasporal bodies solubilized in alkali (fig. 3.4). Thus, these are probably cleavage products resulting from proteolysis, respectively, of the 128- and 135-kDa proteins, a situation similar to that which occurs when the 135-kDa proteins of other subspecies of *B. thuringiensis* are solubilized in the presence of protease. Realization that there are three different proteins of approx-

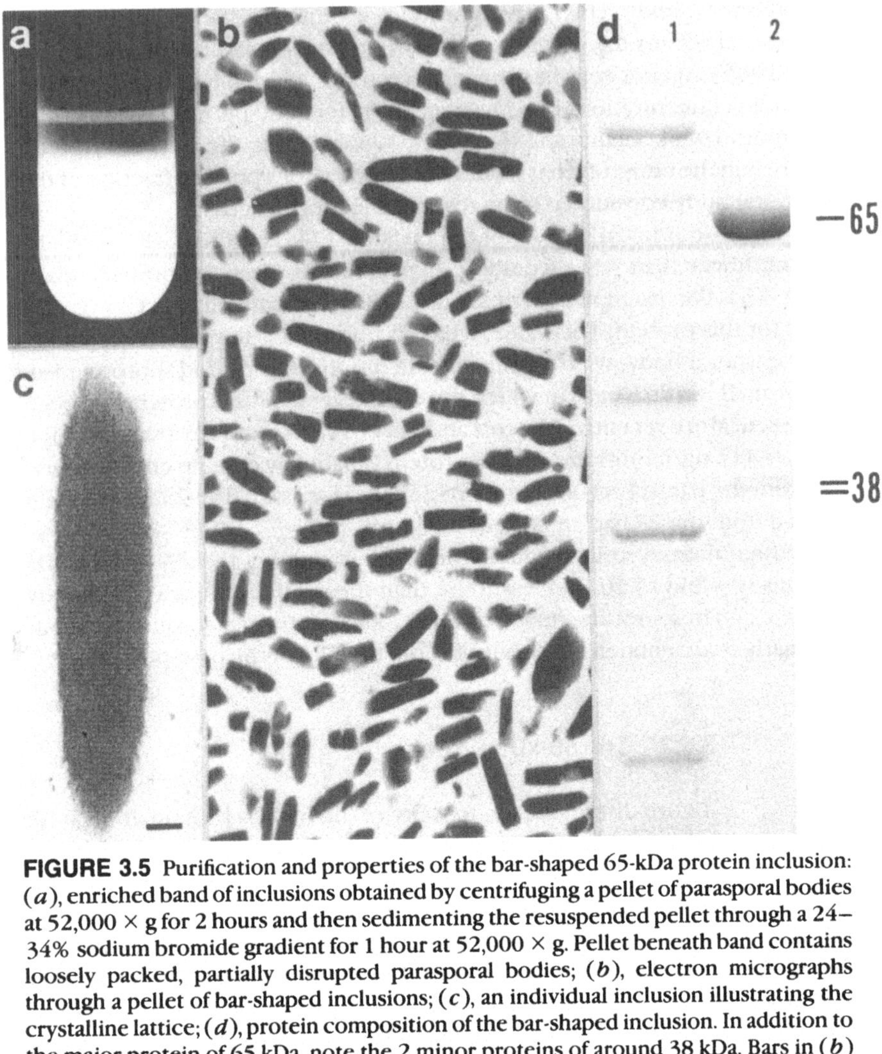

FIGURE 3.5 Purification and properties of the bar-shaped 65-kDa protein inclusion: (*a*), enriched band of inclusions obtained by centrifuging a pellet of parasporal bodies at 52,000 × g for 2 hours and then sedimenting the resuspended pellet through a 24–34% sodium bromide gradient for 1 hour at 52,000 × g. Pellet beneath band contains loosely packed, partially disrupted parasporal bodies; (*b*), electron micrographs through a pellet of bar-shaped inclusions; (*c*), an individual inclusion illustrating the crystalline lattice; (*d*), protein composition of the bar-shaped inclusion. In addition to the major protein of 65 kDa, note the 2 minor proteins of around 38 kDa. Bars in (*b*) and (*c*) represent, respectively, 500 and 100 nm.

imately the same size, each of which possibly, if not probably, has different toxicological properties, is important because protein fractions of the 65-kDa protein purified by column chromatography after solubilization of parasporal bodies in alkali will contain all three proteins. Thus, data on the toxicity of the 65-kDa protein purified by column chromatography probably include the effects of different proteins.

As noted above, the primary 65-kDa protein of the parasporal body is

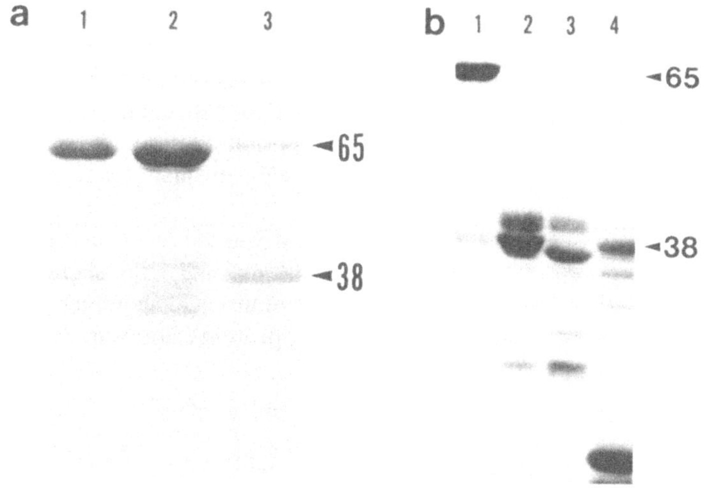

FIGURE 3.6 Analysis of the 65-kDa protein inclusion by SDS-PAGE after solubilization in alkali or digestion with proteases: (*a*), protein composition of (1) intact inclusions, (2) inclusions solubilized in alkali for 1 hour at 37°C prior to electrophoresis, and (3) insoluble residue remaining after solubilization; (*b*), degradation and digestion of the 65-kDa protein by trypsin and α-chymotrypsin of (1) intact inclusion, (2) inclusion solubilized at pH 10.5 and stored at 4°C for 4 weeks, (3) inclusion digested with trypsin, or (4) α-chymotrypsin.

located in a polyhedral, bar-shaped inclusion, which often appears rectangular, or, in transverse sections, with edges that taper. (fig. 3.5). By subjecting parasporal bodies to sonication or high gravitation force, this inclusion can be dislodged and then purified on sodium bromide gradients (Ibarra and Federici 1986b). Highly purified inclusions of this type analyzed by SDS-PAGE without prior solubilization in alkali exhibit a single major protein of 65 kDa and two minor proteins around 38 kDa (fig. 3.5). If solubilized in alkali (50 mM $Na_2CO_3 \cdot HCl$; pH 10.5) and then analyzed by SDS-PAGE, the protein profile is basically the same. Interestingly, the material that does not dissolve in alkali contains remnants of the envelope and a higher concentration of one of the approximately 38-kDa proteins (fig. 3.6a). If solubilized in alkali and allowed to stand at 4°C for several weeks, or if cleaved with trypsin or α-chymotrypsin, the 65-kDa proteins yields a series of peptides in the range of 36–40 kDa (fig. 3.6b). Thus, there are two different proteins of around 38 kDa associated with the 65-kDa inclusion: one, apparently a native protein, bound to the envelope, and the second a cleavage product of the 65-kDa protein. The envelope-bound protein may be the same 38-kDa protein detected at the initiation of parasporal body formation by Lee, Eckblad, and Bulla (1985).

The biochemical properties of the 65-kDa protein have also been studied by Chestukina et al. (1985). By treating parasporal bodies with 100 mM DTT, they were able to solubilize the 27- and 130-kDa proteins. The 65-kDa protein remained in the insoluble residue, but could be dissolved in 0.05 M NaOH. The purity of the 65-kDa protein obtained through this procedure was suitable for amino acid sequencing and permitted these investigators to determine the following N-terminal sequence for the first 16 amino acids of this protein:

Met-Glu-Asn-Xaa-Pro-Leu-Asp-Thr-Leu-Ser-iLeu-Val-Asn-Glu-Thr-Asp

Except for the amino acids designated at positions 3 and 5, this sequence corresponds to the first 16 amino acids of the amino acid sequence deduced from the nucleotide sequence of the 65-kDa protein (Donovan, Dankocsik, and Gilbert 1988).

With respect to toxicity, the bar-shaped, 65-kDa protein inclusion is mosquiticidal, but, as in the case of the 27-kDa inclusion noted above, not nearly as toxic as the intact parasporal body (table 3.3). Ibarra and Federici (1986b) reported an LC_{50} of 43 ng/ml for the 65-kDa inclusion against first instars of *Ae. aegypti*, whereas the intact parasporal body had an LC_{50} of 0.66 ng/ml. The actual toxicity of the 65 kDa is probably higher because the inclusions tested contained envelope and were much smaller than parasporal bodies, presenting the possibility they may not have been ingested as efficiently as the latter. Nevertheless, it is unlikely that elimination of minor contaminants or increasing the amount of protein ingested will show the 65-kDa protein in particulate form to be as toxic as the parasporal body. For example, Insell and Fitz-James (1985) found an LC_{50} of 600 ng/ml against second and third instars of *Ae. aegypti* of a particulate 65-kDa protein fraction precipitated from parasporal bodies solubilized in alkali. More recently, Chilcott and Ellar (1988) also demonstrated that the 65-kDa protein was much less toxic than the intact parasporal body. They obtained LC_{50}s of 0.3 ng/ml and 4 ng/ml, respectively, for the intact parasporal body and purified, precipitated 65-kDa protein against five-day-old larvae of *Ae. aegypti.*

As might be expected, the 65-kDa protein in a solubilized form is less toxic than the inclusion containing this protein. Reported LC_{50} values range from 180 to 720 ng/ml against various instars of *Ae. aegypti* (table 3.3). Although the LC_{50} of 180 ng/ml obtained by Hurley et al. (1985) against the fourth instar indicates a relatively high toxicity, this value likely refers to a mixture of different proteins derived from the 65-kDa protein inclusion and 128/135-kDa proteins. The 65-kDa protein fraction tested was purified by column chromatography from parasporal bodies solubilized in sodium hydroxide at pH 12.

In regard to the toxicity of possible cleavage products of the 65-kDa protein, Cheung and Hammock (1985b) reported an LC_{50} of 6,680 ng/ml for a 31-, 34-, 35-kDa protein fraction against third instars of *Ae. aegypti.* Although this toxicity appears low, using the same assay system they obtained LC_{50} values of 3,630 ng/ml for solubilized parasporal bodies and greater than

25,000 ng/ml for the 25-kDa protein. Thus, the 31- to 35-kDa proteins were almost half as toxic as parasporal bodies and more than three times as toxic as the 25-kDa protein. The origin of 31- to 35-kDa proteins is not known, but given variation in molecular sizes obtained by gel electrophoresis, these proteins are in the size range of the products obtained by proteolytic cleavage of the 65-kDa protein. More recently, Chilcott and Ellar (1988) reported an LC_{50} of 12 ng/ml for the 30- to 35-kDa protein complex (table 3.3). Interestingly, they showed that this complex was toxic to mosquito cells in vitro, but that the 65-kDa protein was not, indicating the latter may be a protoxin.

3.6.4 The 128- and 135-kDa Proteins

Although initial studies focused on the 27- and 65-kDa proteins as being the ones responsible for the mosquiticidal activity of *B.t.i.,* there is now abundant evidence that the 126- and 135-kDa proteins are also substantially mosquiticidal. At this point, however, relatively little information exists regarding the specific toxicity of each of these proteins, and that which does exist is ambiguous even though genes encoding proteins corresponding to these masses have been cloned and sequenced by several groups (table 3.1). Evaluation of the data in the literature on the toxicity of these proteins is complicated by several factors, including the identity of the proteins assayed, the form in which they were assayed (crystalline, solubilized, or precipitated), the host species and stage they were assayed against, and whether the proteins were isolated directly from *B.t.i.* or from other bacterial species in which one of the cloned genes had been expressed. The discussion below will illustrate some of these problems and summarize the data, which at this point basically indicate the 135-kDa protein is the most toxic of the two.

In early assays of these proteins, when it was not clear that there were two proteins in this size range, they were purified by column chromatography after solubilization of parasporal bodies and thus treated as a single protein, that is, the 130-kDa protein (Wu and Chang 1985; Visser et al. 1986). In their study, Visser et al. (1986) reported an LC_{50} of 40 ng/ml against third instars of *Anopheles stephensi* for the 130-kDa protein bound to latex beads. This value clearly places these proteins in the range of significant mosquiticidal toxicity (table 3.3). Wu and Chang (1985) did not find the mixture of these two proteins to be nearly so toxic, but their results are difficult to assess because the data were not analyzed statistically. In any case, the toxicity they determined for the mixture was similar to levels they obtained for the 65-kDa protein; more importantly, both the 130- and 65-kDa proteins tested separately were more toxic than the 27-kDa protein.

In a later study, Chilcott and Ellar (1988), who purified the two high molecular weight proteins by column chromatography and precipitated them for bioassay, reported an LC_{50} of 32 ng/ml for the 130-kDa against five-day-old larvae of *Ae. aegypti.* This toxicity is similar to that reported by Visser et al. (1986).

Within the past two years, the genes encoding the 128- and 135-kDa proteins have been cloned and expressed in several bacterial species (table 3.1), but in most studies little information is provided about the specific toxicity of the product of the cloned gene. Studies by Delecluse et al. (1988) and Ward and Ellar (1988), however, do provide data useful in comparing and evaluating the toxicity of each of these proteins. In essence, Ward and Ellar (1988) found that the product of the gene (pCH 130) encoding a 134-kDa protein was about 10-fold more toxic to *Ae. aegypti* than the product of the gene (pPC 130) encoding a protein of 128-kDa. The LC_{50} values they obtained for these gene products were 5–1 μg/ml (pCH 130 expressed in *E. coli*) and 5–10 μg/ml (pPC 130 expressed in *B. subtilis*). The gene products were purified as inclusions from the bacteria in which they were produced and assayed against five-day-old larvae of *Ae. aegypti.* It is notable that although these proteins were assayed in a particulate form, they were at a minimum 10- to 100-fold less toxic than the 130-kDa protein based on the values reported by Visser et al. (1986) and by Chilcott and Ellar (1988). Potential explanations for these differences are many and include differences in the processing of these proteins in foreign host cells and difficulties in purifying and accurately characterizing the proteins by bioassay.

In the study by Delecluse et al. (1988), a difference in the host range of the cloned 128-kDa gene product was detected. They found that the product of the 128-kDa protein expressed in *E. coli* was highly toxic to larvae of *Ae. aegypti* and *An. stephensi,* but of only low toxicity to *Culex pipiens.* They determined, however, that when the product of ORF 1 (Thorne et al. 1986), a gene that may not even be expressed in *B. t. i.,* was expressed with or added to the 128-kDa protein, the mixture was highly toxic to *Cx. pipiens.*

In summary, it appears at present that the 135-kDa protein is considerably more toxic than the 128-kDa protein, at least to *Ae. aegypti,* and that there may be differences in the specific toxicity of these proteins to different species of mosquitoes. This is not surprising considering that similar findings have been reported for *B. t.* toxins active against lepidopterous insects (Höfte and Whiteley 1989).

3.6.5 Synergistic Interaction of Toxic Proteins

The above data on the four major parasporal body proteins of *B. t. i.* can be summarized by stating that each is mosquiticidal, but none, even in a particulate form, is as toxic as the parasporal body. If only one of the proteins is responsible for mosquiticidal activity and acts alone, it would be expected that this protein on a weight basis would be from two to seven times more toxic than the parasporal body, depending on which protein is designated the toxin. However, the reverse has been found; the purest particulate preparations tested are considerably less toxic than the parasporal body. Differences in the specific activity of major parasporal body proteins do appear to be real (the 65- and 130-kDa proteins are more toxic than the 25-kDa protein), but these differences cannot account for the high toxicity of the parasporal body.

One possible explanation for the parasporal body's high toxicity is that a minor protein, native or a cleaved product, of extremely high toxicity is responsible for mosquiticidal activity. However, there is no evidence that such a protein exists. Thorne et al. (1986) and Garduno et al. (1988) have described a 58-kDa mosquiticidal protein expressed by a cloned *B. t. i.* gene in *E. coli* and *B. subtilis*. Though the gene that encodes this protein is structurally related to the 128-kDa protein gene (Garduno et al. 1988; Delecluse et al. 1988), it is not certain that this gene is expressed in *B. t. i.*

Another possible explanation for the parasporal body's high toxicity is that two or more of its proteins interact synergistically, yielding a higher toxicity than would be expected based on the specific toxicity of individual proteins. Wu and Chang (1985) first presented evidence for such an interaction, and studies by Ibarra and Federici (1986b) and, more recently, Delecluse et al. (1988) and Chilcott and Ellar (1988) also provide evidence of synergistic interactions among two or more proteins as an explanation of the much higher toxicity of the parasporal body in comparison to the individual proteins it contains. Although the toxicity of Wu and Chang's (1985) preparations were low overall, generally in the range of 1,000 ng/ml, using parasporal body proteins separated by column chromatography, they reported that mixtures of the 27-, and 65-and/or 130-kDa proteins were substantially more toxic on a weight basis than any of these proteins tested alone. They suggested that the 27-kDa protein was synergistic for the 65- and 130-kDa proteins. Ibarra and Federici (1986b) also provided evidence for synergism in their study of the 65-kDa protein. They found that the toxicity of the bar-shaped inclusion containing this protein could be correlated with the amount of the 27-kDa protein associated with the inclusion. Inclusion preparations that contained minor quantities of the 27-kDa protein had an LC_{50} of about 9 ng/ml for first instars of *Ae. aegypti*, whereas preparations that contained almost none of this protein had an LC_{50} of 43 ng/ml (fig. 3.7). The LC_{50} of 9 ng/ml is one of the highest toxicities reported for a purified component of the parasporal body, but is still much less toxic than the LC_{50} of 0.66 ng/ml they obtained for the parasporal body. Based on these results, Ibarra and Federici (1986b) suggested the high toxicity of the parasporal body may result from potentiation of the toxicity of the 65- and 130-kDa proteins by the 27-kDa protein.

More recently, Chilcott and Ellar (1988) also provided evidence for synergism between parasporal body proteins. Their data showed that in all cases, addition of the 27-kDa protein to each of the other major parasporal body

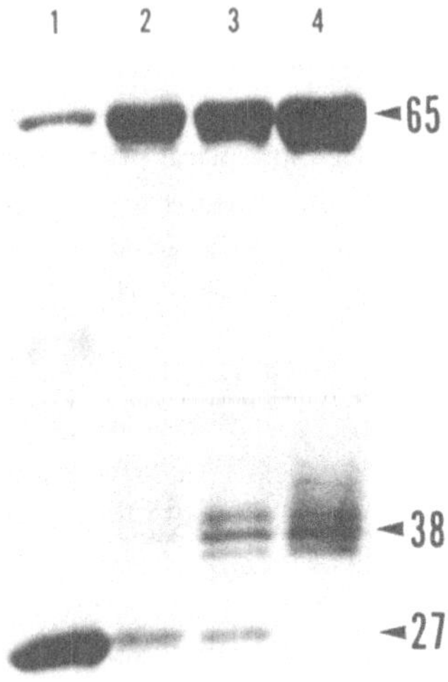

FIGURE 3.7 SDS-PAGE of 65-kDa protein inclusion preparations contaminated with different amounts of the 27-kDa protein. Toxicity was determined through bioassays with first-instar larvae of *Ae. aegypti* as described by Ibarra and Federici (1986b). Gels were stained with silver. Lane 1, intact parasporal body, LC_{50} = 0.66 ng/ml. Lanes 2–4 contain different 65-kDa protein inclusion preparations arranged in order of decreasing levels of toxicity. LC_{50}s for lanes 2–4 were, respectively, 8.83, 10.53, and 43 ng/ml. Note that the lower the amount of 27-kDa protein present in the preparation, the lower the toxicity.

proteins potentiated the toxicity of the latter. The best potentiation/highest toxicity they found with any mixture occurred when they mixed the 65- and 27-kDa proteins at a ratio of 0.75:1.0 (w/w), wherein they obtained an LC_{50} of 2.0 ng/ml against five-day-old larvae of *Ae. aegypti.*

In the study by Delecluse et al. (1988), it was shown that the low toxicity of the 128-kDa protein (or its truncated 65-kDa active portion) to *Cx. pipiens* was enhanced significantly when combined with the product of ORF 1 (the 58-kDa protein of Thorne et al. 1986), which was not toxic to this species when tested separately.

3.7 MOSQUITICIDAL PARASPORAL BODIES OF OTHER SUBSPECIES OF *B. thuringiensis*

Within the past few years, several strains of subspecies of *B. thuringiensis* other than *israelensis* have been reported that produce parasporal bodies containing proteins toxic to mosquitoes. These subspecies include *morrisoni* (H8a,8b), *darmstadiensis* (H10), and *kurstaki* (H3a,3b).

The most toxic isolate of these subspecies is the PG-14 isolate of *Bacillus thuringiensis* subsp. *morrisoni* (fig. 3.8). PG-14 was isolated from a canal in Cebu City, Philippines (Padua, Ohba, and Aizawa 1984), and is as toxic to mosquitoes as *B. t. i.* Studies have shown that the parasporal body of PG-14, like that of *B. t. i.,* is spherical and contains proteins of 27, 65, 128, and 135 kDa assembled into a series of inclusions similar to those found in *B. t. i.* (Ibarra and Federici 1986a). Additionally, the PG-14 27-kDa protein is highly related immunologically to the 27-kDa protein of *B. t. i.,* as are the 65-kDa pro-

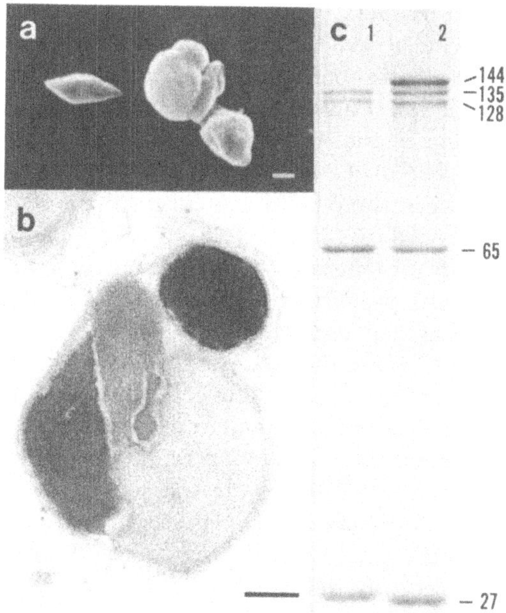

FIGURE 3.8 Characteristics of the mosquiticidal parasporal body of *Bacillus thuringiensis* subsp. *morrisoni* (PG-14): (*a*), scanning electron micrograph of the basically spherical parasporal body and a dislodged bipyramidal crystal. Bar = 200 nm; (*b*), transmission electron micrograph illustrating 4 different inclusions of the parasporal body bound together by an envelope. Bar = 200 nm; (*c*), comparison of the protein composition by SDS-PAGE of the parasporal bodies from *B.t.i.* (1) and *B.t.m.* PG-14 (2). PG-14 contains a protein of 144 kDa not found in *B.t.i.*.

teins of these two isolates (Gill et al., 1987), and the genes encoding the 27-kDa protein differ by only a single base (Galjart, Sivasubramanian, and Federici 1987). Thus, the toxicity of PG-14 to mosquitoes is probably largely due to parasporal body proteins it produces that have toxicological properties similar to those of *B.t.i.* However, an interesting difference between *B.t.i.* and PG-14 is that the parasporal body of the latter contains an additional inclusion of the bipyramidal type composed of a protein or proteins with an apparent molecular size of about 144 kDa. The role this protein plays in mosquiticidal toxicity has not yet been determined.

The mosquiticidal strains of *Bacillus thuringiensis* subsp. *darmstadiensis* were isolated from silkworm litter in Japan (Padua, Ohba, and Aizawa 1980). Preliminary studies of one isolate, 73-E-10-2, have shown that it produces a spherical parasporal body containing two types of inclusions and has a protein complement of about five major proteins with sizes of 25, 49, 70, 125, and 144 kDa (Mikkola et al. 1982; Federici et al. 1987). The spore–parasporal body complex has an LC_{50} of about 110 ng/ml against fourth instars of *Ae. aegypti* (Federici et al. 1987). Kim, Ohba, and Aizawa (1984) purified a 67-kDa protein from a spore–parasporal body complex of this isolate and determined an LC_{50} of 1,680 ng/ml for this protein against four-day-old *Ae. aegypti* larvae, a value about 20 times less toxic than that which they obtained for a 67-kDa protein of *B. t. i.* Additionally, they found no immunological relatedness between the 67-kDa proteins of this isolate and *B. t. i.* The apparent low toxicity of the 67-kDa protein of 73-E-10-2 isolate indicates that other parasporal body proteins must be involved in mosquito toxicity.

The mosquiticidal properties of strain HD1, *Bacillus thuringiensis* subsp. *kurstaki,* are also apparently due to a protein of 67 kDa. Yamamoto and McLaughlin (1981) were able to separate two distinct proteins from the bipyramidal crystals of HD1; P1, a protein of 135 kDa, and P2, a protein of 65 kDa. P1 was only toxic to lepidopterous larvae, whereas P2 was toxic to both lepidopterous larvae and mosquitoes. An LC_{50} of 1,480 ng/ml was determined for the solubilized P2 protein against second instars of *Aedes taeniorhynchus.* Using bioassays and peptide mapping, Yamamoto and McLaughlin (1981) provided sound evidence that P2 was not a cleavage product of P1. More recently, the gene encoding this protein has been cloned and sequenced by Donovan et al. (1988) and Widner and Whiteley (1989). The mass of the P2 protein deduced from the nucleotide sequence is approximately 71 kDa.

Another isolate of *B. thuringiensis* reported to be mosquiticidal is the type-strain of subspecies *kyushuensis* (H11a,11c). This strain produces a parasporal body similar in appearance to that of strain 73-E-10-2, and although toxic to mosquitoes, no information is available regarding the identity or properties of the proteins responsible for the mosquiticidal activity (Ohba and Aizawa 1979).

3.8 DISCUSSION

The further characterization of *B. t. i.*'s parasporal body proteins in regard to their mosquiticidal toxicity will likely continue until the basis for the parasporal body's high toxicity is fully understood. Aside from its academic interest, resolution of the controversy over toxicity has practical implications. If only one protein accounts for mosquiticidal activity, then it should be possible to engineer strains that produce only this protein, thereby improving efficacy and cost-effectiveness. For example, Hurley et al. (1985) reported that the 65-kDa protein accounted for virtually all the mosquiticidal activity of the parasporal body. They found this protein represented about one-seventh of the parasporal body protein, and in pure form was about seven times as toxic. Although most other studies contradict this hypothesis, if true it should be possible to develop strains in which the total amount of parasporal body protein remains the same but consists only of the 65-kDa protein. Such strains would be seven-fold more effective on a weight basis than wild-type *B. t. i.* and would facilitate the development of more effective formulations.

If no single protein is responsible for *B. t. i.*'s mosquiticidal properties, then the specific interactions and functions of each parasporal body protein become of interest, particularly the role of the 27-kDa protein. A useful observation here, perhaps, is that the mosquiticidal isolate PG-14 of subspecies *morrisoni* contains the same four proteins (27, 65, 128, 135 kDa) found in *B. t. i.* The existence of this specific protein mixture in different subspecies isolated from widely separately geographical regions of the world is unlikely to be fortuitous. This suggests that a mixture of proteins is maintained through selection and may be necessary for full toxicity, though perhaps all four proteins are not essential. Another interesting observation in regard to *B. t. i.* and PG-14 is that, in both, an envelope binds the different protein inclusions together. The function of this envelope is not known, but it could be to hold the inclusions together, ensuring that proteins essential for high toxicity are consumed simultaneously.

The combination of several toxins within a single parasporal body is in itself an interesting observation. It is well known that *B. t. i.* has a very broad host range among mosquito and black fly species, a property that may be accounted for by this mixture of toxins. A complex of toxins within a single bacterial strain may also prevent or delay the development of resistance in the host. Similar "strategies" have proven useful in delaying resistance to chemical insecticides in insect populations (Georghiou and Taylor 1986).

Determination of the function of each of the *B. t. i.* parasporal body proteins could also, obviously, prove of more than academic interest. The 128- and 135-kDa proteins, as well as their proteolytic cleavage products, are very similar in size to the typical insecticidal proteins and activated toxins of *B. thuringiensis* subspecies toxic to lepidopterous insects. Comparisons of the nucleotide sequences of the genes that encode toxins active against lepidopterous or dipterous insects indicate the toxins belong to the same gene family (Höfte and Whiteley 1989). Yet the high molecular weight proteins in *B. t. i.* are not toxic to lepidopterous insects and therefore provide useful subjects for studies of toxin specificity.

The proteins of approximately 65 kDa produced by several different subspecies of *B. thuringiensis* are also emerging as a potentially important group for studying toxin specificity. Different proteins of this size have been reported from subspecies ***kurstaki***, *israelensis*, *morrisoni*, *darmstadiensis*, and from strain *tenebrionis* (H8a,8b). The protein from the latter isolate is toxic to beetles (Bernhard 1986), whereas the others are toxic to mosquitoes, or, in the case of ***kurstaki***, lepidopterans also. However, aside from differences in host range, the specific toxicities of these proteins vary widely, although little is known about the molecular basis for these variations.

The 27-kDa protein, though perhaps not highly mosquiticidal, may ultimately turn out to be a very important protein. If this protein does potentiate the toxicity of other parasporal body proteins in *B.t.i.,* it may be capable of enhancing the toxicity of insecticidal proteins from other bacterial isolates. This could lead to more cost-effective strains with higher toxicities per unit weight and broader host ranges. Some concern has been raised in regard to the safety of this protein because of its mammalian and neuromuscular toxicity (Cheung and Hammock 1985b; Cheung et al. 1985). However, this type of toxicity is very probably due primarily to the detergentlike action this protein has on membranes. Because this toxicity has only been found under unnatural routes of exposure, it would not appear to be of any practical significance.

In summary, the discovery of *Bacillus thuringiensis* subsp. *israelensis* and its unique properties have resulted in a novel, environmentally sound insecticide for mosquitoes and black flies, as well as in the identification of a series of proteins, which, either directly or as models, may contribute to the development of even more effective insecticides. The function of the multitoxin complexity of the parasporal body and further delineation of the role that each protein plays in toxicity and synergism are particularly attractive areas for future research.

Acknowledgments

Research by B. A. Federici reported in this paper received partial financial support from the WHO/World Bank/UNDP Special Programme for Research and Training in Tropical Diseases and the University of California Mosquito Control Research Program.

References

Angsuthanasombat, C.; Chungjatupornchai, W.; Kertbundit, S.; Luxananil, P.; Settasatian, C.; Wilairat, P.; and Panyim, S. 1987. Cloning and expression of 130-kd mosquito-larvicidal delta-endotoxin gene of *Bacillus thuringiensis* var. *israelensis* in *Escherichia coli. Mol. Gen. Genet.* 208: 384–389.

Angus, T. A. 1965. Association with protein-crystalline inclusions of *Bacillus sotto ishiwata. Can. J. Microbiol.* 2: 122–131.

Armstrong, J. L.; Rohrmann, G. F.; and Beaudreau, G. S. 1985. Delta endotoxin of *Bacillus thuringiensis* subsp. *israelensis. J. Bacteriol.* 161 (1): 39–46.

Aronson, A. I.; Beckman, W.; and Dunn, P. 1986. *Bacillus thuringiensis* and related insect pathogens. *Microbiol. Rev.* 50: 1–24.

Barjac, H., de. 1978a. Toxicité de *Bacillus thuringiensis* var. *israelensis* pour les larves d'*Aedes aegypti* et d'*Anopheles stephensi. C. R. Acad. Sci.* (Paris) 286D: 1175–1178.

———. 1978b. Une nouvelle variété de *Bacillus thuringiensis* très toxique pour les moustiques: *B. thuringiensis* var. *israelensis* sérotype 14. *C. R. Acad. Sci.* (Paris) 286D: 797–800.

———. 1985. Collection de souches de *Bacillus thuringiensis* and *Bacillus sphaericus.* Paris: Institut Pasteur.

Bernhard, K. 1986. Studies on the delta-endotoxin of *Bacillus thuringiensis* var. *tenebrionis*. *FEMS Microbiol. Lett.* 33: 261–265.

Bourgouin, C.; Delecluse, A.; Ribier, J.; Klier, A.; and Rapoport, G. 1988. A *Bacillus thuringiensis* subsp. *israelensis* gene encoding a 125-kilodalton larvicidal polypeptide is associated with inverted repeat sequences. *J. Bacteriol.* 170: 3575–3583.

Bourgouin, C.; Klier, A.; and Rapoport, G. 1986. Characterization of the genes encoding the haemolytic toxin and the mosquitocidal delta-endotoxin of *Bacillus thuringiensis israelensis*. *Mol. Gen. Genet.* 205: 390–397.

Charles, J.-F., and Barjac, H. de. 1982. Sporulation et cristallogenèse de *Bacillus thuringiensis* var. *israelensis* en microscopie électronique. *Ann. Microbiol.* (Inst. Pasteur) 133A: 425–442.

———. 1983. Action des cristaux de *Bacillus thuringiensis* var. *israelensis* sur l'intestin moyen des larves de *Aedes aegypti* L., en microscopie électronique. *Ann. Microbiol.* (Inst. Pasteur) 134A: 197–218.

Chestukhina, G. G.; Zalunin, I. A.; Kostina, L. I.; Bormatova, M. E.; Klepikova, F. S.; Khodova, O. M.; and Stepanov, V. M. 1985. Structural features of crystal-forming proteins produced by *Bacillus thuringiensis* subspecies *israelensis*. *FEBS Lett.* 190 (2): 345–348.

Cheung, P.Y.K., and Hammock, B. D. 1985a. Micro-lipid-droplet encapsulation of *Bacillus thuringiensis* subsp. *israelensis* δ-endotoxin for control of mosquito larvae. *Appl. Environ. Microbiol.* 50 (4): 984–988.

———. 1985b. Separation of three biologically distinct activities from the parasporal crystal of *Bacillus thuringiensis* var. *israelensis*. *Curr. Microbiol.* 12: 121–126.

Cheung, P.Y.K.; Roe, R. M.; Hammock, B. D.; Judson, C. L.; and Montague, M. A. 1985. The apparent in vivo neuromuscular effects of the δ-endotoxin of *Bacillus thuringiensis* var. *israelensis* in mice and insects of four orders. *Pest. Biochem. Physiol.* 23: 85–94.

Chilcott, C. N., and Ellar, D. J. 1988. Comparative toxicity of *Bacillus thuringiensis* var. *israelensis* crystal proteins in vivo and in vitro. *J. Gen. Microbiol.* 134: 2551–2558.

Chilcott, C. N.; Kalmakoff, J.; and Pillai, J. S. 1983. Characterization of proteolytic activity associated with *Bacillus thuringiensis* var. *israelensis* crystals. *FEMS Microbiol. Lett.* 18: 37–41.

Chunjatupornchai, W.; Höfte, H.; Seurinck, J.; Angsuthanasombat, C.; and Vaeck, M. 1988. Common features of *Bacillus thuringiensis* toxins specific for Diptera and Lepidoptera. *Eur. J. Biochem.* 173: 9–16.

Davidson, E. W., and Yamamoto, T. 1984. Isolation and assay of the toxic component from the crystals of *Bacillus thuringiensis* var. *israelensis*. *Curr. Microbiol.* 11: 171–174.

Delafield, F. P.; Somerville, H. J.; and Rittenberg, S. C. 1968. Immunological homology between crystal and spore protein of *Bacillus thuringiensis*. *J. Bacteriol.* 96: 713–720.

Delecluse, A.; Bourgouin, C.; Klier, A.; and Rapoport, G. 1988. Specificity of action on mosquito larvae of *Bacillus thuringiensis israelensis* toxins encoded by two different genes. *Mol. Gen. Genet.* 214: 42–47.

Donovan, W. P.; Dankocsik, C. C.; and Gilbert, M. P. 1988. Molecular characterization of a gene encoding a 72-kilodalton mosquito-toxic crystal protein from *Bacillus thuringiensis* subsp. *israelensis*. *J. Bacteriol.* 170: 4732–4738.

Donovan, W. P.; Dankocsik, C. C.; Gilbert, M. P.; Gawron-Burke, M. C.; Groat, R. G.; and Carlton, B. C. 1988. Amino acid sequence and entomocidal activity of the P2 crystal protein. *J. Biol. Chem.* 263: 561–567. (Erratum, 1988, *J. Biol. Chem.* 264: 4740).

Federici, B. A.; Ibarra, J. E.; Padua, L. E.; Galjart, N. J.; and Sivasubramanian, N. 1987. Parasporal body of mosquitocidal subspecies of *Bacillus thuringiensis*. In *Biotechnology advances in invertebrate pathology and cell culture*, ed. K. Maramorosch, 115–131. Orlando: Academic Press.

Galjart, N. J.; Sivasubramanian, N.; and Federici, B. A. 1987. Plasmid location, cloning, and sequence analysis of the gene encoding a 27.3-kDa cytolytic protein from *Bacillus thuringiensis* subsp. *morrisoni* (PG-14). *Curr. Microbiol.* 16: 171–177.

Garduno, F.; Thorne, L.; Walfield, A. M.; and Pollock, T. J. 1988. Structural relatedness between mosquitocidal endotoxins of *Bacillus thuringiensis* subsp. *israelensis*. *Appl. Environ. Microbiol.* 54: 277–279.

Georghiou, G. P.; and Taylor, C. E. 1986. Factors influencing the evolution of resistance. In *Pesticide resistance strategies and tactics for management*, 157–169. Proc. Symp. on Management of Resistance to Pesticides. Natl. Acad. Sci., Washington, D.C.

Gill, S. S.; Hornung, S. M.; Ibarra, J. E.; Singh, G. J. P.; and Federici, B. A. 1987. Cytolytic activity and immunological similarity of *Bacillus thuringiensis* subsp. *israelensis* and *morrisoni* (PG-14) toxins. *Appl. Environ. Microbiol.* 53: 1251–1256.

Goldberg, L. J., and Margalit, J. 1977. A bacterial spore demonstrating rapid larvicidal activity against *Anopheles sergentii, Uranotaenia unguiculata, Culex univitattus, Aedes aegypti,* and *Culex pipiens. Mosq. News* 37: 355–358.

Heimpel, A. M. 1967. A critical review of *Bacillus thuringiensis* var. *thuringiensis* Berliner and other crystalliferous bacteria. *Ann. Rev. Entomol.* 12: 287–322.

Höfte, H., and Whiteley, H. R. 1989. Insecticidal crystal proteins of *Bacillus thuringiensis. Microbiol. Rev.* 53: 242–255.

Huber, H. E., and Lüthy, P. 1981. *Bacillus thuringiensis* delta-endotoxin: Composition and activation. In *Pathogenesis of invertebrate microbiol diseases,* ed. E. W. Davidson, 209–233. Totowa, N.J.: Allanheld, Osmun.

Hurley, J. M.; Lee, S. G.; Andrews, R. E., Jr.; Klowden, M. J.; and Bulla, L. A., Jr. 1985. Separation of the cytolytic and mosquitocidal proteins of *Bacillus thuringiensis* subsp. *israelensis. Biochem. Biophys. Res. Commun.* 126 (2): 961–965.

Ibarra, J. E., and Federici, B. A. 1986a. Isolation of a relatively nontoxic 65-kilodalton protein inclusion from the parasporal body of *Bacillus thuringiensis* subsp. *israelensis. J. Bacteriol.* 165 (2): 527–533.

———. 1986b. Parasporal bodies of *Bacillus thuringiensis* subsp. *morrisoni* (PG-14) and *Bacillus thuringiensis* subsp. *israelensis* are similar in protein composition and toxicity. *FEMS Microbiol. Lett.* 34 (1): 79–84.

Iizuka, T.; Faust, R. M.; and Ohba, M. 1983. Comparative profiles of plasmid DNA and morphology of parasporal crystals in four strains of *Bacillus thuringiensis* subsp. *darmstadiensis. Appl. Entomol. Zool.* 18 (4): 486–494.

Insell, J. P., and Fitz-James, P. C. 1985. Composition and toxicity of the inclusion of *Bacillus thuringiensis* subsp. *israelensis. Appl. Environ. Microbiol.* 50 (1): 56–62.

Kim, K. H.; Ohba, M.; and Aizawa, K. 1984. Purification of the toxic protein from *Bacillus thuringiensis* serotype 10 isolate demonstrating a preferential larvicidal activity to the mosquito. *J. Invertebr. Pathol.* 44: 214–219.

Klowden, M. J.; Held, G. A.; and Bulla, L. A., Jr. 1983. Toxicity of *Bacillus thuringiensis* subsp. *israelensis* to adult *Aedes aegypti* mosquitoes. *Appl. Environ. Microbiol.* 46: 312–315.

Krieg, A.; Huger, A. M.; Langenbruch, G. A.; and Schnetter, W. 1983. *Bacillus thuringiensis* var. *tenebrionis:* Ein neuer gegenüber Larven von Coleopteren wirksamer Pathotyp. *Z. Ang. Ent.* 96: 500–508.

Lacey, L. A., and Undeen, A. H. 1986. Microbiol control of black flies and mosquitoes. *Ann. Rev. Entomol.* 31: 265–296.

Lecadet, M.-M.; Chevrier, G.; and Dedonder, R. 1972. Analysis of a protein fraction in the spore coats of *Bacillus thuringiensis:* Comparison with crystal protein. *Eur. J. Biochem.* 25: 349–358.

Lee, S. G.; Eckblad, W.; and Bulla, L. A., Jr. 1985. Diversity of protein inclusion bodies and identification of mosquitocidal protein in *Bacillus thuringiensis* subsp. *israelensis. Biochem. Biophys. Res. Commun.* 126 (2): 953–960.

Lüthy, P., and Ebersold, H. R. 1981. *Bacillus thuringiensis* delta-endotoxin: Histopathology and molecular mode of action. In *Pathogenesis of invertebrate microbial diseases,* ed. E. W. Davidson, 235–267. Totowa, N.J.: Allenheld, Osman.

Mikkola, A. R.; Carlberg, G. A.; Vaara, T.; and Gyllenberg, H. G. 1982. Comparison of inclusions in different *Bacillus thuringiensis* strains: An electron microscope study. *FEMS Microbiol. Lett.* 13: 401–408.

Mulla, M. S.; Federici, B. A.; and Darwazeh, H. A. 1982. Larvicidal efficacy of *Bacillus thuringiensis* ser. H-14 against stagnant-water mosquitoes and its effects on nontarget organisms. *Environ. Entomol.* 11: 788–795.

Ohba, M., and Aizawa, K. 1979. A new subspecies of *Bacillus thuringiensis* possessing 11a:11b flagellar antigenic structure: *Bacillus thuringiensis* subsp. *kyushuensis. J. Invertebr. Pathol.* 33 (3): 387–388.

Padua, L. E.; Ohba, M.; and Aizawa, K. 1980. The isolates of *Bacillus thuringiensis* serotype 10 with a high preferential toxicity to mosquito larvae. *J. Invertebr. Pathol.* 36: 180–186.

———. 1984. Isolation of a *Bacillus thuringiensis* strain (serotype 8a:8b) highly and selectively toxic against mosquito larvae. *J. Invertebr. Pathol.* 44 (1): 12–17.

Pfannenstiel, M. A.; Ross, E. J.; Kramer, V. C.; and Nickerson, K. W. 1984. Toxicity and composition of protease-inhibited *Bacillus thuringiensis* var. *israelensis* crystals. *FEMS Microbiol. Lett.* 21: 39–42.

Ribier, J., and Lecadet, M.-M. 1973. Etude ultrastructurale et cinétique de la sporulation de *Bacillus thuringiensis* var. *berliner* 1715: Remarques sur la formation de l'inclusion parasporales. *Ann. Microbiol.* (Inst. Pasteur) 124A: 311–344.

Schnell, D. J.; Pfannenstiel, M. A.; and Nickerson, K. W. 1984. Bioassay of solubilized *Bacillus thuringiensis* subsp. *israelensis* crystals by attachment to latex beads. *Science* 223: 1191–1193.

Sekar, V. 1986. Biochemical and immunological characterization of the cloned crystal toxin gene of *Bacillus thuringiensis* var. *israelensis. Biochem. Biophys. Res. Commun.* 137 (2): 748–751.

Sen, K.; Honda, G.; Koyama, N.; Nishida, M.; Neki, A.; Sakai, H.; Himeno, M.; and Komano, T. 1988. Cloning and nucleotide sequences of the two 130-kDa insecticidal protein genes of *Bacillus thuringiensis* var. *israelensis. Agric. Biol. Chem.* 52: 873–878.

Sriram, R.; Kamdar, H.; and Jayaraman, K. 1985. Identification of the peptides of the crystals of *Bacillus thuringiensis* var. *israelensis* involved in the mosquito larvicidal activity. *Biochem. Biophys. Res. Commun.* 132 (1): 19–27.

Thomas, W. E., and Ellar, D. J. 1983a. *Bacillus thuringiensis* var. *israelensis* crystal δ-endotoxin: Effects on insect and mammalian cells in vitro and in vivo. *J. Cell Sci.* 60: 181–197.

———. 1983b. Mechanism of action of *Bacillus thuringiensis* var. *israelensis* insecticidal δ-endotoxin. *FEBS Lett.* 154: 362–368.

Thorne, L.; Garduno, F.; Thompson, T.; Decker, D.; Zounes, M.; Wild, M.; Walfield, A. M.; and Pollock, T. J. 1986. Structural similarity between the lepidoptera- and diptera-specific insecticidal endotoxin genes of *Bacillus thuringiensis* subsp. *kurstaki* and *israelensis. J. Bacteriol.* 166 (3): 801–811.

Tungpradubkul, S.; Settasatien, C.; and Panyim, S. 1988. The complete nucleotide sequence of a 130-kDa mosquito-larvicidal delta-endotoxin gene of *Bacillus thuringiensis* var. *israelensis. Nucleic Acids Res.* 16: 1637–1638.

Tyrell, D. J.; Bulla, L. A., Jr.; Andrews, R. E., Jr.; Kramer, K. J.; Davidson, L. I.; and Nordin, P. 1981. Comparative biochemistry of entomocidal parasporal crystals of selected *Bacillus thuringiensis* strains. *J. Bacteriol.* 145 (2): 1052–1062.

Tyrell, D. J.; Davidson, L. I.; Bulla, L. A., Jr.; and Ramoska, W. A. 1979. Toxicity of parasporal crystals of *Bacillus thuringiensis* subsp. *israelensis* to mosquitoes. *Appl. Environ. Microbiol.* 38: 656–658.

Undeen, A. H., and Nagel, W. L. 1978. The effect of *Bacillus thuringiensis* ONR-60A strain (Goldberg) on *Simulium* larvae in the laboratory. *Mosq. News* 38: 524–527.

Visser, B.; van Workum, M.; Dullemans, A.; and Waalwijk, C. 1986. The mosquitocidal activity of *Bacillus thuringiensis* var. *israelensis* is associated with M_r 230,000 and 130,000 crystal proteins. *FEMS Microbiol. Lett.* 30: 211–214.

Waalwijk, C.; Dullemans, A. M.; van Workum, M.E.S.; and Visser, B. 1985. Molecular cloning and the nucleotide sequence of the M_r 28,000 crystal protein gene of *Bacillus thuringiensis* subsp. *israelensis. Nucleic Acids Res.* 13 (22): 8207–8217.

Ward, E. S., and Ellar, D. J. 1988. Cloning and expression of two homologous genes of *Bacillus thuringiensis* subsp. *israelensis* which encode 130-kilodalton mosquitocidal proteins. *J. Bacteriol.* 170: 727–735.

Ward, E. S.; Ridley, A. R.; Ellar, D. J.; and Todd, J. A. 1986. *Bacillus thuringiensis* var. *israelensis* δ-endotoxin: Cloning and expression of the toxin in sporogenic and asporogenic strains of *Bacillus subtilis. J. Mol. Biol.* 191: 13–22.

Widner, W. R., and Whiteley, H. R. 1989. Two highly related insecticidal crystal proteins of *Bacillus thuringiensis* subsp. *kurstaki* possess different host range specificities. *J. Bacteriol.* 171: 965–974.

Wu, D., and Chang, F. N. 1985. Synergism in mosquitocidal activity of 26 and 65 kDa proteins from *Bacillus thuringiensis* subsp. *israelensis* crystal. *FEBS Lett.* 190: 232–236.

Yamamoto, T.; Iizuka, T.; and Aronson, J. N. 1983. Mosquitocidal protein of *Bacillus thuringien-*

sis subsp. *israelensis:* Identification and partial isolation of the protein. *Curr. Microbiol.* 9: 279–284.

Yamamoto, T., and McLaughlin, R. E. 1981. Isolation of a protein from the parasporal crystal of *Bacillus thuringiensis* var. *kurstaki* toxic to the mosquito larvae, *Aedes taeniorhynchus. Biochem. Biophys. Res. Commun.* 103 (2): 414–421.

Yamamoto, T.; Watkinson, I. A.; Kim, L.; Sage, M. V.; Stratton, R.; Akanda, N.; Li, Y.; Ma, D. P.; and Poe, B. A. 1988. Nucleotide sequence of the gene coding for a 130-kDa mosquitocidal protein of *Bacillus thuringiensis* var. *israelensis. Gene* 66: 107–120.

4

Mechanism of Action of *Bacillus thuringiensis israelensis* Parasporal Body

CHRIS N. CHILCOTT
BARBARA H. KNOWLES
DAVID J. ELLAR
FRANCIS A. DROBNIEWSKI

4.1 INTRODUCTION

The *Bacillus thuringiensis* subsp. *israelensis* (*B. t. i.*) parasporal body is a gut poison, and the midgut epithelium of affected larvae is considered to be its initial site of action (de Barjac 1978). In *B. t. i.*-treated mosquito larvae, midgut epithelial cells swell and burst, causing severe damage to the gut wall (de Barjac 1978; Charles and de Barjac 1983; Lahkim-Tsror et al. 1983). The general characteristics of poisoning of *B. t. i.*-treated mosquito larvae (4 μg/ml) are cessation of feeding within one hour, reduced activity by two hours, extreme sluggishness by four hours, and general paralysis by six hours. Singh, Schouest, and Gill (1986) showed that the effect on the posterior midgut ultrastructure was detected as early as one hour, when the epithelial striated border was damaged. This was followed by swelling of epithelial organelles. Midgut circular and longitudinal muscles were also damaged, as indicated by their swelling and separation from the basement lamina. Six hours after *B. t. i.* treatment the midgut wall was severely ruptured, with the peritrophic membrane and basement lamina broken. Of these lesions, disruption of microvilli and epithelial swelling in the midgut coincide with cessation of feeding. The sluggish behavior and paralysis of insects only occurred in advanced stages of parasporal body poisoning.

B. t. i. parasporal body consists of at least four major proteins of 27 kDa, 65 kDa, and a doublet of 130 kDa (fig. 4.1), all of which are mosquiticidal (Visser et al. 1986; Ibarra and Federici 1986; Ward et al. 1986; Ward and Ellar 1988). Purification of the parasporal body proteins using fast-protein liquid chromatography (Chilcott and Ellar 1988), followed by acid precipitation of these proteins, have shown that the 65-kDa protein has the highest mosquiticidal activity (table 4.1). It was also shown by feeding mosquito larvae mixtures of 27-, 65-, or 130-kDa proteins that when a mixture of 27- and

FIGURE 4.1 SDS-polyacrylamide gel of *B. t. i.* parasporal body proteins.

130-kDa proteins was used there was increased mosquiticidal activity compared to when they were fed individually to the larvae. Wu and Chang (1985) reported that when 27- and 65-kDa or 27- and 130-kDa proteins were fed to mosquito larvae the toxicity was greater than if either had been fed alone to the larvae. Hurley, Bulla, and Andrews (1987) found that a mixture of 27- and 65-kDa proteins showed increased activity. These observations suggest a synergistic interaction between the parasporal body proteins in their toxicity to mosquito larvae.

Reports have shown that alkali-solubilized *B. t. i.* parasporal body is cytolytic to insect and mammalian cells (Thomas and Ellar 1983a; Chilcott, Kalmakoff, and Pillai 1985; Gill and Hornung 1987; Chilcott and Ellar 1988), hemolytic and toxic when injected into suckling mice (Thomas and Ellar 1983a), and neurotoxic (Chilcott, Kalmakoff, and Pillai 1984; Singh and Gill

TABLE 4.1.
Toxicity of Purified *B.t.i.* Parasporal Proteins against *Aedes aegypti* Larvae

Preparation	LC_{50} (μg/ml)
Native parasporal bodies	0.32
Solubilized parasporal bodies[a]	1.0
130-kDa protein	32.0
65-kDa protein	4.0
27-kDa protein	115.0

SOURCE: Chilcott and Ellar 1988.
[a] *B. t. i.* parasporal bodies solubilized in 50 mM Na_2CO_3.HCl pH 9.5 containing 10 mM dithiothreitol, at 37° C for 60 minutes, followed by incubation in 50 mM Na_2CO_3, pH 11.3, at 37° C for 60 minutes. Both supernatants were pooled and the protein precipitated with 12% citric acid at pH 4.5.

TABLE 4.2.
Toxicity of Purified *B.t.i.* Parasporal Body Proteins to Insect Cell Lines

Cell Line	LC50 (μg/ml)		
	25 kDa	30–35 kDa	53 kDa
Aedes aegypti	2.8	125	—[a]
Anopheles gambiae	3	79	35
Culex quinquefasciatus	3	90	34
Choristoneura fumiferana	34	—[b]	—
Heliothis zea	68	—	—
Spodoptera frugiperda	11	—	115

SOURCE: Chilcott and Ellar, 1988
[a]No toxicity at 150 μg/ml.
[b]No toxicity at 100 μg/ml.

1985). The major cytolytic protein of the *B.t.i.* parasporal body is the 25-kDa protein derived from proteolysis of the 27-kDa protein (Thomas and Ellar 1983a; Armstrong, Rohrmann, and Beaudreau 1985; Chilcott and Ellar 1988). Thomas and Ellar (1983a) reported that *Aedes albopictus* cells incubated with alkali-soluble parasporal body protein (5 μg/ml) showed lysis in 30–40 minutes, but *Choristoneura fumiferana* cells incubated with 25 μg/ml lysed in 120 minutes. A 53-kDa protein derived from the 130-kDa protein and the 30- to 35-kDa proteins derived from the 65-kDa protein are much less toxic to insect cell lines (table 4.2) (Chilcott and Ellar 1988). The 25-kDa protein is hemolytic (Yamamoto, Iizuka, and Aronson 1983; Ward et al. 1986), while the 65- and 130-kDa proteins are not hemolytic (Visser et al. 1986; Hurley et al. 1985). Mayes et al. (1989) demonstrated that the factor responsible for mammalian toxicity of intraperitoneally administered *B.t.i.* soluble proteins was the 25-kDa toxin.

Many theories have been proposed to explain the mode of action of the *Bacillus thuringiensis* lepidopteran toxic parasporal bodies (table 4.3). Since the primary effect of *B.t.* lepidopteran and dipteran toxins is on the midgut of susceptible larvae, which show the same histopathological changes, it may be that some of the theories for *B.t.* lepidopteran toxins are also applicable to the *B.t.i.* parasporal body. The theories that have been proposed for *B.t.i.* are that it has a detergentlike activity, neurotoxicity, or causes colloid-osmotic lysis. Several authors have reported the neurotoxicity of *B.t.i.* (Chilcott, Kalmakoff, and Pillai 1984; Cheung et al. 1985, 1987; Singh and Gill 1985; Singh, Schouest, and Gill 1986). This is consistent with the observation that the *B.t.i.* 27-kDa toxin can lyse all cells containing suitable phospholipid receptors (see section 4.2.1). Isolated neurons or the intact neuromuscular system of insects to which the toxin has been administered would be expected to be killed by permeabilization of their plasma membranes. How-

TABLE 4.3.
Theories for *Bacillus thuringiensis* Toxin Mechanism of Action

Target of Action	Reference
Plasma membrane (Ionophore)	Angus 1968 Nickerson and Schnell 1983
Mitochondria	Travers, Faust, and Reichelderfer 1976
Plasma membrane (General breakdown)	Lüthy and Ebersold 1981 Nishiitsusuji-Uwo, Endo, and Himeno 1979
Goblet cell K^+ pump	Griego, Moffett, and Spence 1979 Harvey and Wolfersberger 1979
Plasma membrane (*B.t.i.*) (detergentlike action)	Thomas and Ellar 1983b
Neuromuscular system	Chilcott, Kalmakoff, and Pillai 1984 Cheung et al. 1985, 1987 Singh, Schouest, and Gill 1986
Na^+ and/or K^+ transport	Himeno et al. 1985 Gupta et al. 1985 Sacchi et al. 1986
Plasma membrane (colloid-osmotic lysis)	Knowles and Ellar 1987 Haider and Ellar 1987 Drobniewski and Ellar 1988a

ever, since the insect nervous system is not exposed to the toxin until the gut border has broken down, it would not be able to reach the nervous system. Thomas and Ellar (1983b) suggested that the interaction of the toxin with specific plasma membrane receptors causes a detergentlike rearrangement of the lipids, leading to disruption of membrane integrity and eventual cytolysis. An elaboration of the original theory proposed by Thomas and Ellar (1983b) is described in detail in section 4.2.3. Mode-of-action studies have so far been confined to the 25-kDa protein of the parasporal body, because it was initially described as the principal toxin and because of its wide range of target cells in vitro. However, as it has been reported that the 65- and 130-kDa proteins are more mosquiticidal than the 25-kDa protein, studies should be carried out on the mechanism of action of these proteins to increase our understanding of the toxicity of the *B.t.i.* parasporal body to mosquito and black fly larvae. Section 4.2 describes the possible mode of action of the 25-kDa protein.

4.2 MECHANISM OF ACTION

Studies on the mode of action of the *B.t.i.* parasporal body have been carried out using in vitro systems such as insect or mammalian tissue culture cells or erythrocytes (Thomas and Ellar 1983a; Davidson and Yamamoto 1984; Chilcott, Kalmakoff, and Pillai 1985). The methods used to investigate the mode of action are leakage of ions or other compounds from cells (Knowles and Ellar 1987; Drobniewski and Ellar 1988a), neutral red staining (Chilcott and Ellar 1988), or exclusion of the dye trypan blue (Thomas and Ellar 1983a).

4.2.1 Receptors

The first step in toxicity consists of binding of the toxin to specific receptors on the cell surface, but the nature of these binding sites is still uncertain. Ellar et al. (1985) showed that the 25-kDa protein was the only parasporal body protein that inserted into lipid bilayers. Thomas and Ellar (1983b) reported that lipids extracted from *Ae. albopictus* cells inactivated the alkali-soluble parasporal body protein. Their experiments with lipid dispersions and multilamellar liposomes showed that the toxin bound to phosphatidyl choline, sphingomyelin, and phosphatidyl ethanolamine provided these lipids contained unsaturated fatty acids. Phosphatidyl serine bound the toxin less efficiently, and phosphatidyl inositol, cardiolipin, cerebroside, and cholesterol did not bind the toxin. Gill, Singh, and Hornung (1987) confirmed a specific interaction between the 24/25-kDa toxins and certain unsaturated phospholipids. The net charge on the liposomes did not appear to be a determining factor in toxin-lipid interactions at pH 7.3 as there was no difference between phosphatidyl choline liposomes containing stearylamine or dicetylphosphate (Thomas and Ellar 1983a), but at higher pH (pH 10) the protein only bound to positively charged liposomes (Knowles et al. n.d.). It was noted that the toxic effect of the 25-kDa protein on red blood cells (Knowles et al. n.d.) and Malpighian tubules (Maddrell et al. 1989) was decreased when the pH was raised from 7 to 10. The 25-kDa protein from the *B.t.i.* parasporal body released ^{86}Rb from labeled synthetic phosphatidyl choline liposomes containing unsaturated fatty acids but not from phosphatidyl choline liposomes containing saturated fatty acids (Drobniewski and Ellar 1988b) and released glucose from unsaturated phosphatidyl choline liposomes (Knowles et al. n.d.). Despite the evidence that phospholipids bind the 25-kDa toxin, the possibility still exists that in cells sensitive to the toxin an additional receptor may be present. Mosquito cell lines, for example, are more sensitive to the 25-kDa toxin than lepidopteran cell lines (table 4.2; Chilcott and Ellar 1988), although both types of cells contain suitable receptor phospholipids in their membranes. Gill and Hornung (1987) found that dog kidney cells were much less sensitive than *Ae. albopictus* cells to the 25-kDa toxin. One explanation for this is that the phospholipids in nondipteran cells are less accessible to the toxin than in mosquito cells (see section 4.3 for a discussion of this theory). However, as has been suggested in the case of *Staphylococcus aureus* α-toxin (Maharaj and Fackrell 1980) and diphtheria toxin (Olsnes et al. 1985), two classes of plasma membrane receptors may exist. Chilcott (1983) observed that *Aedes aegypti* cells treated with

Streptomyces griseus protease or α-amylase were not susceptible to the 25-kDa toxin, suggesting that in these cells, in addition to the phospholipid receptors, a glycoprotein receptor may be involved in binding of the toxin to the cells. A glycoprotein is thought to be the receptor for *B. t.* subsp. *kurstaki* 130-kDa P1 toxin (Knowles and Ellar 1986).

Some support for the possibility that the 25-kDa toxin may bind to cell-surface components other than phospholipids has come from recent studies with mutant forms of the toxin generated by site-directed mutagenesis (Ward, Ellar, and Chilcott 1988; Ellar et al. n.d.). One such mutant, in which a glutamic-acid residue at position 204 was replaced by alanine, was found to have altered its ability to bind to unsaturated phosphatidyl choline liposomes. Nevertheless, this mutant toxin retained the ability to lyse *Ae. aegypti* and *Anopheles gambiae* cells, albeit with a higher LC_{50}. Most significant, however, was the finding that in contrast to the wild-type toxin, the mutant protein failed completely to lyse cells of the *Choristoneura fumiferana* lepidopteran cell line. At this stage no firm conclusions are justified, but one interpretation of these results would be that the cytolytic action of the 25-kDa toxin is enhanced by the presence of mosquito cell–specific surface receptors in addition to the ubiquitous membrane phospholipids. The former may orient the molecule(s) at the hydrophobic membrane surface prior to a second step in which the toxin penetrates the hydrophobic lipid bilayer. In the case of nondipteran cells that lack a specific receptor, a direct interaction of a higher concentration of toxin with membrane phospholipids may bypass the requirement for a receptor. Conceivably, the replacement of glutamate 204 by alanine in the mutant may affect the ability of the protein to interact directly with phospholipids, thus precluding insertion into lepidopteran membranes that lack a specific receptor. Lysis of mosquito cells is still possible via the receptor-mediated pathway.

4.2.2 Toxin Structure and Membrane Insertion

Detailed information on the three-dimensional structure of the *B. t.* parasporal body is a prerequisite for a full understanding of the mechanism by which it initiates lysis. This is likely to come from X-ray crystallographic studies of individual parasporal body proteins that are currently in progress in a number of laboratories. Preliminary X-ray diffraction analysis of crystals of the 25-kDa toxin of *B. t. i.* suggested that a major structural component of the protein is parallel or antiparallel helical bundles (McPherson et al. 1987). It has recently been proposed that *B. t. i.* toxins may be glycoproteins and that it is the sugar part of the molecule that may be responsible for toxin specificity (Pfannenstiel et al. 1987; Muthukumar and Nickerson 1987). In contrast to these results, we have observed that *B. t. i.* 25-kDa toxin has identical properties of toxicity and specificity whether produced by *B. t. i.*, *Bacillus*

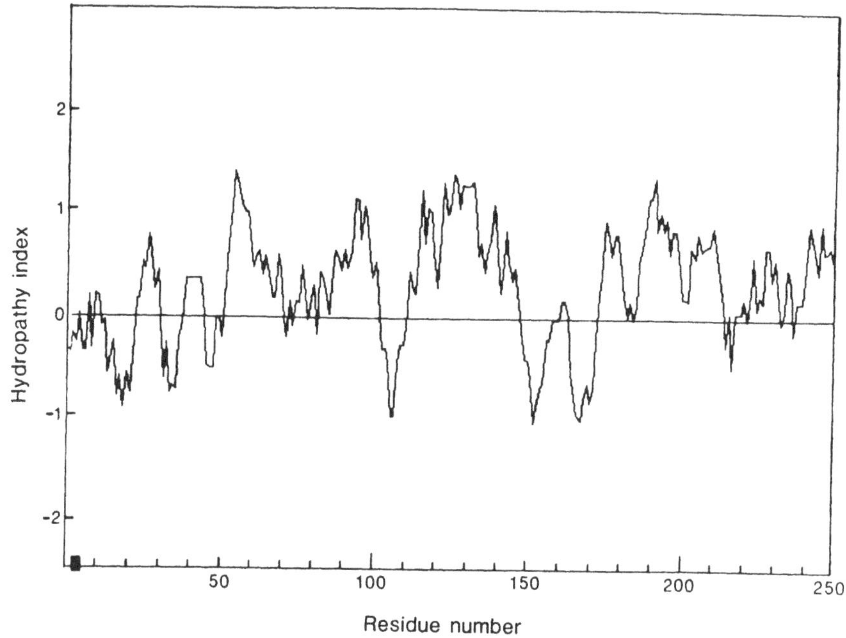

FIGURE 4.2 Hydropathy plot of the *B. t. i.* 27-kDa protein. Positive values represent hydrophobicity, and negative values represent hydrophilicity. Reproduced from Ward and Ellar 1986.

subtilis, or *Escherichia coli* (Ward et al 1986; Knowles and Ellar 1987; Drobniewski and Ellar 1988a; Chilcott and Ellar 1988), even though the latter organism is known to not glycosylate proteins. We have been unable to detect the presence of covalently bound sugar in preparations of purified *B. t. i.* toxins (Knowles, unpublished observation).

In the meantime, the structures of those parasporal bodies whose primary sequences have been determined can be analyzed with the aid of a number of computational models that have been designed to predict likely three-dimensional conformations. One of the most widely known of these models (Kyte and Doolittle 1982) scans a primary sequence, progressively evaluating the hydrophobicity and hydrophilicity of the protein. A hydropathy scale is constructed by assigning to each of the 20 amino acids a value that reflects its relative hydrophobicity/hydrophilicity. The computer program then scans through the protein sequence from NH_2 to COOH terminus, continuously determining the average hydropathy of a moving segment relative to a universal midline (hydropathy plot).

Figure 4.2 shows such a plot for the *B. t. i.* 27-kDa toxin (Ward and Ellar 1986), with positive values corresponding to hydrophobicity and negative values representing hydrophilicity. The overall hydropathy of the molecule and the presence of a number of hydrophobic sequences of membrane-spanning dimensions are both consistent with a protein having a high affinity for the apolar environment of biological membranes. Proteins normally resident in membranes and other proteins known to be capable of inserting into membranes frequently display a good correlation between the large peaks on the hydrophobic side of the midpoint line and the portions of their sequences known to penetrate or span the lipid bilayer. These apolar sequences are often the correct length (18–25 amino acids) to form a transmembrane α-helix, but other secondary structural features, such as extended antiparallel β-sheets (β-barrels), could also form transmembrane domains.

A number of other proteins that bind to or insert into membranes appear to have transmembrane segments that are amphiphilic helices. These are characterized by a concentration of charged or hydrophilic residues on one surface of a helix, with another surface entirely composed of hydrophobic side chains. Analysis of the 26–amino acid cytolytic polypeptide, melittin, has shown it to be highly amphiphilic; and the same feature has been seen in several other cytolytic peptides (Argiolas and Pisano 1985) and surface-seeking proteins (Eisenberg et al. 1984; Roise et al. 1986), suggesting that cytolysis and/or membrane penetration may be related to this amphiphilic structural feature. Upon penetration into a membrane, it would be energetically unfavorable for the hydrophilic face of such an amphiphilic helix to remain in contact with the apolar lipid acyl chains. One way of avoiding this is for two or more transmembrane amphiphilic helices to aggregate together, with their hydrophobic surfaces facing the bilayer lipid and their hydrophilic side chains facing each other. A recent study of the conformation of melittin in lipid membranes by Raman spectroscopy (Vogel and Jahnig 1986) has suggested that individual melittin monomers aggregate together in this way within the lipid bilayer to form tetramers with a central polar pore or hole. For amphiphilic helices to disrupt membrane integrity, transmembrane segments may not, however, be necessary. Thus, an amphiphilic helix is also capable of associating with the membrane parallel to the surface in such a way that its hydrophilic face interacts with charged residues on the surface (lipid headgroups and carbohydrates), while its hydrophobic face penetrates into the apolar region of the outer monolayer of the bilayer.

Ward, Ellar, and Chilcott (1988) employed computer predictive methods to produce a model of the possible secondary structure of the *B.t.i.* 27-kDa toxin. However, since these methods are designed to predict the conformation of proteins in an aqueous environment, the resulting model may not accurately depict the toxin configuration in a membrane. Figure 4.3 represents an attempt to model the toxin structure as it exists in a lipid bilayer. In producing this model we have therefore arbitrarily ignored certain predictions contained in the more complex conformational structure prediction of Ward, Ellar, and Chilcott (1988). In this simpler model (fig. 4.3) we have also

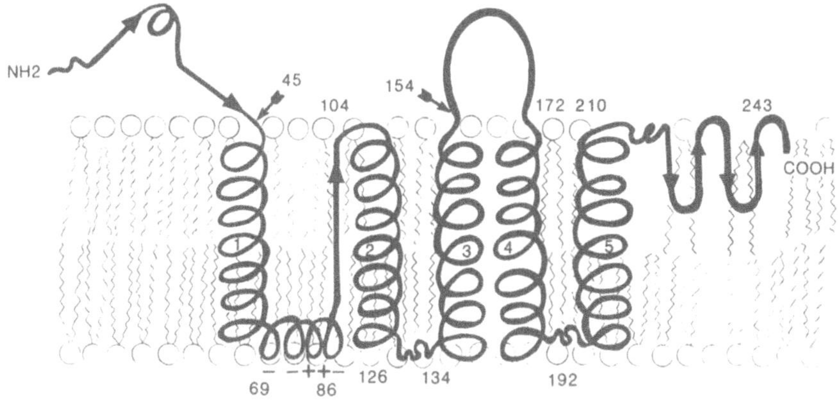

FIGURE 4.3 Schematic diagram showing one predicted secondary structure for the 27-kDa protein as it might fold in a lipid bilayer.

taken into account results from experiments in which we treated liposomes that had bound the 27-kDa toxin with proteinase K (Knowles et al. n.d.). Two large peptides (8.5 kDa and 9 kDa) were protected from digestion, with N-termini corresponding to residues 45 and 154 of the 248 residue toxin. This result suggests that the bulk of this toxin is inserted into the lipid phase, consistent with its high hydrophobicity. Conceivably, only a segment at the N-terminus (residues 1–44) and a region containing lysine 154 may be exposed in the aqueous phase after insertion. Although shown in figure 4.3 as an extramembranous loop, the location of the latter region with respect to the membrane is not known with certainty. Site-directed mutagenesis studies (Ward, Ellar, and Chilcott 1988) suggested that replacement of lysine 154 with uncharged alanine abolished the ability of the toxin to bind phospholipids under the conditions used. Ward, Ellar, and Chilcott (1988) suggested that lysine 154 has a key location in the overall tertiary conformation of the toxin and possibly also in the interaction of the toxin with phospholipids. Their predicted model, based on conformational analysis of proteins in aqueous solution, placed this residue at the N-terminal end of a short 10-residue α-helix. Figure 4.3 shows residues 50–69 folding into a hydrophobic α-helix of sufficient length to span the bilayer. The amphipathic nature of the segment from residues 72–86 is shown as a possible surface-seeking domain. The extended apolar β-segment from residues 87–104 could form a second membrane-spanning element. Analysis of the sequence from residues 105–210 suggests four more membrane-spanning α-helical segments. Taken together these predictions suggest five, and possibly six, potential transmembrane segments, and their locations agree reasonably well with the hydropathy plot in figure 4.2. It should be stressed, however, that

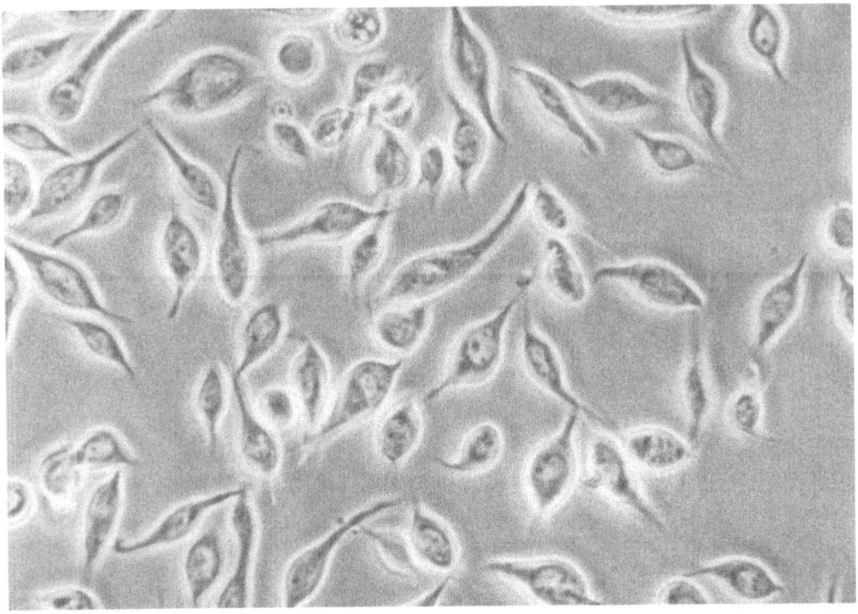

a

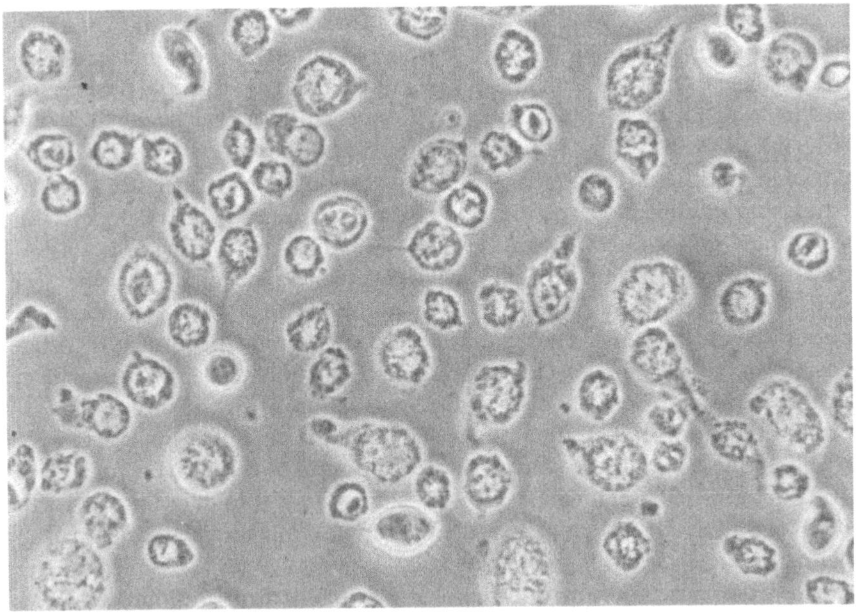

b

FIGURE 4.4 Cytopathic effect of *B.t.i.* 25-kDa protein on insect cell lines: (*a*), *Choristoneura fumiferana* cells treated with 25 µl/ml 50 mM sodium carbonate/HCl pH 10.5 for 60 minutes; (*b*), *C. fumuferana* cells treated with 25 µg/ml *B.t.i.* parasporal body protein soluble in 50 mM sodium carbonate/HCl pH 10.5 for 60 min-

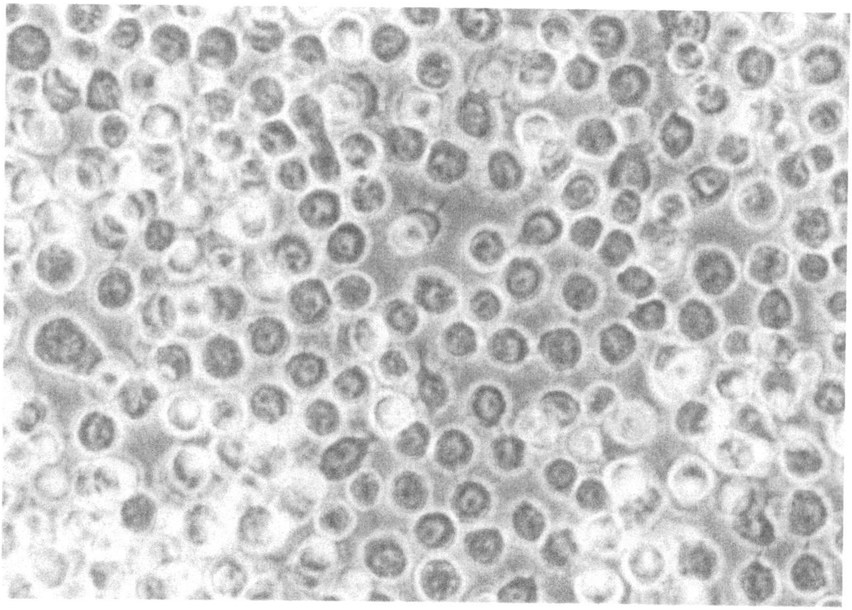

c

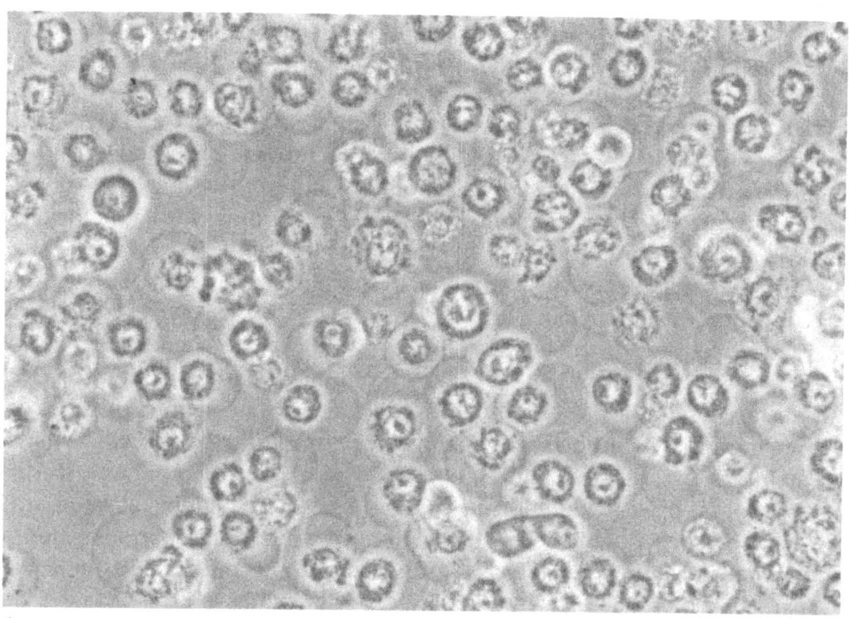

d

FIGURE 4.4 *(Continued)*
utes; (*c*), *Ae. aegypti* cells treated with 10 μl/ml 50 mM sodium carbonate/HCl, pH 10.5 for 30 minutes; (*d*), *Ae. aegypti* cells treated with 10μg/ml *B.t.i.* parasporal body protein soluble in 50 mM sodium carbonate/HCl pH 10.5 for 30 minutes.

alternative structures are possible. Nevertheless, a recent detailed predictive analysis, which included all the known classes of *B.t.* toxins except the 27-kDa toxin (Hodgman and Ellar n.d.), shows a remarkable conservation of six potential transmembrane helical segments in the N-terminal 250 residues of ***all*** the toxins. In this respect the highly hydrophobic 27-kDa toxin shows a striking similarity to the N-terminal domain of all the other *B.t.* toxins.

Figure 4.3 does not attempt to model the intramolecular interactions that may be involved in stabilizing the 27-kDa toxin-membrane association or creating a transmembrane pore. If the 27-kDa toxin pore requires the aggregation of several toxin monomers (see section 4.2.4), this may occur at the membrane surface prior to insertion or through oligomerization of monomers after insertion into the bulk lipid phase. It is likely that both these processes would be dependent on intermolecular interactions involving specialized regions of the 27-kDa protein.

4.2.3 Colloid–Osmotic Lysis Theory

The experiments described in this section were all carried out using the *B.t.i.* parasporal body proteins soluble in 50 mM sodium carbonate/HCl pH 10.5 (Thomas and Ellar 1983a), activated with *Pieris brassicae* gut extract (Knowles et al. 1984). This preparation contains predominantly the 25-kDa protein, but also some 30- and 35-kDa material. These latter proteins have only weak cytolytic activity in comparison to the 25-kDa active portion of the 27-kDa toxin (table 4.1). However, in view of the suggestion of synergism between *B.t.i.* proteins, it should be noted that when the experiments are carried out using cloned 27-kDa protein purified from inclusions made in *B. subtilis* (Ward et al. 1986) and activated with gut extract, the results are the same as those using carbonate-soluble *B.t.i.* parasporal body protein (Drobniewski and Ellar 1988a).

Our theory for the mechanism of action of the 25-kDa cytolytic toxin of *B.t.i.* (Knowles and Ellar 1987) is that after binding to phospholipids in the plasma membrane (section 4.2.1) the toxin generates small holes or pores in the membrane. These holes will disrupt the permeability barrier of the plasma membrane, leading to equilibration of ions across the membrane, disturbing the colloid-osmotic equilibrium, and resulting in a net influx of water into the cell. The consequent cell swelling will cause a further disruption of membrane integrity and eventual lysis. An advantage of this theory is that it leads to several predictions that can be tested experimentally.

1. Toxin-treated cells will swell before they lyse. This was observed by Thomas and Ellar (1983a) and is shown in figure 4.4.
2. Osmotic protectants that are too large to penetrate the toxin-

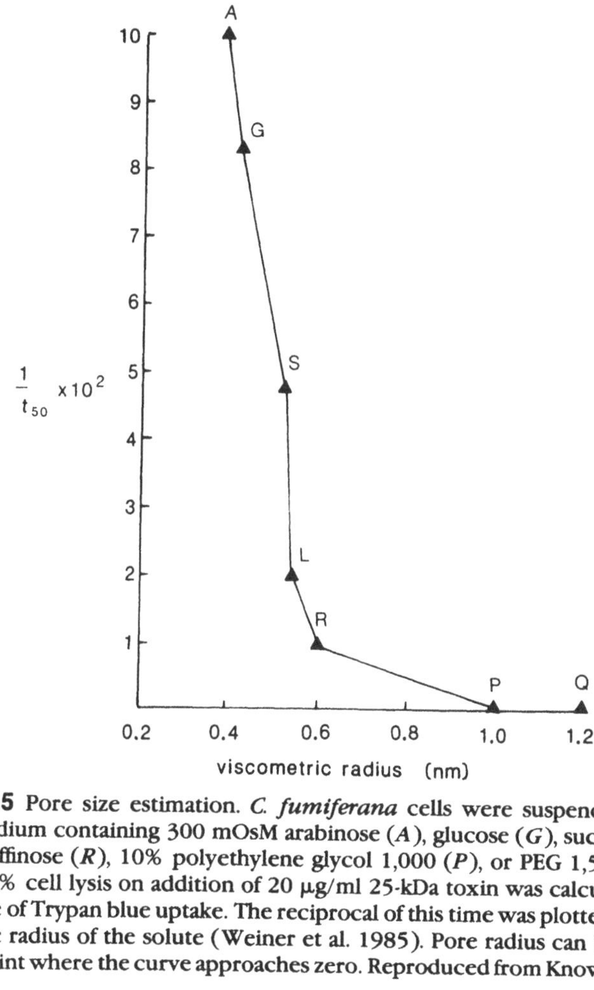

FIGURE 4.5 Pore size estimation. *C. fumiferana* cells were suspended in tissue culture medium containing 300 mOsM arabinose (*A*), glucose (*G*), sucrose (*S*), lactose (*L*), raffinose (*R*), 10% polyethylene glycol 1,000 (*P*), or PEG 1,500 (Q). The time for 50% cell lysis on addition of 20 μg/ml 25-kDa toxin was calculated from a time course of Trypan blue uptake. The reciprocal of this time was plotted against the viscometric radius of the solute (Weiner et al. 1985). Pore radius can be estimated from the point where the curve approaches zero. Reproduced from Knowles and Ellar 1987.

induced holes should inhibit or delay cytolysis. The minimum size of molecule inhibiting lysis completely provides an estimate of the size of the hole (Weiner et al. 1985). We have found that lysis is partially inhibited by sucrose and raffinose, and is completely inhibited by polyethylene glycol (M_r 1,000) (fig 4.5). Our results suggest a pore radius of 0.6–1.0 nm, a value that is consistent with a mechanism of colloid-osmotic lysis: toxins thought to act by this mechanism have pore sizes of 1.2 nm (*Pseudomonas aeruginosa* cytotoxic protein, Weiner et al. 1985) or 1–1.5 nm (*Staphylococcus aureus* α-toxin,

Bhakdi, Tranum-Jensen, and Sziegoleit 1985); while toxins that make larger holes in the membrane, such as streptolysin O (30 nm, Bhakhdi, Tranum-Jensen, and Sziegoleit 1985), cause immediate lysis with no prior swelling since macromolecules can leak out.

3. Although the initial toxic lesion must be too small to allow cellular macromolecules to leak out of the cell, the subsequent cell swelling will further disrupt the membrane, allowing progressively larger molecules to leak out. Thus, the theory would predict that the effect of the toxin would be to allow molecules of increasing size to leak out of the cell with time. In order to test this, we loaded cells with radioactive markers of different size and followed their rate of efflux from toxin-treated cells (fig. 4.6). The results showed that the order of efflux was ^{86}Rb (M_r 86) > ^{3}H-uridine (M_r < 1,000) > $^{51}CrO_4$ (M_r 3,000). This result is also consistent with the estimated pore size; $^{51}CrO_4$, which could not penetrate a 1-nm pore, should only leak out of the cell through holes formed in the damaged membrane as a secondary result of osmotic swelling.

We believe that this theory can account for the observed effects of *B. t. i.* 25-kDa toxin. For example, neurotoxicity would be a result of disrupting the permeability barrier of nerve cells and hence destroying their ability to function. Singh, Schouest, and Gill (1986) speculated that the histopathological effect of *B. t. i.* on the muscle cells of *Ae. aegypti* larvae may be due to breakdown of the cells' osmotic barrier. It is interesting to note that injecting *B. t. i.* parasporal body solubilized in 50 mM sodium carbonate pH 10.5 (i.e., mainly 25-kDa toxin) into housefly larvae produced the same histopathological effect on muscle cells as feeding mosquito larvae with the whole *B. t. i.* parasporal bodies (Singh, Schouest, and Gill 1986). This raises the possibility that the 25-kDa protein is responsible for both phenomena. The important fact remains, however, that this effect on muscle cells is secondary to the gut damage, since the toxin is unable to reach the muscle cells until the gut barrier is breached. Gill, Singh, and Hornung (1987) also proposed that the 25-kDa toxin forms a pore in membranes, and Maddrell et al. (1989) showed that dextran (M_r 5,000) significantly delayed the effect of the toxin on Malpighian tubules, supporting the notion that the toxin acts by colloid-osmotic lysis.

Recently, a direct demonstration that *B. t. i.* 25-kDa toxin forms a pore in membranes was obtained by Knowles et al. (1989) using planar lipid bilayers. The toxin formed a cation-selective channel that showed rapid opening and closing, and divalent cations increased the probability of the closed state. The channels showed multiple conductance states and complex gating kinetics (Knowles and Blatt, unpub. data) reminiscent of those formed by *Staphy-*

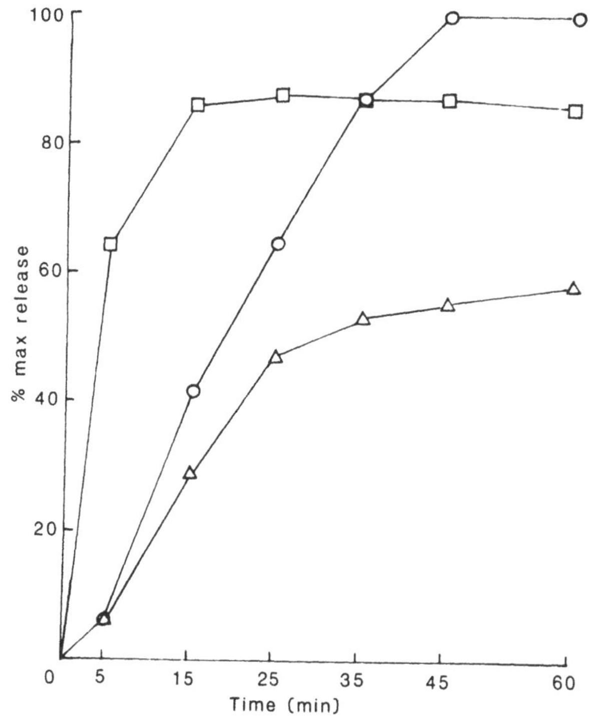

FIGURE 4.6 Release of internal markers. *C. fumiferana* cells were loaded with ^{86}Rb (□), ^{3}H-uridine (○) or $^{51}CrO_4$ (△). (method as described by Thelestam and Mollby 1975). Twenty-five-kDa toxin was added at time 0, and release of label at each time point was expressed as percentage of maximal release. Reproduced from Knowles and Ellar 1987.

lococcus aureus δ-toxin (Mellor, Thomas, and Sansom 1988). The latter toxin was proposed to form a pore consisting of six toxin molecules (Mellor, Thomas, and Sansom 1988).

4.2.4 Toxin Oligomerization

The model proposed in section 4.2.2 raised the question of whether one molecule of 25-kDa toxin is able to form a hole in the cell membrane or whether several molecules must be present to generate a pore. Many pore-forming cytolytic toxins are known to form their toxic lesion by generation of a toxin oligomer in the plasma membrane. An example of this is *Staphylococcus aureus* α-toxin (Bhakhdi, Tranum-Jensen, and Sziegoleit 1985). In the case of *B. t. i.* 25-kDa toxin, SDS-polyacrylamide gel electropho-

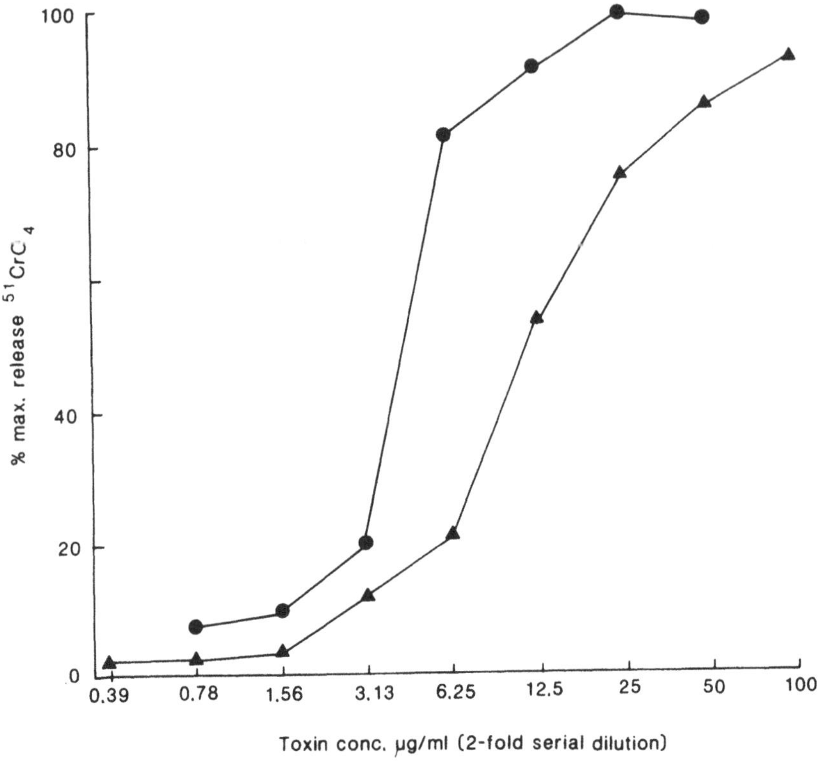

FIGURE 4.7 Dose-response curve. *Ae. aegypti* cells were loaded with $^{51}CrO_4$ as in fig. 4.6. Labeled cells were incubated with *B.t.i.* 25-kDa toxin (●) or melittin (▲) for 60 minutes and the percentage of maximal release determined.

resis of toxin-liposome and toxin-erythrocyte membrane preparations suggested that toxin oligomerization may be taking place in the membrane (Drobniewski and Carroll, unpublished observation). Detergents bind to isolated membrane proteins by virtue of the ability of their apolar portions to compensate for the lipid acyl chains. In this context it is interesting to note that the pure 27-kDa toxin readily forms dimers and trimers in solutions containing sodium dodecyl sulfate.

Maddrell et al. (1988) observed a lag in the inhibition of fluid secretion when isolated Malpighian tubules were exposed to *B.t.i.* 25-kDa toxin and, on the basis of this and other observations, suggested that lytic pore formation depended upon the association of a critical number of monomers at one membrane location. Similarly, lateral diffusion of toxin molecules prior to oligomerization may explain the sigmoid dose-response curve obtained for *B.t.i.* 25-kDa toxin on *Ae. aegypti* cells (fig. 4.7). The same result was obtained using the 25-kDa toxin cloned in *B. subtilis* (Drobniewski, unpublished observation). A sigmoid dose-response curve was also obtained by Johnson and Davidson (1984), although they did not comment on its significance. The fact that a similar dose-response curve was observed for melittin (fig. 4.7) is particularly interesting since melittin appears to aggregate into tetramers upon insertion into lipid bilayers, and kinetic analyses of dose-response curves for this cytolytic peptide show that the permeability increases in proportion to the fourth power of the melittin concentration (Vogel and Jahnig 1986). Taken together, the results discussed here support the proposal that the toxic lesion generated by *B.t.i.* 25-kDa toxin in plasma membranes is a pore consisting of an oligomer of toxin molecules.

4.3 DISCUSSION

Studies on the mechanism of action of *B.t.i.* parasporal body have been confined to the 25-kDa protein; however, the other parasporal body proteins are toxic to mosquito larvae, and the interaction between the parasporal body proteins in toxicity and their mechanism of action is under investigation. Histopathological studies do not necessarily give us an indication of mechanism of action since Percy and Fast (1983) suggested that the ultrastructural changes observed in midgut cells treated with *B.t.* parasporal body protein are indicators of cell injury and/or necrosis and are not attributable to a specific toxic action of the parasporal body protein. Studies using insect tissue culture or larval midgut cells will give us a better understanding of the mechanism of action of these toxins. The results described above provide us with some ideas about how *B.t.i.* 25-kDa cytolytic toxin acts. However, there are still unanswered questions. For instance, although *B.t.i.* 25-kDa toxin is cytolytic to all lepidopteran cell lines tested and is lethal to *P. brassicae* larvae on injection into the hemolymph, it is not toxic when administered *per os* (Thomas and Ellar 1983a). There are several explanations for this observation. Either the toxin is inactivated by gut enzymes or is neutralized by binding to lipids in the food or peritrophic membrane before it reaches the midgut cells, or the midgut cells themselves are resistant to the toxin. Preliminary results (Knowles, unpublished observation) suggested that isolated gut cells from *P. brassicae* were less sensitive to the 25-kDa toxin than isolated muscle cells, supporting the latter possibility. An explanation for this might be that, although the phospholipid receptors are undoubtedly present in the midgut cells, they may be inaccessible to the toxin due to steric or charge hindrance from a layer of carbohydrate formed by glycoconjugates at the cell surface. Experimental support for this theory comes from results obtained in a study on erythrocytes (Drobniewski, Knowles, and Ellar 1987). It was observed that sheep red blood cells were insensitive to the 25-kDa toxin compared to human red blood cells (Thomas and Ellar 1983a). However, treatment of sheep red blood cells with neuraminidase to remove the sialic acid from surface glycoconjugates rendered these cells as susceptible to 25-kDa toxin as human red blood cells (Drobniewski, Knowles, and Ellar 1987).

The hypothesis for our model is that the 25-kDa protein binds to a specific plasma membrane receptor, most likely a phospholipid. In the case of mosquito cells, we have evidence to suggest that there is an additional specific receptor, possibly a glycoconjugate. The primary action of the toxin is to insert into the plasma membrane and generate small pores. It is likely that the pores are composed of toxin oligomers. The creation of these pores will lead to colloid-osmotic lysis; that is, an equilibration of ions through the pore resulting in a net inflow of ions, accompanied by influx of water, cell swelling, and eventual cell lysis. We propose that the same mechanism of action can account for the effects of other *B. t.* toxins with very different structures and insect specificities (Knowles and Ellar 1987; Haider and Ellar 1987; Drobniewski and Ellar 1988a).

Acknowledgments

Research carried out in Cambridge and referred to in this chapter was conducted with the aid of grants from the Agriculture and Food Research Council, the Science and Engineering Research Council, Shell Agricultural Chemicals Company, and E. I. Dupont de Nemours and company. BHK is a Royal Society 1983 University Research Fellow.

References

Angus, T. A. 1968. Similarity of effect of valinomycin and *Bacillus thuringiensis* parsporal protein in larvae of *Bombyx mori. J. Invertebr. Pathol.* 11: 145–146.

Argiolas, A., and Pisano, J. J. 1985. Bombolitins: A new class of mast cell degranulation peptides from the venom of the bumblebee *Megabombus pennsylvanicus. J. Biol. Chem.* 260: 1437–1444.

Armstrong, J. L.; Rohrmann, G. F.; and Beaudreau, G. S. 1985. Delta endotoxin of *Bacillus thuringiensis* subsp. *israelensis. J. Bacteriol.* 161 (1): 39–46.

Barjac, H., de. 1978. Etude cytologique de l'action de *Bacillus thuringiensis* var. *israelensis* sur larvas de moustiques. *C. R. Acad. Sci.* (Paris) 286b:1629–1632.

Bhakdi, S.; Tranum-Jensen, J.; and Sziegoleit, A. 1985. Structure of streptolysin O in target membranes. In *Bacterial protein toxins,* ed. J. Alouf, F. Fehrenbach, J. Frier, and J. Jeljaszewicz, 173–180. London: Academic Press.

Charles, J.-F., and Barjac, H. de. 1983. Action des cristaux de *Bacillus thuringiensis* var. *israelensis* sur l'intestin moyen des larves de *Aedes aegypti* L., en microscopie électronique. *Ann. Microbiol.* (Inst. Pasteur) 134A: 197–218.

Cheung, P.Y.K.; Buster, D.; Hammock, B. D.; Roe, R. M.; and Alford, A. R. 1987. *Bacillus thuringiensis* var. *israelensis* δ-endotoxin: Evidence of neurotoxic action. *Pest. Biochem. Physiol.* 27: 42–49.

Cheung, P.Y.K.; Roe, R. M.; Hammock, B. D.; Judson, C. L., and Montague, M. A. 1985. The apparent in vivo neuromuscular effects of the δ-endotoxin of *Bacillus thuringiensis* var. *israelensis* in mice and insects of four orders. *Pest. Biochem. Physiol.* 23: 86–94.

Chilcott, C. N. 1983. *Bacillus thuringiensis* var. *israelensis* delta endotoxin. Ph.D. diss., University of Otago, Dunedin, New Zealand.

Chilcott, C. N., and Ellar, D. J. 1988. Comparative toxicity of *Bacillus thuringiensis* var. *israelensis* crystal proteins in vivo and in vitro. *J. Gen. Microbiol.* 134: 2551–2558.

Chilcott, C. N.; Kalmakoff, J.; and Pillai, J. S. 1984. Neurotoxic and haemolytic activity of a protein isolated from *Bacillus thuringiensis* var. *israelensis* crystals. *FEMS Microbiol. Lett.* 25: 259–263.

———. 1985. Cytotoxicity of two proteins isolated from *Bacillus thuringiensis* var. *israelensis* crystals to insect and mammalian cell lines. *FEMS Microbiol. Lett.* 26: 83–87.

Davidson, E. W., and Yamamoto, T. 1984. Isolation and assay of the toxic component from the crystals of *Bacillus thuringiensis* var. *israelensis. Curr. Microbiol.* 11: 171–174.

Drobniewski, F. A., and Ellar, D. J. 1988a. Investigation of the membrane lesion induced in vitro by two mosquitocidal δ-endotoxins of *Bacillus thuringiensis. Curr. Microbiol.* 16: 195–199.

———. 1988b. Toxin-membrane interactions of *Bacillus thuringiensis* δ-endotoxins. *Biochem. Soc. Trans.* 16: 39–40.

Drobniewski, F. A.; Knowles, B. H.; and Ellar, D. J. 1987. Non-specific ionic effects on the cytolytic and hemolytic properties of *Bacillus thuringiensis* δ-endotoxins. *Curr. Microbiol.* 15: 295–299.

Eisenberg, D.; Schwarz, E.; Komaromy, M.; and Wall, R. 1984. Analysis of membrane and surface protein sequences with the hydrophobic moment plot. *J. Mol. Biol.* 179: 125–142.

Ellar, D. J.; Knowles, B. H.; Carroll, J. C.; Horsnell, J. M.; Haider, M. Z.; Ahmad, W.; Nicholls, C. N.; Armstrong, G.; and Hodgman, C. n.d. Genetic and biochemical studies of the mechanism of action of *Bacillus thuringiensis* entomocidal δ-endotoxins. In *Fourth European workshop.* See Knowles et al. n.d.

Ellar, D. J.; Thomas, W. E.; Knowles, B. H.; Ward, S.; Todd, J.; Drobniewski, F.; Lewis, J.; Sawyer, T.; Last, D.; and Nicholls, C. 1985. Biochemistry, genetics, and mode of action of *Bacillus thuringiensis* δ-endotoxins. In *Molecular biology of microbial differentiation,* ed. J. A. Hoch and P. Setlow, 230–240. Washington, D.C.: American Society for Microbiology.

Gill, S. S., and Hornung, J. M. 1987. Cytolytic activity of *Bacillus thuringiensis* proteins to insect and mammalian cell lines. *J. Invertebr. Pathol.* 50: 16–25.

Gill, S. S.; Singh, G. J. P.; and Hornung, J. M. 1987. Cell membrane interaction of *Bacillus thuringiensis* subsp. *israelensis* cytolytic toxin. *Infect. Immun.* 55: 1300–1308.

Griego, V. M.; Moffett, D. F.; and Spence, K. D. 1979. Inhibition of active K^+ transport in the tobacco hornworm (*Manduca sexta*) midgut after ingestion of *Bacillus thuringiensis* endotoxin. *J. Insect Physiol.* 25: 283–288.

Gupta, B. L.; Dow, J. A. T.; Hall, T. A.; and Harvey, W. R. 1985. Electron probe X-ray microanalysis of the effects of *Bacillus thuringiensis* var. *kurstaki* crystal protein insecticide on ions in an electrogenic K^+-transporting epithelium in the larval midgut in the lepidopteran, *Manduca sexta,* in vitro. *J. Cell. Sci.* 74: 137–152.

Haider, M. Z., and Ellar, D. J. 1987. Analysis of the molecular basis of insecticidal specificity of *Bacillus thuringiensis* crystal δ-endotoxin. *Biochem. J.* 248: 197–201.

Harvey, W. E., and Wolfersberger, M. G. 1979. Mechanism of inhibition of active potassium transport in isolated midgut of *Manduca sexta* by *Bacillus thuringiensis* endotoxin. *J. Exp. Biol.* 83: 293–304.

Himeno, M.; Koyama, N.; Funato, T.; and Komano, T. 1985. Mechanism of action of *Bacillus thuringiensis* insecticidal delta-endotoxin on insect cells in vitro. *Agric. Biol. Chem.* 49: 1461–1468.

Hodgman, T. C., and Ellar, D. J. n.d. Models for the structure and function of the *Bacillus thuringiensis* δ-endotoxins determined by compilational analysis. *Sequence.* In press.

Hurley, J. M.; Bulla, L. A.; and Andrews, R. E. 1987. Purification of the mosquitocidal and cytolytic proteins of *Bacillus thuringiensis* subsp. *israelensis. Appl. Environ. Microbiol.* 53: 1316–1321.

Hurley, J. M.; Lee, S. G.; Andrews, R. E., Jr.; Klowden, M. J.; and Bulla, L. A., Jr. 1985. Separation of the cytolytic and mosquitocidal proteins of *Bacillus thuringiensis* subsp. *israelensis. Biochem. Biophys. Res. Commun.* 126: 961–965.

Ibarra, J. E., and Federici, B. A. 1986. Isolation of a relatively nontoxic 65-kilodalton protein inclusion from the parasporal body of *Bacillus thuringiensis* subsp. *israelensis. J. Bacteriol.* 165 (2): 527–533.

Johnson, D. E., and Davidson, L. I. 1984. Specificity of cultured insect tissue cells for bioassay of entomocidal protein from *Bacillus thuringiensis. In vitro* 20: 66–70.

Knowles, B. H.; Blatt, M. R.; Tester, M.; Horsnell, J. M.; Carroll, J.; Menestrina, G.; and Ellar, D. J. 1989. A cytolytic toxin from *Bacillus thuringiensis* var. *israelensis* forms cation-selective channels in planar lipid bilayers. *FEBS Lett.* 244: 259–262.

Knowles, B. H.; Carroll, J. G.; Horsnell, J. M.; and Ellar, D. J. n.d. Interactions of a cytolytic toxin from *Bacillus thuringiensis* var. *israelensis* with liposomes and membranes. In *Fourth European workshop on bacterial protein toxins,* ed. R. Rappuoli, J. Aluof, J. Frier, F. Fehrenbach, T. Wadstrom, and B. Witholt. Stuttgart: Gustav Fischer Verlag. In press.

Knowles, B. H., and Ellar, D. J. 1986. Characterisation and partial purification of a plasma membrane receptor for *Bacillus thuringiensis* var. *kurstaki* lepidopteran-specific δ-endotoxin. *J. Cell Sci.* 83: 89–101.

———. 1987. Colloid-osmotic lysis is a general feature of the mechanism of action of *Bacillus thuringiensis* δ-endotoxins with different insect specificity. *Biochim. Biophys. Acta* 924: 509–518.

Knowles, B. H.; Thomas, W. E.; and Ellar, D. J. 1984. Lectin-like binding of *Bacillus thuringiensis* var. *kurstaki* lepidopteran-specific toxin is an initial step in insecticidal action. *FEBS Lett.* 168: 197–202.

Kyte, J., and Doolittle, R. F. 1982. A simple method for displaying the hydropathic character of a protein. *J. Mol. Biol.* 157: 105–132.

Lahkim-Tsror, L.; Pascar-Gluzman, C.; Margalit, J.; and Barak, Z. 1983. Larvicidal activity of *Bacillus thuringiensis* subsp. *israelensis* serovar H14 in *Aedes aegypti:* Histopathological studies. *J. Invertebr. Pathol.* 41: 104–116.

Laurent, P., and Charles, J-F. 1984. Action comparee des cristaux solubilses des serotypes H-14 de *Bacillus thuringiensis* sur des cultures des *Aedes aegypti. Ann. Microbiol.* (Inst. Pasteur) 135A: 473–484.

Lüthy, P., and Ebersold, H. R. 1981. *Bacillus thuringiensis* delta-endotoxin histopathology and molecular mode of action. In *Pathogenesis of invertebrate microbial diseases,* ed. E. W. Davidson, 235–268. Totowa, N.J.: Allanheld, Osmum.

McPherson, A.; Jurnak, F.; Singh, G.J.P.; and Gill, S. S. 1987. Preliminary X-ray diffraction analysis of crystals of *Bacillus thuringiensis* toxin, a cell membrane disrupting protein. *J. Mol. Biol.* 195: 755–757.

Maddrell, S.H.P.; Lane, N. J.; Harrison, J. B.; Overton, J. A.; and Moreton, R. B. 1988. The initial stages in the action of an insecticidal δ-endotoxin of *Bacillus thuringiensis* var. *israelensis* on the epithelial cells of the Malpighian tubules of the insect *Rhodnius prolixus. J. Cell Sci.* 90: 131–144.

Maddrell, S.H.P.; Overton, J. A.; Ellar, D. J.; and Knowles, B. H. 1989. Action of activated 27000 M_r toxin from *Bacillus thuringiensis* var. *israelensis* on Malpighian tubules of the insect, *Rhodnius prolixus. J. Cell Sci.* 94: 601–608.

Maharaj, I., and Fackrell, H. B. 1980. Rabbit erythrocyte band 3: A receptor for staphylococcal alpha toxin. *Can. J. Microbiol.* 26: 524–531.

Mayes, M. E.; Held, G. A.; Lau, C.; Seely, J. C.; Roe, R. M.; Dauterman, W. C.; and Kawanishi, C. Y. 1989. Characterization of the mammalian toxicity of the crystal polypeptides of *Bacillus thuringiensis* subsp. *israelensis. Fundamental and Applied Toxicology* 13: 310–322.

Mellor, I. R.; Thomas, D. H.; and Sansom, M.S.P. 1988. Properties of ion channels formed by *Staphylococcus aureus* δ-toxin. *Biochim. Biophys. Acta* 942: 280–294.

Muthukumar, G., and Nickerson, K. W. 1987. The glycoprotein toxin of *Bacillus thuringiensis* subsp. *israelensis* indicates a lectinlike receptor in the larval mosquito gut. *Appl. Environ. Microbiol.* 53: 2650–2655.

Nickerson, K. W., and Schnell, D. J. 1983. Toxicity of cyclic peptide antibiotics to larvae of *Aedes aegypti. J. Invertebr. Pathol.* 42: 407–409.

Nishiitsutsuji-Uwo, J.; Endo, Y.; and Himeno, M. 1979. Mode of action of *Bacillus thuringiensis* δ-endotoxin: Effect on TN 368 cells. *J. Invertebr. Pathol.* 34: 267–275.

Olsnes, S.; Carvagal, E.; Sundan, A.; and Sandvig, K. 1985. Evidence that membrane phospholipids and protein are required for binding of diptheria toxin in Vero cells. *Biochim. Biophys. Acta* 846: 334–341.

Percy, J., and Fast, P. G. 1983. *Bacillus thuringiensis* crystal toxin: Ultrastructure studies of its effect on silkworm midgut cells. *J. Invertebr. Pathol.* 41: 86–98.

Pfannenstiel, M. A.; Muthukumar, G.; Couche, G. A.; and Nickerson, K. W. 1987. Amino sugars in the glycoprotein toxin from *Bacillus thuringiensis* subsp. *israelensis. J. Bacteriol.* 169: 796–801.

Roise, D.; Horvath, S. J.; Tomich, J. M.; Richards, J. H.; and Schatz, G. 1986. A chemically synthesized pre-sequence of an imported mitochondrial protein can form an amphiphilic helix and perturb natural and artifical phospholipid bilayers. *EMBO J.* 5: 1327–1334.

Sacchi, V. F.; Parenti, P.; Hanozet, G. M.; Giordana, B.; Luthy, P.; and Wolfersberger, M. G. 1986. *Bacillus thuringiensis* toxin inhibits K^+-gradient-dependent amino acid transport across the brush border membrane of *Pieris brassicae* midgut cells. *FEBS Lett.* 204: 213–218.

Singh, G.J.P., and Gill, S. S. 1985. Myotoxic and neurotoxic activity of *Bacillus thuringiensis* var. *israelensis* crystal toxin. *Pest. Biochem. Physiol.* 24: 406–414.

Singh, G. J. P.; Schouest, L. P., Jr.; and Gill, S. S. 1986. Action of *Bacillus thuringiensis* subsp. *israelensis* δ-endotoxin on the ultrastructure of the house fly larvae neuromuscular system in vitro. *J. Invertebr. Pathol.* 47: 155–166.

Thelestam, M., and Mollby, R. 1975. Sensitive assay for detection of toxin-induced damage to the cytoplasmic membrane of human diploid fibroblasts. *Infect. Immun.* 12: 225–232.

Thomas, W. E., and Ellar, D. J. 1983a. *Bacillus thuringiensis* var. *israelensis* crystal δ-endotoxin: Effects on insect and mammalian cells in vitro and in vivo. *J. Cell Sci.* 60: 181–197.

———. 1983b. Mechanism of action of *Bacillus thuringiensis* var. *israelensis* insecticidal δ-endotoxin. *FEBS Lett.* 154: 362–368.

Travers, R. S.; Faust, R. M.; and Reichelderfer, C. F. 1976. Effects of *Bacillus thuringiensis* var. *kurstaki* δ-endotoxin on isolated lepidopteran mitochondria. *J. Invertebr. Pathol.* 28: 351–356.

Visser, B.; van Workum, M.; Dullemans, A.; and Waalwijk, C. 1986. The mosquitocidal activity of *Bacillus thuringiensis* var. *israelensis* is associated with M_r 230,000 and 130,000 crystal proteins. *FEMS Microbiol. Lett.* 30: 211–214.

Vogel, H., and Jahnig, F. 1986. The structure of melittin in membranes. *Biophys. J.* 50: 573–582.

Ward, E. S., and Ellar, D. J. 1986. *Bacillus thuringiensis* var. *israelensis* δ-endotoxin: Nucleotide sequence and characterization of the transcripts in *Bacillus thuringiensis* and *Escherichia coli*. *J. Mol. Biol.* 191: 1–11.

———. 1988. Cloning and expression of two homologous genes of *Bacillus thuringiensis* subsp. *israelensis* which encode 130-kilodalton mosquitocidal proteins. *J. Bacteriol.* 170: 727–735.

Ward, E. S.; Ellar, D. J.; and Chilcott, C. N. 1988. Single amino acid changes in the *Bacillus thuringiensis* var. *israelensis* δ-endotoxin affect the toxicity and expression of the protein. *J. Mol. Biol.* 202: 527–535.

Ward, E. S.; Ridley, A. R.; Ellar, D. J.; and Todd, J. A. 1986. *Bacillus thuringiensis* var. *israelensis* δ-endotoxin: Cloning and expression of the toxin in sporogenic and asporogenic strains of *Bacillus subtilis*. *J. Mol. Biol.* 191: 13–22.

Weiner, R. N.; Schneider, E.; Haest, C. W. M.; Deuticke, B.; Benz, R.; and Frimmer, M. 1985. Properties of leak permeability induced by a cytotoxic protein from *Pseudomonas aeruginosa* (PACT) in rat erythrocytes and black lipid membranes. *Biochim. Biophys. Acta* 820: 173–182.

Wu, D., and Chang, F. N. 1985. Synergism in mosquitocidal activity of 26 and 65 kDa proteins from *Bacillus thuringiensis* subsp. *israelensis* crystals. *FEBS Lett.* 190: 232–236.

Yamamoto, T.; Iizuka, T.; and Aronson, J. N. 1983. Mosquitocidal protein of *Bacillus thuringiensis* subsp. *israelensis*: Identification and partial isolation of the protein. *Curr. Microbiol.* 9: 279–284.

5

Genetics of *Bacillus thuringiensis israelensis*

VAITHILINGAM SEKAR

5.1 INTRODUCTION

Bacillus thuringiensis (*B. t.*) is a gram-positive, endospore-forming soil bacterium that has the unique ability to produce proteinaceous, generally bipyramidal, crystalline inclusions during sporulation (Bulla, Kramer, and Davidson 1977). These crystal proteins (δ-endotoxins) exhibit selective toxicity to many lepidopteran insect pests (Dulmage 1979). Over 20 subspecies of *B. t.* have been isolated to date and classified according to their flagellar immunological characteristics (de Barjac 1981). Crystal inclusions of one such subspecies, *B. t.* subsp. *israelensis* (*B. t. i.*), differ from others by virtue of their irregular morphology and also due to selective toxicity to larvae of dipteran insects such as mosquitoes (Goldberg and Margalit 1977). The flagellar immunology of *B. t. i.* is also quite distinct from that of other *B. t.* strains, and hence this organism is the only member of the serotype H14 (de Barjac 1981). Due to its unique insecticidal property, *B. t. i.* has a significant commercial importance in the control of human disease vectors in various parts of the world.

In order to use it as an effective mosquiticidal agent, extensive genetic characterization of *B. t. i.* would be very useful. However, until now most of the genetic studies performed on this bacterium have dealt only with the molecular biology of its crystal toxin. The objective of this review is, therefore, not only to evaluate the current findings on the genetics of its δ-endotoxin, but also to present brief details about some of the genetic methods that are applicable to *B. t. i.* By introducing readers to these simple yet effective methods, it is hoped that an increased interest in gaining a better genetic understanding of this organism is generated among researchers in this area.

5.2 GENETIC EXCHANGE SYSTEMS

To further develop its value as a vector control agent, a more detailed knowledge of the genetics of *B. t. i.* is essential. This could be achieved by developing dependable DNA exchange systems such as transformation, transduction, and conjugation.

5.2.1 Transformation

Efficient transformation of *Bacillus subtilis* protoplasts with plasmid DNA was reported some years ago (Chang and Cohen 1979). This method involved the generation of protoplasts by treating early log-phase cells with lysozyme followed by the addition of plasmid DNA in the presence of polyethyleneglycol (PEG), which induced the uptake of foreign DNA by the protoplasts. Subsequently, the cell walls were regenerated by plating the protoplasts on osmotically stable media. Finally, the transformants were scored by allowing the regenerated cells to grow on appropriate selective media. Following the above report, PEG-induced protoplast transformation of several *Bacillus* species, including *B.t.*, has been reported (Martin, Lohr, and Dean 1981; Alikhanian et al. 1981; Fischer, Lüthy, and Schweitzer 1984). It appears from these reports that a few *B.t.* subspecies are transformable at rather low frequencies, while others are virtually nontransformable. Among these, Fischer, Lüthy, and Schweitzer (1984) were the only authors to attempt to transform *B.t.i.* These authors tried to introduce pC194, a 2.9-kb chloramphenicol-resistance plasmid from *Staphylococcus aureus,* into 11 different serovars of *B.t.*, including *B.t.i.*. In several subspecies (*aizawai, alesti, galleriae,* and *thuringiensis*), a rather low efficiency of transformation (10^{-8} to 10^{-7}) was observed. Although a significantly higher transformation frequency of approximately 10^{-4} was found in *B.t.* subsp. ***kurstaki*** strain HD1, in another ***kurstaki*** strain (strain 4D11A, a plasmid-free variant) the frequency was approximately 10^{-7}. Thus, it appears that there is considerable variation in the transformation frequencies not only between several *B.t.* subspecies, but also among strains of the same subspecies. In the case of *B.t.i.*, however, no transformants were obtained even after screening approximately 10^9 protoplasts. Similar to these results, we have also been unable to introduce plasmid vectors such as pBC16, pUB110, pBD64, and pHV23 into a plasmid-free *B.t.i.* strain by various PEG-induced protoplast transformation methods (V. Sekar and B. J. Brown, unpublished observations). This may be due to inherent difficulties in the uptake and/or the maintenance of foreign DNA molecules in *B.t.i.* Loprasert, Pantuwatana, and Bhumiratana (1986) recently reported on an efficient method for the protoplast transformation of *B.t.i.* These authors have introduced plasmids pC194 and pHV33 into a plasmid-free, lysozyme-sensitive *B.t.i.* strain, O-016. In addition, they have demonstrated that these plasmids could then be transferred to other *B.t.i.* strains via cell mating.

5.2.2 Transduction

The possibility of genetic characterization of *B.t.* by transduction is currently being developed primarily by Thorne and his colleagues (Thorne 1978; Barsomian, Robillard, and Thorne 1984; Ruhfel, Robillard, and Thorne 1984) and by Boman and collaborators (Landen, Heierson, and Boman 1981; Heierson, Landen, and Boman 1983). Stepanova and Azizbekyan (1983) have been able to demonstrate intervarietal transduction between *B.t.* subspecies *galleriae* and *israelensis* by a generalized transducing phage (CP-55) obtained from the laboratory of Thorne. Using another set of soil-derived temperate bacteriophages, TP-13 and TP-18, Thorne and his associates have shown that mapping of large segments of the chromosome of several *B.t.* strains, including *B.t.i.*, is indeed possible. While TP-13, a broad host-range temperate phage, which infected 18 of 19 *B.t.* subspecies tested, is useful in scanning large segments of chromosome (Perlak, Mendelsohn, and Thorne 1979), a narrower host-range phage, TP-18, which infected 9 of 21 *B.t.* subspecies and had a head size seven times smaller than that of TP-13, was effective for ordering markers that were too closely linked (Barsomian, Robillard, and Thorne 1984). Barsomian, Robillard, and Thorne (1984) were able to map three groups of linked markers by using two-, three- and four-factor crosses mediated by these two phages in *B.t.* subsp. *aizawai.*

Another broad host-range temperate phage, CP-51, which is active in generalized transduction of chromosomal markers of *Bacillus cereus, B.t.*, and a nonvirulent *Bacillus anthracis* strain, was isolated by Thorne (1978). Ruhfel, Robillard, and Thorne (1984) showed that CP-51 could also be used to transduce plasmid molecules such as pBC16 and pC194 into strains of *B. cereus, B. anthracis,* and *B.t.* at frequencies as high as 10^{-5}. This finding may prove very useful if a readily transformable *Bacillus* strain that could serve as an intermediate host for CP-51 is found. Recombinant plasmids constructed in vitro or those carried by strains outside the host range of CP-51 could then be introduced into *B.t.i.* or other *B.t.* strains via CP-51 and the intermediate host.

5.2.3 Plasmid Transfer

Another approach to understand the genetics of especially the plasmid molecules found in most *B.t.* strains is the recently discovered plasmid transfer system by cell mating (Gonzalez, Brown, and Carlton 1982). The mechanism of this highly efficient plasmid transfer found in *B.t.* appears to be conjugationlike in that it requires cell-to-cell contact and is resistant to DNase treatment (Chapman and Carlton 1985). In this method, the donor and recipient strains are grown together for a few generations in nutrient broth, and the mixture is plated onto an appropriate medium that is selective for the recipient cells. The transcipients (recipient cells that have acquired one or more plasmids from the donor) are scored by analyzing the plasmid profiles of resulting recipient colonies by the slot-lysis agarose-gel electrophoresis method of Eckhardt (1978) with the modifications suggested by Gonzalez and Carlton (1982). By the cell-mating method, both large and small plasmids were found to transfer equally well between a few *B.t.* strains and into *B. cereus* (Gonzalez and Carlton 1982; Gonzalez, Brown, and Carlton 1982). A recombinant plasmid from *B. subtilis* (containing a cloned *B.t.* toxin gene) transferred readily into *B.t.* subsp. ***kurstaki, israelensis,*** and *sotto* (Klier, Bourgouin, and Rapoport 1983). Lereclus, Menou, and Lecadet (1983) were able to isolate by plasmid transfer a transposonlike structure that was found to be associated with the crystal toxin gene of a *B.t.* subsp. ***kurstaki*** strain using pAMβ1 plasmid of *Streptococcus faecalis* as a conjugative vector. By taking advantage of the plasmid transfer system, Gonzalez and Carlton (1984) were able to demonstrate the transfer of a 75-megadalton (MDa) plasmid, previously implicated in the crystal toxin production of *B.t.i.* (Gonzalez and Carlton 1982; Ward and Ellar 1983), into a plasmid-free, acrystalliferous (Cry^-) recipient. The transfer of this plasmid converted the recipient to a Cry^+ strain, indicating that the 75-MDa plasmid is both necessary and sufficient for the toxin synthesis.

5.3 PLASMIDS AND CRYSTAL TOXIN PRODUCTION

5.3.1 Plasmids and Plasmid Curing Analysis

The presence of plasmids of different sizes has been reported for most *B.t.* strains by several investigators. Similar to these results, it was observed that *B.t.i.* also had an array of plasmids (Clark and Dean 1983; Ward and Ellar 1983; Gonzalez and Carlton 1984). The relative ease with which Cry^- mutants were spontaneously obtained suggested that the crystal toxin gene of *B.t.i.* may be plasmid-borne. In order to further elucidate the exact location of the δ-endotoxin gene, mutants are generated by growing the wild-type strains under conditions that would induce plasmid curing (e.g., growth at 42°C and/or in the presence of mutagenic agents such as ethidium bromide). Plasmid profiles of the cured derivatives, spontaneous Cry^- variants and Cry^+ strains are analyzed by slot-lysis agarose-gel electrophoresis. By comparing their plasmid profiles, correlation between any specific plasmid(s) and crystal production is obtained. By the application of curing techniques, locations of the crystal toxin gene of several other *B.t.* subspecies have been reported (Carlton and Gonzalez 1984; Sekar 1986b).

5.3.2 Location of the δ-endotoxin Gene

Clark and Dean (1983) employed the above techniques to identify any possible involvement of specific plasmid(s) in crystal toxin production in *B.t.i.* Among several plasmids detected, these authors noticed the loss of a large plasmid in most Cry$^-$ mutants. Gonzalez and Carlton (1984) examined the complex plasmid array of *B.t.i.* and several of its partially plasmid-cured mutants that are either Cry$^+$ or Cry$^-$. The type-strain (HD-567, a re-isolate of ONR-60A) was found to contain eight plasmids with approximate sizes of 3.3, 4.2, 4.9, 10.6, 68, 75, 105, and 135 MDa, as well as a plasmidlike linear DNA of approximately 10 MDa. The 75-MDa plasmid was absent consistently in all the 15 partially cured Cry$^-$ mutants. In plasmid transfer experiments using a plasmid-free Cry$^-$ *B.t.i.* strain as a recipient (see section 5.2.3), it was found that the transfer of the 75-MDa plasmid alone is both necessary and sufficient to convert the recipient into Cry$^+$ strain. Southern blot analysis (using ^{32}P-labeled 75-MDa plasmid as a probe) of an unusual variant that did not contain the 75-MDa plasmid but still remained Cry$^+$ revealed that sequences homologous to this plasmid were still present but possibly integrated into the chromosome. It was also observed that the 75-MDa plasmid could recombine with the 68-MDa plasmid, to which it was partially homologous.

In an independent study, Ward and Ellar (1983) also assigned the 75-MDa plasmid (referred to as "72-MDa plasmid" by these authors) an essential role in the crystal toxin production. Analysis of over 100 isolates cured of one or more plasmids revealed that loss of the 75-MDa plasmid was invariably accompanied by the loss of insecticidal δ-endotoxin production, implicating the involvement of the 75-MDa plasmid in the crystal toxin production in *B.t.i.* Observations by Himeno et al. (1985) confirmed the above findings.

The results of other investigators—correlating a low-M_r plasmid (Faust et al. 1983) or a plasmid DNA migrating underneath the chromosome (Kamdar and Jayaraman 1983)—are questionable due to shortcomings in the methods used by these authors.

5.4 CLONING OF THE CRYSTAL TOXIN GENE(S)

The molecular cloning of the crystal toxin gene of *B.t.* subsp. *kurstaki* was reported several years ago by Schnepf and Whiteley (1981), but the finding that there is no nucleotide sequence homology between the *B.t.i.* and (cloned) *B.t. kurstaki* toxin genes (Kronstad, Schnepf, and Whiteley 1983) has left researchers without any suitable DNA probe for the cloning of the *B.t.i.* toxin gene until recently. Within the last few years, however, several reports have appeared on the cloning of the crystal toxin gene of *B.t.i.*

Sekar and Carlton (1985) have described a transformant of *Bacillus megaterium*, VB131, containing a 6.3-kb *Xba* I segment (obtained from purified 75-MDa plasmid) cloned into a *B. cereus* tetracycline-resistance plasmid vector, pBC16. The chimeric plasmid, pVB131, was introduced into *B. subtilis* by competence transformation. Both the *B. megaterium* and the *B. subtilis* recombinant strains produced irregular, proteinaceous crystalline inclusions during sporulation. Sporulated recombinant *B. megaterium* and *B. subtilis* cells, as well as purified crystal preparations from these strains, were found to be highly toxic to second-instar larvae of *Aedes aegypti*. Specific hybridization of ^{32}P-labeled pVB131 DNA with the 75-MDa plasmid of Cry^+ strains was observed, but no hybridization with other plasmids or chromosomal DNA of either Cry^+ or Cry^- strains of *B.t.i.* In addition, it was shown by western blot analysis that the recombinant pVB131 DNA encodes for the 130-kDa polypeptide and not the 28- or 65-kDa peptides of the crystal toxin of *B.t.i.* (Sekar 1986a). The complete DNA sequence of this gene has been determined (Yamamoto et al. 1988). The 130-kDa protein has also been cloned by other research groups in recent years (Bourgouin, Klier, and Rapoport 1986; Tungpradubkul, Settasatien, and Panyim 1988; Sen et al. 1988). A second 130-kDa protein with mosquiticidal properties was described by Visser et al. (1986). This protein has been cloned by Bourgouin, Klier, and Rapoport (1986); Ward and Ellar (1987) and Sen et al. (1988). Bourgouin, Klier, and Rapoport (1986) observed that the initially identified 130-kDa protein (type 1) was highly toxic (LC_{50} = 10–25 ng/ml) to larvae of *Ae. aegypti*. Although the second 130-kDa protein (type 2) was found to be mosquiticidal, the LC_{50} value is not available at present. Interestingly, the type 2 130-kDa protein was found to be identical to the type 1 protein in a region of the C-terminal 467 amino acids (Ward and Ellar 1987; Sen et al. 1988).

Ward, Ellar, and Todd (1984) presented evidence for the cloning of the crystal toxin gene of *B.t.i.* from a purified 75-MDa plasmid preparation. A 9.7-kb *Hin*dIII fragment containing the crystal toxin gene was inserted into the *Escherichia coli* cloning vector, pUC12. The recombinant plasmid pIP174 was introduced into *E. coli*. Concentrated protein preparations obtained from the recombinant strain were found to be extremely cytolytic to *Aedes albopictus* cells in culture. Although the recombinant strain was toxic to larvae of *Ae. aegypti*, no LC_{50} value was reported. It was identified by immunoprecipitation that the product of the cloned gene is the 28-kDa peptide of the *B.t.i.* crystal toxin.

Following a procedure similar to that of Ward, Ellar, and Todd (1984) for cloning and screening of the recombinants, Waalwijk et al. (1985) have also cloned the 28-kDa protein of *B.t.i.* crystal toxin. These authors have constructed a recombinant plasmid, p425, containing a 9.7-kb *Hin*dIII fragment (obtained from total plasmid DNA of *B.t.i.*) cloned into pBR322. The location of the crystal toxin gene within the *Hin*dIII fragment was identified by further subcloning. Nucleotide sequence analysis of the cloned crystal toxin gene revealed an open reading frame with a coding capacity of 249 amino acids (M_r 27,340). The presence of all essential regulatory elements were observed in the nucleotide sequence of the cloned 28-kDa peptide. Despite their presence, only a limited expression of this peptide was seen in *E. coli*. It was postulated that this could be due to poor recognition of *B.t.i.* promoter sequences by *E. coli* RNA polymerase. Bourgouin, Klier, and Rapoport (1986) have reported the cloning of the 28-kDa protein. This gene has also been cloned by McLean and Whiteley (1987). Extract of this clone was found to be somewhat toxic (i.e., LC_{50} = 170 ng/ml) to larvae of *Ae. aegypti*.

Thorne et al. (1986) have reported the cloning of mosquiticidal protein(s) from *B.t.i.* The recombinant clone constructed by these authors, pSY408, was shown by nucleotide sequencing analysis to contain two adjacent ORFs resembling a transcriptional operon. The large ORF had a capacity to code for a protein of approximately 72 kDa. Since the stop codon for the second ORF was not within the sequenced region, the coding capacity of this ORF is unknown at present. When pSY408 was introduced into *B. subtilis* and into a Cry^- *B.t.i.* strain, the recombinant plasmid was found to encode for a peptide of approximately 58 kDa. The recombinant *B. subtilis* strain exhibited no hemolytic activity but had a weak larvicidal activity. N-terminal amino acid sequences corresponding to these two ORFs did not resemble that of either the 28-kDa peptide (Waalwijk et al. 1985) or the 65-kDa peptide (Chestukhina et al. 1985). Since the partial restriction map of pVB131 (Sekar and Carlton 1985), which has been shown to code for the 130-kDa peptide of *B.t.i.* crystal toxin (Sekar 1986a), did not show any homology to that of pSY408, it is likely that this cloned gene also does not code for a truncated product of the 130-kDa peptide. Hence, the exact nature of the gene product of pSY408 remains unknown at this point.

5.5 GENETICS AND BIOCHEMISTRY OF THE CRYSTAL TOXIN

Unlike the alkali-solubilized crystal proteins of most *B.t.* strains, which contain only one major polypeptide of 130 kDa when analyzed by polyacrylamide-gel electrophoresis, the presence of a series of polypeptides ranging in size from 20 kDa to 130 kDa has been observed for *B.t.i.* crystal inclusions by several investigators (Tyrell et al. 1981; Dean 1984; Insell and Fitz-James 1985). It was not clear until recently whether these polypeptides are products of multiple genes or are proteinase digestion products of one large precursor. It appears at present that both of the conditions do contribute to the complexity of the *B.t.i.* crystal inclusions. Recent molecular cloning studies reveal that the *B.t.i.* toxin contains products from at least four different genes, two coding for the 130-kDa peptides (i.e., types 1 and 2), one for the 58-kDa peptide, and one for the 28-kDa peptide. Involvement of additional genes is also possible. In addition, it has been established by current biochemical and cloning experiments that at least one precursor (the 28-kDa peptide) is proteolytically processed to a smaller peptide (the 25-kDa peptide), which is relatively protease resistant (Armstrong, Rohrmann, and Beaudreau 1984; Visser et al. 1986).

Ingestion of the crystal toxin of *B. t. i.* by larvae of susceptible insects such as mosquitoes results in severe toxicity and rapid mortality of these organisms. Nontarget animals, such as other insect species (e.g., *Lepidoptera*) and several higher animals, appear to be quite unaffected when fed with this toxin. However, when solubilized crystal protein is injected into nontarget animals, including mammals, extensive cytolytic and hemolytic activity is exhibited by the crystal toxin of *B. t. i.* (Thomas and Ellar 1983a, 1983b). Assignment of these two different activities to specific peptides has generated an intense debate among researchers in this field (Yamamoto, Iizuka, and Aronson 1983; Thomas and Ellar 1983b; Kim, Ohba, and Aizawa 1984; Lee, Eckblad, and Bulla 1985; Hurley et al. 1985; Ibarra and Federici 1986; Sekar 1986a). Among the several peptides of the *B. t. i.* crystal toxin, only a few (130 kDa, 65 kDa, 58 kDa, and 28 kDa) are considered to be the major components. It appears very clear from recent findings that the 28-kDa peptide and its proteinase-digestion product, 25-kDa peptide, are the primary cytolytic/hemolytic principles of the *B. t. i.* toxin (Hurley et al. 1985; Ward, Ellar, and Todd 1984). The actual mosquiticidal component(s) of the δ-endotoxin is still a matter of controversy. Although it was generally believed until recently that the 28-kDa protein is also the major mosquiticidal component (Thomas and Ellar 1983b; Yamamoto, Iizuka, and Aronson 1983; Sriram, Kamdar, and Jayaraman 1985), it is increasingly becoming evident that this is not the case (Kim, Ohba, and Aizawa 1984; Lee, Eckblad, and Bulla 1985; Cheung and Hammock 1985; Visser et al. 1986; Sekar 1986a). Hence, the mosquiticidal activity must reside within one or more of the other major peptides. Purified 65-kDa peptide has been shown to elicit larvicidal activity (Kim, Ohba, and Aizawa 1984; Lee, Eckblad, and Bulla 1985). But the LC_{50} value obtained from the treatment of neonate *Ae. aegypti* larvae with crystal inclusions containing only the 65-kDa peptide is 43 ng/ml; this value is 50–70 times higher than that obtained for total *B. t. i.* crystals (Ibarra and Federici 1986). It has also been shown (Ibarra and Federici 1986) that the 65-kDa peptide could be processed by proteases to generate a fragment of 38 kDa. This protease-digestion product may possibly be the same as the weakly mosquito-toxic 35-kDa peptide described by Cheung and Hammock (1985). These results indicate that the 65-kDa peptide and its protease-digestion product(s) may not be the prime candidates for the mosquiticidal activity of the *B. t. i.* toxin.

Mosquito larvae (second instar, *Ae. aegypti*) treated with the cloned type 1 130-kDa protein (Sekar and Carlton 1985; Sekar 1986a) showed a

100% mortality at 25 ng/ml, which is somewhat comparable (5- to 10-fold lower) in toxicity to that of the total *B. t. i.* toxin, suggesting that this peptide quite possibly is the primary candidate responsible for the larvicidal activity of the *B. t. i.* crystal inclusions. Other recent evidence from molecular cloning studies (Bourgouin, Klier, and Rapoport 1968; Sen et al. 1988) also suggests that the type 1 and type 2 130-kDa peptides are likely the mosquiticidal components of the *B. t. i.* Still, the observed increased potency of the intact *B. t. i.* crystal inclusions over any of its individual components may very well be due to the result of synergistic interactions (Wu and Chang 1985) between the 130-kDa peptide and other peptides of reduced larvicidal activity (i.e., 65-kDa and 28-kDa peptides).

5.6 CONCLUSIONS

In this review I have briefly outlined some details about a few genetic techniques that are applicable to *B. t. i.* Also included is a summary of the current findings on the genetics of the crystal toxin of this mosquito pathogen. Although attempts to transform protoplasts with plasmid DNA have not been very successful, efficient DNA exchange methods such as transduction and plasmid transfer by cell mating have been developed recently for *B. t. i.* In addition, the exact location of the δ-endotoxin gene(s) of this organism has been identified by several molecular approaches. Cloning and characterization of at least four different toxin genes, responsible for the cytolytic and mosquiticidal activities, have been accomplished. Involvement of one or more additional gene(s) also seems likely. The participation of multiple genes in the crystal toxin production has brought an unexpected complexity to the genetics of *B. t. i.* toxin. Although details about the biological activity spectrum of some of the major peptides are becoming clear now, more information is needed to resolve this problem thoroughly. By the effective use of various genetic and molecular biological methods, we can certainly expect in the near future a wealth of information about the biology of its crystal toxin and about the basic genetics of this effective mosquito pathogen.

Acknowledgments

Assistance and encouragement offered by B. Sekar during the preparation of this manuscript is gratefully acknowledged. I thank M. Clamer, L. Brereton, and T. Jones for expert technical assistance. Thanks are also due to Dr. J. M. Gonzalez, Jr., for critical reading of this manuscript.

References

Alikhanian, S. I.; Ryabchenko, N. F.; Bukanov, N. O.; and Sakanyan, V. S. 1981. Transformation of *Bacillus thuringiensis* subsp. *galleria* protoplasts by plasmid pBC16. *J. Bacteriol.* 146: 7–9.

Armstrong, J. L.; Rohrmann, G. F.; and Beaudreau, G. S. 1985. Delta endotoxin of *Bacillus thuringiensis* subsp. *israelensis*. *J. Bacteriol.* 161 (1): 39–46.

Barjac, H. de. 1981. Identification of H-serotypes of Bacillus thuringiensis. In *Microbial control of pests and plant diseases, 1970–1980,* ed. H. D. Burgess, 35–43. London: Acad. Press.

Barsomian, G. D.; Robillard, N. J.; and Thorne, C. B. 1984. Chromosomal mapping of *Bacillus thuringiensis* by transduction. *J. Bacteriol.* 157: 746–750.

Bourgouin, C.; Klier, A.; and Rapoport, G. 1986. Characterization of the genes encoding the haemolytic toxin and the mosquitocidal delta-endotoxin of *Bacillus thuringiensis israelensis*. *Mol. Gen. Genet.* 205: 390–397.

Bulla, L. A.; Kramer, K. J.; and Davidson, L. I. 1977. Characterization of the entomocidal parasporal crystal of *Bacillus thuringiensis*. *J. Bacteriol.* 130: 375–383.

Carlton, B. C., and Gonzalez, J. M., Jr. 1984. The genetics and molecular biology of *Bacillus thuringiensis*. In *The molecular biology of the bacilli,* ed. D. Dubnau, 2: 211–249. New York: Acad. Press.

Chang, S., and Cohen, S. N. 1979. High frequency transformation of *Bacillus subtilis* protoplasts by plasmid DNA. *Mol. Gen. Genet.* 168: 111–115.

Chapman, J. S., and Carlton, B. C. 1985. Conjugal plasmid transfer in *Bacillus thuringiensis*. In *Plasmids in bacteria,* ed. B. C. Clewell, D. A. Jackson, and A. Hollaender, 453–467. New York: Plenum.

Chestukhina, G. G.; Zalunin, I. A. Kostina, L. I.; Bormatova, M. E.; Klepikova, F. S.; Khodova, O. M.; and Stepanov, V. M. 1985. Structural features of crystal-forming proteins produced by *Bacillus thuringiensis* subspecies *israelensis*. *FEBS Lett.* 190 (2): 345–348.

Cheung, P.Y.K., and Hammock, B. D. 1985. Separation of three biologically distinct activities from the parasporal crystal of *Bacillus thuringiensis* var. *israelensis*. *Curr. Microbiol.* 12: 121–126.

Clark, B. D., and Dean, D. H. 1983. A high molecular weight plasmid is associated with toxicity in *Bacillus thuringiensis* var. *israelensis*. *Abstr. Ann. Mtg. ASM,* 121, H-9.

Clark, B. D.; Perlak, F. J.; Chu, C. Y.; and Dean, D. H. 1984. The *Bacillus thuringiensis* genetic system. In *Comparative pathobiology,* ed. T. C. Cheng, 7: 155–171. New York: Plenum.

Dean, D. H. 1984. Biochemical genetics of the bacterial-insect control agent *Bacillus thuringiensis:* Basic principles and prospects for genetic engineering. In *Biotechnology and genetic engineering reviews,* ed. G. E. Russell, 2: 341–363. Newcastle upon Tyne: Intercept.

Dulmage, H. T. 1979. Genetic manipulation of pathogens: Selection of different strains. In *Genetics in relation to insect management,* ed. M. A. Hoy and J. J. McKelvey, Jr., 116–127. Rockefeller Foundation Working Papers.

Eckhardt, T. 1978. A rapid method for the identification of deoxyribonucleic acid in bacteria. *Plasmid* 1: 584–588.

Faust, R. M.; Abe, K.; Held, G. A.; Iizuka, T.; Bulla, L. A.; and Meyers, C. L. 1983. Evidence for plasmid-associated crystal toxin production in *Bacillus thuringiensis* subsp. *israelensis*. *Plasmid* 9: 98–103.

Fischer, H. M.; Lüthy, P.; and Schweitzer, S. 1984. Introduction of plasmid pC194 into *Bacillus thuringiensis* by protoplast transformation and plasmid transfer. *Arch. Microbiol.* 139: 213–217.

Goldberg, L. J., and Margalit, J. 1977. A bacterial spore demonstrating rapid larvicidal activity against *Anopheles sergentii, Uranotaenia unguiculata, Culex univitattus, Aedes aegypti,* and *Culex pipiens*. *Mosq. News* 37: 355–358.

Gonzalez, J. M., Jr.; Brown, B. J.; and Carlton, B. C. 1982. Transfer of *Bacillus thuringiensis* plasmids coding for δ-endotoxin among strains of *B. thuringiensis* and *B. cereus*. *Proc. Natl. Acad. Sci.* 79: 6951–6955.

Gonzalez, J. M., Jr., and Carlton, B. C. 1980. Plasmid transfer in *Bacillus thuringiensis*. In *Genetic Exchange,* ed. U. N Streips, S. H. Goodgal, W. R. Guild, and G. A. Wilson, 85–95. New York: Marcel Dekker.

________. 1982. Patterns of plasmid DNA in crystalliferous strains of *Bacillus thuringiensis. Plasmid* 3: 92–98.

________. 1984. A large transmissible plasmid is required for crystal toxin production in *Bacillus thuringiensis* variety *israelensis. Plasmid* 11: 28–38.

Heierson, A.; Landen, R.; and Boman, H. G. 1983. Transductional mapping of nine linked chromosomal genes in *Bacillus thuringiensis. Mol. Gen. Genet.* 192: 118–123.

Himeno, M.; Ikeda, M.; Sen, K.; Koyama, N.; Komano, T.; Yamamoto, H.; and Nakayama, I. 1985. Plasmids and insecticidal activity of delta-endotoxin crystals from *Bacillus thuringiensis* var. *israelensis. Agric. Biol. Chem.* 49: 573–580.

Hurley, J. M.; Lee, S. G.; Andrews, R. E., Jr.; Klowden, M. J.; and Bulla, L. A., Jr. 1985. Separation of the cytolytic and mosquitocidal proteins of *Bacillus thuringiensis* subsp. *israelensis. Biochem. Biophys. Res. Commun.* 126: 961–965.

Ibarra, J. E., and Federici, B. A. 1986. Isolation of a relatively nontoxic 65-kilodalton protein inclusion from the parasporal body of *Bacillus thuringiensis* subsp. *israelensis. J. Bacteriol.* 165 (2): 527–533.

Insell, J. P.; and Fitz-James, P. C. 1985. Composition and toxicity of the inclusion of *Bacillus thuringiensis* subsp. *israelensis. Appl. Environ. Microbiol.* 50 (1): 56–62.

Kamdar, H., and Jayaraman, K. 1983. Spontaneous loss of a high molecular weight plasmid and the biocide of *Bacillus thuringiensis* var. *israelensis. Biochem. Biophys. Res. Commun.* 110: 477–482.

Kim, K. H.; Ohba, M.; and Aizawa, K. 1984. Purification of the toxic protein from *Bacillus thuringiensis* serotype 10 isolate demonstrating a preferential larvicidal activity to the mosquito. *J. Invertebr. Pathol.* 44: 214–219.

Klier, A.; Bourgouin, C.; and Rapoport, G. 1983. Mating between *Bacillus subtilis* and *Bacillus thuringiensis* and transfer of cloned crystal genes. *Mol. Gen. Genet.* 191: 257–262.

Kronstad, J. W.; Schnepf, H. E.; and Whiteley, H. R. 1983. Diversity of locations for *Bacillus thuringiensis* crystal protein genes. *J. Bacteriol.* 160: 95–102.

Landen, R.; Heierson, A.; and Boman, H. G. 1981. A phage for generalized transduction in *Bacillus thuringiensis* and mapping of four genes for antibiotic resistance. *J. Gen. Microbiol.* 123: 49–59.

Lee, S. G.; Eckblad, W.; and Bulla, L. A., Jr. 1985. Diversity of protein inclusion bodies and identification of mosquitocidal protein in *Bacillus thuringiensis* subsp. *israelensis. Biochem. Biophys. Res. Commun.* 126 (2): 953–960.

Lereclus, D.; Menou, G.; and Lecadet, M. M. 1983. Isolation of a DNA sequence related to several plasmids from *Bacillus thuringiensis* after a mating involving the *Streptococeus faecalis* plasmid pAMB1. *Mol. Gen. Genet.* 191: 307–313.

Loprasert, S.; Pantuwatana, S.; and Bhumiratana, A. 1986. Transfer of pBC16 and pC194 into *Bacillus thuringiensis* subsp. *israelensis. J. Invebr. Pathol.* 48: 325–334.

McLean, K. M., and Whiteley, H. R. 1987. Expression in *Escherichia coli* of a cloned crystal protein gene of *Bacillus thuringiensis* subsp. *israelensis. J. Bacteriol.* 169: 1017–1023.

Martin, P. A. W.; Lohr, J. R.; and Dean, D. H. 1981. Transformation of *Bacillus thuringiensis* protoplasts by plasmid deoxyribonucleic acid. *J. Bacteriol.* 145: 980–983.

Perlak, F. J.; Mendelsohn, C. L.; and Thorne, C. B. 1979. Converting bacteriophage for sporulation and crystal formation in *Bacillus thuringiensis. J. Bacteriol.* 140: 699–706.

Ruhfel, R. E.; Robillard, N. J.; and Thorne, C. B. 1984. Interspecies transduction of plasmids among *Bacillus anthracis, B. cereus,* and *B. thuringiensis. J. Bacteriol.* 157: 708–711.

Schnepf, H. E., and Whiteley, H. R. 1981. Cloning and expression of the *Bacillus thuringiensis* crystal protein gene in *Escherichia coli. Proc. Natl. Acad. Sci.* 78: 2893–2897.

Sekar, V. 1986a. Biochemical and immunological characterization of the cloned crystal toxin gene of *Bacillus thuringiensis* var. *israelensis. Biochem. Biophys. Res. Commun.* 137 (2): 748–751.

________. 1986b. Location of the delta-endotoxin gene of *Bacillus thuringiensis* var. *aizawai. Curr. Micobiol.* 14: 301–304.

Sekar, V., and Carlton, B. C. 1985. Molecular cloning of the delta-endotoxin gene of *Bacillus thuringiensis* var. *israelensis. Gene* 33: 151–158.

Sen, K.; Honda, G.; Koyama, N.; Nishida, M.; Neki, A.; Sakai, H.; Himeno, M.; and Komano, T. 1988. Cloning and nucleotide sequences of the two 130-kDa insecticidal protein genes of *Bacillus thuringiensis* var. *israelensis. Agric. Biol. Chem.* 52: 873–878.

Sriram, R.; Kamdar, H.; and Jayaraman, K. 1985. Identification of the peptides of the crystals of *Bacillus thuringiensis* var. *israelensis* involved in the mosquito larvicidal activity. *Biochem. Biophys. Res. Commun.* 132 (1): 19–27.

Stepanova, T. V., and Azizbekyan, R. R. 1983. Transduction of strains of *Bacillus thuringiensis* serotype 14. *Dokl. Acad. Nauk.* (USSR) 271: 735–737.

Thomas, W. E., and Ellar, D. J. 1983a. *Bacillus thuringiensis* var. *israelensis* crystal δ-endotoxin: Effects on insect and mammalian cells in vitro and in vivo. *J. Cell Sci.* 60: 181–197.

———. 1983b. Mechanism of action of *Bacillus thuringiensis* var. *israelensis* insecticidal δ-endotoxin. *FEBS Lett.* 154: 362–368.

Thorne, C. B. 1978. Transduction in *Bacillus thuringiensis. Appl. Microbiol.* 35: 1109–1115.

Thorne, L.; Garduno, F.; Thompson, T.; Decker, D.; Zounes, M.; Wild, M.; Walfeld, A. M.; and Pollock, T. J. 1986. Structural similarity between the lepidoptera- and diptera-specific insecticidal endotoxin genes of *Bacillus thuringiensis* subsp. *kurstaki* and *israelensis. J. Bacteriol.* 166 (3): 801–811.

Tungpradubkul, S.; Settasatien, C.; and Panyim, S. 1988. The complete nucleotide sequence of a 130-kDa mosquito-larvicidal delta-endotoxin gene of *Bacillus thuringiensis* var. *israelensis. Nucleic Acids Res.* 16: 1637–1638.

Tyrell, D. J.; Bulla, L. A., Jr.; Andrews, R. E., Jr.; Kramer, K. J.; Davidson, L. I.; and Nordin, P. 1981. Comparative biochemistry of entomocidal parasporal crystals of selected *Bacillus thuringiensis* strains. *J. Bacteriol.* 145 (2): 1052–1062.

Visser, B.; van Workum, M.; Dullemans, A.; and Waalwijk, C. 1986. The mosquitocidal activity of *Bacillus thuringiensis* var. *israelensis* is associated with M_r 230,000 and 130,000 crystal protein. *FEMS Microbiol. Lett.* 30: 211–214.

Waalwijk, C.; Dullemans, A. M.; van Workum, M. E. S.; and Visser, B. 1985. Molecular cloning and nucleotide sequence of the M_r 28,000 crystal protein gene of *Bacillus thuringiensis* subsp. *israelensis. Nucleic Acids Res.* 13 (22): 8207–8217.

Ward, E. S., and Ellar, D. J. 1983. Assignment of the δ-endotoxin gene of *Bacillus thuringiensis* var. *israelensis* to a specific plasmid by curing analysis. *FEBS Lett.* 158: 45–49.

———. 1987. Nucleotide sequence of a *Bacillus thuringiensis* var. *israelensis* gene encoding a 130 kDa delta-endotoxin. *Nucleic Acids Res.* 15: 7195.

Ward, E. S.; Ellar, D. J.; and Todd, J. A. 1984. Cloning and expression in *Escherichia coli* of the insecticidal δ-endotoxin gene of *Bacillus thuringiensis* var. *israelensis. FEBS Lett.* 175: 377–382.

Ward, E. S.; Ridley, A. R.; Ellar, D. J.; and Todd, J. A. 1986. *Bacillus thuringiensis* var. *israelensis* δ-endotoxin: Cloning and expression of the toxin in sporogenic and asporogenic strains of *Bacillus subtilis. J. Mol. Biol.* 191: 13–22.

Wu, D., and Chang, F. N. 1985. Synergism in mosquitocidal activity of 26 and 65 kDa proteins from *Bacillus thuringiensis* subsp. *israelensis* crystals. *FEBS Lett.* 190: 232–236.

Yamamoto, T.; Iizuka, T.; and Aronson, J. N. 1983. Mosquitocidal protein of *Bacillus thuringiensis* subsp. *israelensis:* Identification and partial isolation of the protein. *Curr. Microbiol.* 9: 279–284.

Yamamoto, T.; Watkinson, I. A.; Kim, L.; Sage, M. V.; Stratton, R.; Akande, N.; Li, Y.; Ma, D. P.; and Roe, B. A. 1988. Nucleotide sequence of the gene coding for a 130-kDa mosquitocidal protein of *Bacillus thuringiensis* var. *israelensis. Gene* 66: 107–120.

6

Cloning of *Bacillus thuringiensis israelensis* Mosquito Toxin Genes

THOMAS M. BOYLE
DONALD H. DEAN

6.1 INTRODUCTION

The survey of the literature for the original version of this review was completed in May 1986. A delay in publication, which now seems fortuitous, has allowed time for a more complete picture of the *B.t.i.* toxins to unfold. Section 6.2 describes the status three years ago; essentially all of the toxin proteins were recognized at the time, with the exception of the 127-kDa toxin, which was hidden by the 135-kDa toxin. However, since the genetic analysis was not clear and there was the overriding opinion that *B.t.i.* had only one δ-endotoxin, or at least one main δ-endotoxin, the genetic picture was poorly defined. At this moment, we have a somewhat broader picture of the genes encoding *B.t.i.* toxins. Section 6.3 gives what we hope is the complete view of the toxins of *B.t.i.*

By early 1986, four papers had reported the cloning of *Bacillus thuringiensis* subsp. *israelensis* (*B.t.i.*) mosquiticidal toxin genes (Ward, Ellar, and Todd 1984; Sekar and Carlton 1985; Waalwijk et al. 1985; Thorne et al. 1986b). A comparison of these four papers showed the cloning of three unique DNA fragments, each apparently coding for different toxic proteins. This information added more complexity to an already complex search, which had focused on locating and cloning the *B.t.i.* δ-endotoxin gene as if it were a single entity. More recent reports have led to a clearer understanding of the true nature of *B.t.i.* insecticidal crystal proteins and their mosquiticidal activity. At this time, it appears that the *B.t.i.* mosquito toxin(s) represents a more complex and perhaps more interesting system than the δ-endotoxin of other *B. thuringiensis* varieties.

The cloning of genes from other *B. thuringiensis* varieties was facilitated by two types of data about the nature of the insect toxin genes. One type of study involved the determination of the toxin gene location (usually on a plasmid). This was done initially by plasmid-curing experiments (Gonzalez and Carlton 1980). A second type of investigation that proved invaluable to

B.t. cloning was the extensive biochemical study done on the various crystals. Identification of the sizes and numbers of distinct proteins present and the clear identification of the toxic protein component (Bulla, Kramer, and Davidson 1977; Huber et al. 1981) simplified the search and made the identification of the cloned toxin relatively straightforward. Separation and purification techniques for the toxins allowed specific antibodies to be made (Krywienczyk, Dulmage, and Fast 1978) for use in the cloning procedure. Ultimately, the cloning of various toxin genes confirmed the validity of the research that had preceded it.

In the case of *B.t.i.*, similar investigations to those mentioned above were done. However, difficulty in moving the plasmids of *B.t.i.* by mating and difficulty in performing a good plasmid-screening technique to visualize the entire range of plasmids delayed the identification of the location of the crystal genes (Clark et al. 1984). Eventually, a clear association was shown between the presence of a 112-kb plasmid and crystal production, mosquiticidal activity, and the presence of predominant crystal proteins (Gonzalez and Carlton 1984; Kamdar and Jayaraman 1983; Ward and Ellar 1983; Clark and Dean 1983).

Unfortunately, the biochemical identification of the so-called *B.t.i.* δ-endotoxin was much less successful. The composition and specific activities of the *B.t.i.* crystal were not completely resolved. The seemingly conflicting data about the crystal proteins revealed the difficulty of working with the *B.t.i.* crystal and reflected some of the disparity in the current *B.t.i.* toxin cloning papers.

A more complete explanation of the *B.t.i.* crystal and its biochemistry is presented by Federici, Lüthy, and Ibarra (chapter 3). However, some information about the crystal proteins is presented here to help interpret the results of the four papers reporting the cloning of the *B.t.i.* toxin gene. The information that is most important for determining cloning strategy or for evaluating cloning reports is the number, size, and specific activity of the unique proteins of the crystal that are involved in mosquiticidal activity. At least four groups of proteins have been reported to be responsible for the primary mosquiticidal activity: the 28/25-kDa protein (Yamamoto et al. 1983; Armstrong, Rohrmann, and Beaudreau 1985); the 35-, 34-, 31-kDa protein group (Cheung and Hammock 1985); the 67/65-kDa protein (Hurley et al. 1985; Kim, Ohba, and Aizawa 1984); and the 230- and 130-kDa proteins (Visser et al. 1986). All of these are much less toxic than the reported LC_{50} values for intact crystal (Insell and Fitz-James 1985; Pfannenstiel et al. 1984; Ibarra and Federici 1986a; and Yamamoto, Iizuka, and Aronson 1983), but are comparable to solubilized crystal (Cheung and Hammock 1985; Hurley et al. 1985; Visser et al. 1986). In addition, direct support for the idea of synergism among several of these proteins is presented in a paper by Wu and Chang (1985). They report a very low activity for well-separated crystal proteins when tested individually, but substantial increases in activity when the 26- and 65-kDa proteins are combined and also when the 25- and 130-kDa proteins are combined.

With this degree of confusion in the identification of the mosquiticidal proteins, we have little biochemical guidance in the problem of cloning mosquito toxin genes of *B. t. i.* It could be assumed that only one of the above reports is correct in its assertion of identifying the primary mosquiticidal protein of the crystal, but it could also be assumed that several of the proteins are involved in the mosquiticidal activity. In the first case, cloning and expression of a single protein will yield toxicity equal to *B. t. i.*; in the second case, several genes must be cloned and expressed to recover full *B. t. i.* activity.

A final type of biochemical information that is of great importance in analyzing cloned *B. t. i.* toxin genes is data on amino acid composition and sequence. Unfortunately, prior to 1986 this type of information was available only on one of the crystal proteins, the 25-kDa protein, which was analyzed by Armstrong, Rohrmann, and Beaudreau (1985) and by Davidson and Yamamoto (1984). These papers also report the 25-kDa protein to be a proteolytic cleavage fragment of the 28- to 26-kDa protein.

6.2 EARLY CONFUSION IN CLONING *B.t.i.* TOXIN PROTEIN GENES

The first published report of the cloning of a *B. t. i.* toxic activity was by Ward, Ellar, and Todd (1984). They chose to use *Escherichia coli* as a host for the *B. t. i.* DNA and an in vitro transcription-translation system to assay for the recombinants containing the toxic activity. Plasmids of a partially cured strain of *B. t. i.* were collected and used as the source of DNA. An *E. coli* transcription-translation system was first used to determine potential restriction enzymes for use in the cloning of the gene. A 26-kDa protein was produced by the system using a *Hin*d III total digest for the plasmids. This protein was precipitated using antibodies raised in rabbits against the 26-kDa protein purified by preparative gel electrophoresis. Because of this information, a *Hin*d III digest of the plasmid DNA was used for the cloning of the *B. t. i.* toxin gene. The *Hin*d III digest was ligated to a *Hin*d III digest of pUC12 and used to transform *E. coli* JM101. Recombinant plasmid DNA was pooled and used in the in vitro transcription-translation system to identify clones producing the 26-kDa protein. Two recombinants, pIP173 and pIP174, were identified as making an immunoprecipitable 26-kDa protein. Each clone contained pUC12 with a 9.7-kb insert. A restriction map of an internal *Eco*RI (the gene-containing) fragment of this insert, given in a later publication (Ellar et al. 1985), is shown in figure 6.1. The clone containing pIP174 and a control of *E. coli* JM101 containing pUC12 (alone) were used to determine cytotox-

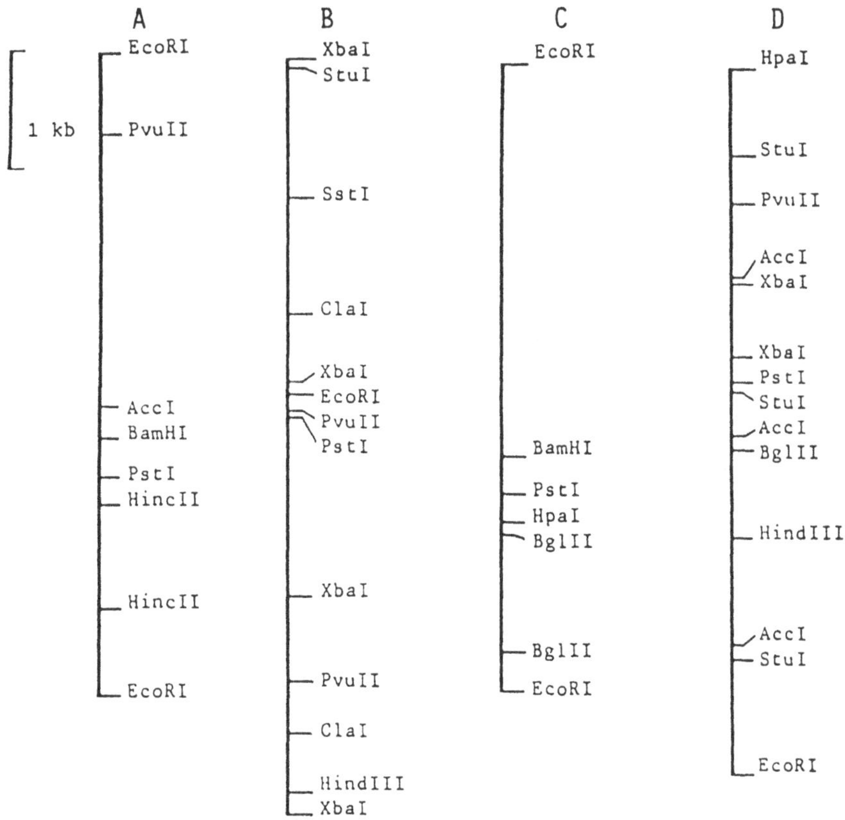

FIGURE 6.1 Restriction maps of the cloned *B.t.i.* DNA fragments containing toxin genes: (*A*) *B.t.i.* DNA fragment of pIPEco5 (Ellar et al. 1985); (*B*) *B.t.i.* DNA fragment of pVB131 (Sekar and Carlton 1985); (*C*) internal *Eco*RI fragment of *B.t.i.* DNA in p425 (Waalwijk et al. 1985); (*D*) *B.t.i.* DNA fragment of pSY368 (Walfield et al. 1986).

icity and mosquiticidal activity of this gene. Protein equivalent to 6 ml of the original *E. coli* culture produced typical cytotoxic effects on *Aedes albopictus* cells after 4 hours. Equivalent protein from the control showed no effect. Cytotoxicity was neutralized by the above-mentioned antiserum as well as by preincubation with sonicated preparations of phospholipids previously reported as cell membrane receptors for the *B.t.i.* δ-endotoxin. *Aedes aegypti* larvae were reported to be killed in 4 hours when offered the equivalent of 25 ml of a 16-hour culture of the clone containing pIP174 resuspended in distilled water. Controls using an equivalent amount of bacteria were unaffected. Because the *E. coli* promoter used in cloning the gene was from the β-galactosidase gene, the control of the 26-kDa protein gene in *E. coli* 174 was tested using a β-galactosidase inducer. This failed to raise the cytotoxicity, therefore suggesting that the gene is not using the promoter of pUC12 but its own promoter.

An article by Sekar and Carlton (1985) also reported the cloning of the *B.t.i.* toxin gene. In this paper, the authors chose *Bacillus megaterium* as the host for cloning the *B.t.i.* gene. Purified 112-kb plasmid from a *B.t.i.* strain carrying only this plasmid was used as the source of the gene. An *Xba*I partial digest of this DNA was ligated to an *Xba*I total digest of plasmid vector pBC16 collected from *B. megaterium* VT1650. The ligated DNA was used to protoplast transform *B. megaterium* 1660. Twenty-eight tetracycline-resistant transformants contained inserts of 0.8 to 15 kb. One isolate of these 28, VB131, also produced phase-refractile parasporal inclusions. Only recombinant VB131 was found to be mosquiticidal in a plate bioassay using *Ae. aegypti.* The inclusions isolated from VB131 caused 100% mortality in 36 hours at a concentration of 25 ng/ml. This assay was performed on 20 second-instar *Ae. aegypti* larvae in 25 ml plastic cups containing 20 ml deionized water. Solubilized crystals from VB131 showed cross-reactivity to rabbit antiserum raised against *B.t.i.* crystal protein in a double diffusion plate.

Crystals of the *B.t.i.* strain that had been the donor of the 112-kb plasmid were used as a comparison for mosquiticidal activity. The data given showed similar activity using a concentration of 5 ng/ml. The authors reported that at similar concentrations the *B.t.i.* crystals acted about five times faster.

Digestion of pVB131 yielded pBC16 and three *Xba*I fragments, calculated as a total insert of 6.3 kb. A restriction map of this insert can be found in figure 6.1. *Bacillus subtilis* strain MI112 was transformed with purified pVB131 plasmid DNA using a competent DNA transformation protocol. The transformant SB131 also produced parasporal inclusions and showed mosquiticidal activity similar to VB131.

A southern blot hybridization procedure using nick-translated pVB131 showed no homology between this insert and any other normal plasmids of *B.t.i.* Subcloning of the individual *Xba*I fragments was done, placing each fragment by itself in *B. megaterium.* None of the subclones produced crystals or had mosquiticidal activity.

The cloning and nucleotide sequence of the 28-kDa crystal protein of *B.t.i.* was reported by Waalwijk et al. (1985). The authors chose *E. coli* as the host for the *B.t.i.* gene and, as Ward, Ellar, and Todd (1984) had done, used an in vitro transcription-translation system to identify *B.t.i.* genes. Plasmid DNA from *B.t.i.* IPS82 was used as the source of the toxin gene. Samples of this DNA were digested with several restriction enzymes, including *Hin*d III, and ligated to linearized pBR322. This DNA was used to transform competent *E. coli* RRI cells. Recombinant DNA was tested in the transcription-translation system using (^{35}S)-methionine to label polypeptides and immunoprecipitation with antiserum raised against whole solubilized *B.t.i.* crystal. A recombinant plasmid designated p425 produced a 28-kDa protein as well as some 25-kDa protein, both of which were precipitated by the antibody. The plasmid p425, which originated from the *Hind* III digest, contained a 9.7-kb *Hind* III insert. Upon further subcloning, the gene was located on an internal *Eco*RI fragment. Some restriction data on this fragment is given in figure 6.1. A nick-translated probe of p425 was used to monitor the production of mRNA for the gene during growth and sporulation. A weak signal from an 800–900 nucleotide mRNA was seen during vegetative growth. The mRNA was most highly transcribed during early sporulation. S-1 nuclease mapping data showed that only the mRNA samples, taken at 4 and 7 hours after the onset of sporulation, protected by 320-bp probe made over a part of the *Eco*RI fragment.

Nucleotide sequencing was done on a portion of the insert approximately 1,000-bp long. Only one open reading frame large enough to specify the 28-kDa protein was revealed by the sequence. The coding capacity of this reading frame is 249 amino acids. The nucleotide sequence showed the presence of 3 in-frame stop codons, an ATG initiation codon, and an apparent Shine-Dalgarno sequence, AAGGAG. The data also revealed a low GC content of 33.6% and a preference for T or A in the third position of the codons. Regarding transcription, the sequence data showed some homology between the various *B. subtilis* −10 and −35 consensus sequences and the sequences in the 5′ proximal end of the cloned gene, but the spacing between the sequences was incorrect. A double hairpin was shown in the sequence following the stop codons. The deduced amino acid sequence predicts a protein of 27,340 daltons. This sequence also agrees with the amino acid composition of a 25-kDa toxin protein-purified trypsin protease k digestion of the *B.t.i.* crystal (Armstrong, Rohrmann, and Beaudreau 1985). The deduced amino acid sequence from residue 30–59 match the NH_2-terminal analysis of this same 25-kDa protein with the exception of a single amino acid (residue 57) where Met is predicted and Ser is found (Armstrong, Rohrmann, and Beaudreau 1985).

An abstract by Thorne et al. (1986a) included a report of cloning a gene of *B.t.i.* that coded for a mosquiticidal polypeptide in *E. coli* and *B. subtilis.* A copy of the complete article reporting the cloning of the gene was kindly sent to us after the meeting (Walfield et al. 1986) along with a copy of the title article comparing the *B.t.i.* gene with a toxin gene of *B. thuringiensis* subsp. *kurstaki* (Thorne et al. 1986b). Both papers contained information about the cloning of a *B.t.i.* insecticidal crystal protein gene and its subsequent characterization. The authors used *E. coli* as the initial host for the *B.t.i.* genes. Purified plasmid DNA from *B.t.i.* strain ONR-60A was used as the source of the toxin genes. The purified plasmids were partially digested with restriction enzyme *Sau*3A to produce 10- to 20-kb overlapping fragments. Vector DNA, λ L47.1, was covalently circularized, digested with *Bam*HI, CIP treated and ligated to the *B.t.i.* partial digests. The resulting DNA was packaged in vitro and transfected into *E. coli* Q359. Resulting plaques were screened in situ using rabbit antiserum raised against both intact and solubilized purified *B.t.i.*

crystals. Fifteen reactive clones were plaque-purified. An *E. coli* maxicell system was used to determine the proteins coded for by the recombinant *B.t.i.* inserts. Of the immunoreactive clones, three were reported to produce unique polypeptides that were immunoprecipitable.

SDS-PAGE analysis of these unique polypeptides showed sizes of 67 ±3 kDa for isolate F2, 92 ±4 and 35 ±2 kDa for isolate C2-1, and 95 ±4 and 59 ±3 kDa for isolate F1. These recombinant phage-infected *E. coli* were tested for mosquiticidal activity against third- and fourth-instar *Ae. aegypti* larvae. Unlysed cells of these clones showed no toxicity. Crude lysates of these clones showed unreproducible levels of toxicity. Protein precipitates using ammonium sulfate at 10% of saturation were usable for toxicity determinations. The precipitates, which were insoluble after dialysis in water, were added to third- and fourth-instar larvae of *Ae. aegypti.* Clone F1 (reported as the most reproducibly toxic) caused a toxicity of 90% in 45 hours at a concentration of 100 μg/ml. The control proteins showed a 10% mortality in comparison.

Subcloning was done from the 11-kb F1 insert using vector pUC13. Toxicity of the subclones was determined by using whole cells. An internal *Eco*RI fragment of 8,500 bp was subcloned and found to retain toxicity. Further subcloning produced a smaller *Hpa*I to *Eco*RI fragment, designated pSY368, of approximately 6,000 bp and was also found to retain toxicity in both orientations. The LC_{50} for pSY368 *E. coli* was 2,000 μg/ml of lyophilized cells. The restriction map of this *Hpa*I to *Eco*RI fragment is given in figure 6.1. This *Hpa*I to *Eco*RI fragment was subsequently placed into *B. subtilis* using pUB110.

The process was accompanied by a spontaneous deletion, which removed the *B.t.i.* DNA from *Hpa*I to just beyond the rightmost *Bal*I site. This new plasmid was designated pSY408. The *B. subtilis* containing pSY408 was also found to be toxic with an LC_{50} of approximately 100 μg/ml of lyophilized cells. Preliminary tests of partially purified toxin from this *B. subtilis* showed no hemolysis to human red blood cells. This *B. subtilis* clone produced an immunoreactive protein of 58 kDa. Plasmid pSY408 was transferred into a *B.t.i.* strain, which had been previously cured of the 72-MDa plasmid. Mosquito toxicity was restored to 10% of the wild-type activity, and the cloned gene was expressed as a 58-kDa protein. This protein is believed by the authors to be a processed product because the gene codes for polypeptides as large as 72 kDa in an in vitro *E. coli* transcription and translation system.

Nucleotide sequencing was performed on about 3,750 nucleotides of the cloned fragment. Two open reading frames were discovered oriented in the same direction. The first had a coding capacity of 72 kDa. The second had a coding capacity of at least 26 kDa, but continued beyond the boundary of sequence analysis. The nucleotide sequence and the deduced amino acid sequence revealed other information about the gene and its similarity to the Lepidoptera-specific toxin gene of *B. thuringiensis* subsp. *kurstaki.* Two regions of homology upstream of the structural portion of ORF 1 were found between the *B. t. i.* DNA and the *B. t. k.* genes control region. No DNA homology was found between the two genes in their structural areas, but similarities were found between the deduced amino acid sequences of both genes and in their patterns of hydrophobic and hydrophilic regions. Both of these toxins began with a hydrophilic region, followed by a hydrophobic segment of about 32 amino acids, then, approximately 200 residues downstream, there was a sharp transition from very hydrophobic to very hydrophilic amino acids, a property associated with many insecticidal crystal proteins.

The sequence data also showed three potential translation start points for ORF 1, one of which was preceded by a GGAGG sequence consistent with the consensus ribosome binding sequence. A potential start point for ORF 2 was also found. It was also preceded by a possible ribosome binding site in a position that left the two reading frames separated by as few as 64 bases. Using DNA hybridization, the authors found the F1 cloned insert to be derived from the 70- to 75-MDa plasmid. Finally, it is reported by the authors that antibody raised against the 26-kDa protein, purified by SDS-PAGE gels, cross-reacted with the protein produced by their clone as well as other components of the crystal. It is noteworthy that there is no major crystal component of either 72-kDa or 58-kDa in the crystal, and it is not clear at the time of this publication whether the genes on this cloned DNA, albeit mosquiticidal, are actually crystal proteins.

In summary of the first section, we have prepared a figure that draws together the reported cloned genes to that point in time (fig. 6.1 a, b, c, d). A close examination of figure 6.1 shows that the various attempts to clone the *B. t. i.* δ-endotoxin resulted in the discovery of three distinct DNA fragments that encode mosquito toxicity. Dissimilarity in restriction patterns of these three distinct fragments, as well as the nucleotide sequence data available on portions of the fragments (Waalwijk et al. 1985; Walfield et al. 1986), shows that the genes coded for by these fragments are not the same and do not appear to be closely related.

The restriction data and the size and immunoreactivity reported for the proteins made by pIPEco5 (Ward, Ellar, and Todd 1984) and p425 (Waalwijk et al. 1985) clearly show these DNA fragments to be identical. The 27.3-kDa protein coded for by this DNA fragment has also been shown to be almost completely identical to the amino acid composition and N-terminal sequence data of the 25-kDa protein, which Armstrong, Rohrmann, and Beaudreau (1985) and Davidson and Yamamoto (1984) have analyzed. The genes cloned by Ward, Ellar, and Todd (1984) and by Waalwijk et al. (1985) therefore account for the 28/26-kDa protein of the *B. t. i.* crystal. One piece of data that is unfortunately not available in these publications is the actual LC_{50} of the purified cloned protein. This information would be quite enlightening considering the diversity of the various reports about its activity. It is also clear that the toxic activity of this protein is not expressed well in *E. coli.*

The fragment of DNA cloned by Sekar and Carlton (1985) not only makes a parasporal inclusion, but is sufficiently toxic to show that it most likely contains one or more of the other crystal proteins previously implicated in toxicity. Unfortunately, since no data have been presented on the size of immunological reactivity of the product(s) made by this cloned fragment, we can only speculate as to which proteins it might be making. The size of the fragment and its failure to maintain activity when the individual *Xba*I fragments were subcloned causes us to consider that it could code for any one or more of the crystal proteins up to and including the 145-kDa protein.

The paper by Walfield et al. (1986) reports three isolates, each making proteins in *E. coli* maxicells that are recognized by crystal-specific antibodies. A search of the literature on the *B.t.i.* crystal proteins reveals at least one report showing a crystal protein close to the same size as each of the proteins made by the three clones above; that is, 92, 67, 55, and 35 kDa. Clone F1, whose initial immunoreactive products were given as 95 and 59 kDa, was eventually shown with an open reading frame capable of coding for a 72-kDa protein and was actually found to make a product of 58 kDa in *B. subtilis.* This information suggests that the F1 gene could be the source of any of several reported crystal proteins. However, nucleotide sequence data show that it is not the 28-kDa protein, as discussed above, and it is not large enough to encode the 135-kDa protein. Its toxic activity and the fact that it makes a 58-kDa protein in *B.t.i.* suggest the possibility that it could be the source of the 67/65-kDa protein; however, the authors reporting this gene have no evidence at this time to show that this is the case. In fact, this protein may actually be another smaller protein of the size 55–53 kDa (Yamamoto, Iizuka, and Aronson 1983; Pfannenstiel et al. 1984), which have not previously been associated with toxic activity.

6.3 CURRENT PICTURE OF *B.t.i.* TOXIN PROTEIN GENES

The first paper that began to draw together pieces of the puzzle was by Bourgouin, Klier, and Rapoport (1986). They reported cloning the 28-kDa cytolytic toxin gene and found it to be hemolytic but not mosquiticidal in *E. coli.* The most important aspect of this paper, however, was the cloning of the 130-kDa crystal protein gene. The cloning attempt was not straightforward and involved first cloning a fragment that was not active against mosquitoes, but encoded a protein that cross-reacted against crystal protein antisera. This cloned gene (clone pCP1) hybridized against two *Eco*RI fragments of the 72-MDa plasmid. The authors correctly reasoned that the other fragment contained the *B.t.i.* δ-endotoxin gene. They constructed a gene bank, which yielded two other cloned fragments. One of the clones (clone pRX7, which was not exactly the same as pCB1) was not toxic to mosquitoes, but encoded a protein that cross-reacted to crystal protein antisera; the other clone (pRX8) encoded a 130-kDa protein that was very active against mosquitoes. Toxicity against *Culex pipiens* was $LC_{50} = 1$ μg/ml in terms of *E. coli* protein. This compares to 2 μg/ml for solubilized crystal protein for *B.t.i.* and 2 ng/ml for whole crystal. The success of this tour de force effort is a credit to the confidence and energy of this French team from the Institut Pasteur.

Several other interesting facts were reported in this paper (Bourgouin, Klier, and Rapoport 1986). The 130-kDa protein from the pRX8 clone was also active against *Ae. aegypti,* but less so than against *Cx. pipiens.* This is contrary to the specificity ratio of *B.t.i.* crystal and thus created a paradox, which was to be resolved in later papers by other authors. Bourgouin, Klier, and Rapoport (1986) proposed the theory that the 72-MDa plasmid contains at least one toxin gene (type 1; e.g., the pRX8 cloned gene) and another type of gene, which contained an extensive region of homology (type 2; e.g., the pRX7 cloned gene). This turned out to be a correct prediction. The authors observed that the cloned gene of pRX8 contained the same restriction enzyme pattern as that described by Sekar and Carlton (1985). They also noticed that ORF 1 and ORF 2 (Thorne et al. 1986b) were also included on the cloned insert of pRX8 (but not on pRX7 or pCB1). This observation linked these two mosquito toxin genes in juxtaposition on the 72-MDa plasmid.

A brief report by Sekar (1986) confirmed that the protein component expressed by the clone VB131, from the previous paper of Sekar and Carlton (1985), was the 130-kDa toxin protein. This missing piece of data confirmed the work of Bourgouin, Klier, and Rapoport (1986).

Independently, the Thai group at Mahidol University in Bangkok cloned the 130-kDa gene in *E. coli* (Angsuthanasombat et al. 1987). Starting with the pure 72-MDa plasmid as the target DNA, 800 candidate clones were obtained. Thirty-two of these were selected by hybridization with stationary phase-enriched stable mRNA, and 17 of these hybridized anti–δ-endotoxin antisera. Five of these that gave the strongest signal were toxic to insects. A partial DNA sequence of the 5′ end of the cloned gene revealed homologies to the lepidopteran insecticidal crystal protein genes (see Höfte and Whiteley 1989), but no homologies to the 28-kDa or 72-kDa genes of *B.t.i.* The complete nucleotide sequence of this gene was published by some of the same authors (Tungpradubkul, Settasatien, and Panyim) in 1988.

Meanwhile, Ward and Ellar (1987) also published a nucleotide sequence of a 130-kDa mosquiticidal gene from *B.t.i.* This gene sequence was clearly different from that of Angsuthanasombat et al. (1987), and the restriction enzyme map was clearly different from that of Sekar and Carlton (1985) and Bourgouin, Klier, and Rapoport (1986). Since the paper was only a brief note listing the sequence and no other data was given, it was difficult to assess the significance of this new gene.

The paradox was quickly resolved by a publication from a collaborative effort between Japanese groups at Kyoto University and Tokyo University reporting the cloning and sequencing of two 130-kDa genes from *B.t.i.* (Sen et al. 1988). One gene encoded 1,180 amino acids (sufficient for a 134.4-kDa protein), and the other encoded 1,136 amino acids (sufficient for a 127.8-kDa protein). The ratio of successful clones to potential candidates (7:1,000), despite using high molecular weight *B.t.i.* plasmids as the target DNA, also indicated the relative difficulty of cloning these genes. The 3′ terminal regions of these two genes were identical, but the 5′ terminal regions were nonhomologous except for limited tracts of conserved amino acids. The 134.4-kDa gene differed by four amino acids from the sequence reported by Ward and Ellar (1987), and by three amino acids from the sequence reported by Tungpradubkul, Settasatien, and Panyim (1988).

Two other research groups reported the DNA sequence of the 127.8-kDa protein gene in 1988, Chungjatupornchai et al. (1988) from Plant Genetic Systems in Belgium and Yamamoto et al. (1988) from Shell Development Corporation. The latter researchers cloned the actual gene cloned by Sekar and Carlton (1985). Their reported sequences differed from the 127.8-kDa protein gene of Sen et al. (1988) by 1 and 97 amino acids respectively. The 127.8-kDa mosquiticidal protein is more active against *Ae. aegypti* than *Cx. pipiens,* explaining the unusual insect spectrum observed with the original *E. coli* clone of the 134.4-kDa gene of Bourgouin, Klier, and Rapoport (1986). Another reason for the specificity of action is explained by the relatively greater *Aedes* activity of the ORF 1 gene product of Thorne et al. (1986b) as demonstrated by Delecluse et al. (1988).

One major *B.t.i.* crystal protein remained to be cloned, the 65- or 72-kDa protein, and that was also reported in late 1988 by the group at Ecogen, Inc. (Donovan, Dankocsik, and Gilbert 1988). This gene was successfully cloned in *B. megaterium,* where it was expressed as a crystal. The gene did not express detectable levels of protein in *E. coli.* DNA sequence analysis revealed significant homologies (33%) between certain regions of the 72-kDa protein gene and regions of the P2 toxin of *B. thuringiensis* (Yamamoto and McLaughlin 1981; Donovan et al. 1988), which is active against both lepidopterans and dipterans, and the coleopteran toxin from *B. thuringiensis* subsp. *tenebrionis* and *san diego* (Herrnstadt et al. 1986; Sekar et al. 1987). No significant homologies were observed between the ORF 1 protein gene of Thorne et al. (1986b) or the 28-kDa protein gene (Waalwijk et al. 1985).

6.4 DISCUSSION

B.t.i. is a bacterium that has evolved an elaborate set of toxins to create an additional ecological niche as a mosquito pathogen. We presume its major ecological niche is as a soil microorganism. Because of its enormous importance as a microbial pesticide with slight activity on nontarget organisms, there has been a flurry of activity to define the important factors in de-

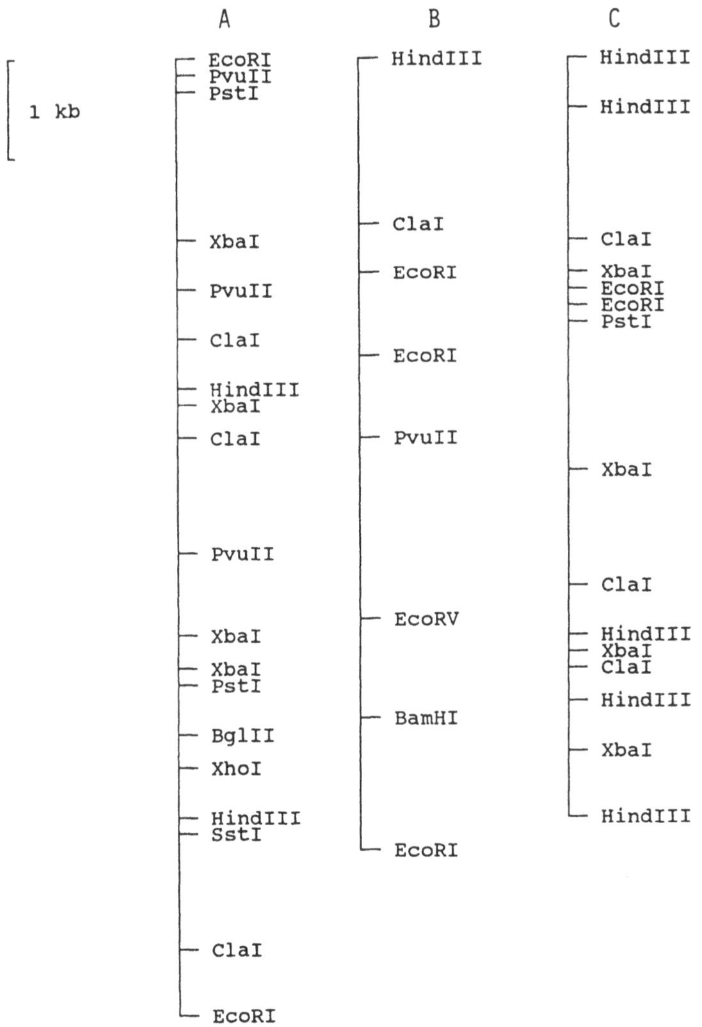

FIGURE 6.2 Restriction maps of cloned *B.t.i.* DNA fragments containing toxin genes: (*A*) *B.t.i.* DNA fragment of pRX8 containing the *cryIVA* and *cryIVC* genes (Bourgouin, Klier, and Rapoport 1986); (*B*) part of the *B.t.i.* DNA fragment of pEG216 containing the *cryIVD* and *cytA* genes (Donovan et al. 1988); (*C*) *B.t.i.* DNA fragment of pBGH3 containing the *cryIVB* gene (Sen et al. 1988).

termining the mechanism of mosquiticidal activity. Molecular cloning has played an important role in this process. At this moment, all of the known major crystal proteins involved in mosquiticidal activity have been cloned and sequenced; also, there is an additional mosquiticidal protein gene, the 72-kDa protein toxin gene, or ORF 1 gene, which apparently encodes a non-crystal toxin.

What have we learned about these toxins from these genetic studies? We have learned that two of the major proteins are very similar in overall structure to the insecticidal crystal proteins of lepidopteran activity. They are large protoxins of approximately 130 kDa, which are processed by proteolytic action to become active cytolytic toxins (Chungjatupornchai et al. 1988). There are certain structural features in common between the *B.t.i.* insecticidal crystal proteins and other *B. thuringiensis* insecticidal crystal proteins, namely, short tracts of conserved amino acids. These are of unknown function, but may play structural as well as functional roles. A recent review by Höfte and Whiteley (1989) addresses these conservative tracts of amino acids. Another feature of this review is the adoption of a nomenclature for the insecticidal crystal protein genes of *B. thuringiensis*. The pertinent names of genes from *B.t.i.* are as follows: the 134.4-kDa crystal protein gene is *cryIVA;* the 127.8-kDa crystal protein gene is *cryIVB;* the 72-kDa gene, or ORF 1 protein (of unknown cellular location) gene, is *cryIVC;* the 65/72-kDa crystal protein gene is *cryIVD;* and the 28-kDa crystal protein gene is *cytA.* The latter is not considered to be of the same type as the other mosquiticidal protein genes by Höfte and Whiteley (1989) because it is hemolytic and structurally quite different. Restriction maps of *B.t.i.* DNA containing these five toxin genes are shown in fig. 6.2.

The final question (and the first one) is also the question leading to the necessity to clone the mosquiticidal genes of *B.t.i.:* Which gene is the real δ-endotoxin? The answer is now clearer. There are many toxins, and they all may be considered δ-endotoxins by the original definition (Heimpel 1967), except the ORF 1 gene product, which is apparently not a crystal protein. None of the toxins alone has been shown to contain the total mosquiticidal activity of *B.t.i.,* so perhaps we should have expected a synergistic action of the several toxins from the beginning.

What is left for the future and what will be the role of molecular cloning? Now that the genetic nature of some of the mosquiticidal protein genes is known to the level of the nucleotide sequence, genes may be introduced into organisms that are more persistent in the environment of the mosquito than *B.t.i.* (chapter 7). This has already begun in the introduction of mosquiticidal genes into blue-green algae (Tandeau de Marsac, de la Torre, and Szulmajster 1987) and *B. sphaericus* (Bourgouin et al. 1989). Genetic knowledge will also lead to further genetic engineering and improvement of the proteins as durable pesticides that are competitive with chemical pesticides.

References

Angsuthanasombat, C.; Chungjatupornchai, W.; Kertbundit, S.; Luxananil, P.; Settasatian, C.; Wilairat, P.; and Panyim, S. 1987. Cloning and expression of 130-kd mosquito-larvicidal delta-endotoxin gene of *Bacillus thuringiensis* var. *israelensis* in *Escherichia coli. Mol. Gen. Genet.* 208: 384–389.

Armstrong, J. L.; Rohrmann, G. F.; and Beaudreau, G. S. 1985. Delta-endotoxin of *Bacillus thuringiensis* subsp. *israelensis. J. Bacteriol.* 161 (1): 39–46.

Bourgouin, C.; Delecluse, A.; de la Torre, F.; and Szulmajster, J. 1989. Transfer and expression of the toxin protein gene of *Bacillus sphaericus* into *Bacillus thuringiensis israelensis.* Paper read at the Fifth International Conference on Genetics and Biotechnology of Bacilli. Asilomar, Calif.

Bourgouin, C.; Klier, A.; and Rapoport, G. 1986. Characterization of the genes encoding the haemolytic toxin and the mosquiticidal delta-endotoxin of *Bacillus thuringiensis israelensis. Mol. Gen. Genet.* 205: 390–397.

Bulla, L. A.; Kramer, K. J.; and Davidson, L. I. 1977. Characterization of the entomocidal parasporal crystal of *Bacillus thuringiensis. J. Bacteriol.* 130: 375–383.

Cheung, P. Y. K., and Hammock, B. D. 1985. Separation of three biologically distinct activities from the parasporal crystal of *Bacillus thuringiensis* var. *israelensis. Curr. Microbiol.* 12: 121–126.

Chungjatupornchai, W.; Höfte, H.; Seurinck, J.; Angsuthanasombat, C.; and Vaeck, M. 1988. Common features of *Bacillus thuringiensis* toxins specific for Diptera and Lepidoptera. *Eur. J. Biochem.* 173: 9–16.

Clark, B. D., and Dean, D. H. 1983. A high molecular weight plasmid is associated with toxicity in *Bacillus thuringiensis* var. *israelensis. Abstr. Ann. Mtg. ASM,* 121, H-91.

Clark, B. D.; Perlak, F. J.; Chu, C. Y.; and Dean, D. H. 1984. The *Bacillus thuringiensis* genetic systems. In *Comparative physiology,* ed. T. C. Cheng, 7: 155–171. New York: Plenum.

Davidson, E. W., and Yamamoto, T. 1984. Isolation and assay of the toxic component from the crystals of *Bacillus thuringiensis* var. *israelensis. Curr. Microbiol.* 11: 171–174.

Delacluse, A.; Bourgouin, C.; Klier, A.; and Rapoport, G. 1988. Specificity of action on mosquito larvae of *Bacillus thuringiensis israelensis* toxins encoded by two different genes. *Mol. Gen. Genet.* 214: 42–47.

Donovan, W. P.; Dankocsik, C. C.; and Gilbert, M. P. 1988. Molecular characterization of a gene encoding a 72-kilodalton mosquito-toxic crystal protein from *Bacillus thuringiensis* subsp. *israelensis. J. Bacteriol.* 170: 4732–4738.

Donovan, W. P.; Dankocsik, C. C.; Gilbert, M. P.; Gawron-Burke, M. C.; Groat, R. G.; and Carlton, B. C. 1988. Amino acid sequence and entomocidal activity of the P2 crystal protein. *J. Biol. Chem.* 263: 561–567. (Erratum, 1988 *J. Biol. Chem.* 264: 4740).

Ellar, D. J.; Thomas, W. E.; Knowles, B. H.; Ward, S.; Todd, J.; Drobniewski, F.; Lewis, J.; Sawyer, T.; Last, D.; and Nichols, C. 1985. Biochemistry, genetics, and mode of action of *Bacillus thuringiensis* δ-endotoxin. In *Molecular biology of microbial differentiation,* ed. J. A. Hoch and P. Setlow, 230–240. Washington, D.C.: American Society for Microbiology.

Gonzalez, J. M., Jr., and Carlton, B. C. 1980. Patterns of plasmid DNA in Crystalliferous and Acrystalliferous strains of *Bacillus thuringiensis. Plasmid* 3: 92–98.

———. 1984. A large transmissible plasmid is required for crystal toxin production in *Bacillus thuringiensis* variety *israelensis. Plasmid* 11: 28–38.

Heimpel, A. M. 1967. A critical review of *Bacillus thuringiensis* var. *thuringiensis* Berliner and other crystalliferous bacteria. *Ann. Rev. Entomol.* 12: 287–322.

Herrnstadt, C.; Soares, G. G.; Wilcox, E. R.; and Edwards, D. L. 1986. A new strain of *Bacillus thuringiensis* with activity against coleopteran insects. *Bio/Technology* 4: 305–308.

Höfte, H., and Whiteley, H. R. 1989. Insecticidal crystal proteins of *Bacillus thuringiensis. Microbiol. Rev.* 53: 242–255.

Huber, H. E.; Lüthy, P.; Ebersold, H. R.; and Cordier, J. L. 1981. The subunits of the parasporal crystal of *Bacillus thuringiensis:* Size and linkage and toxicity. *Arch. Microbiol.* 129: 14–18.

Hurley, J. M.; Lee, S. G.; Andrews, R. E., Jr.; Klowden, M. J.; and Bulla, L. A., Jr. 1985. Separation of the cytolytic and mosquitocidal proteins of *Bacillus thuringiensis* subsp. *israelensis. Biochem. Biophys. Res. Commun.* 126: 961–965.

Ibarra, J. E., and Federici, B. A. 1986a. Isolation of a relatively nontoxic 65-kilodalton protein inclusion from the parasporal body of *Bacillus thuringiensis* subsp. *israelensis. J. Bacteriol.* 165 (2): 527–533.

———. 1986b. Parasporal bodies of *Bacillus thuringiensis* subsp. *morrisoni* (PG-14) and *Bacillus thuringiensis* subsp. *israelensis* are similar in protein composition and toxicity. *FEMS Microbiol. Lett.* 34 (1): 79–84.

Insell, J. P., and Fitz-James, P. C. 1985. Composition and toxicity of the inclusion of *Bacillus thuringiensis* subsp. *israelensis. Appl. Environ. Microbiol.* 50 (1): 56–62.

Kamdar, H., and Jayaraman, K. 1983. Spontaneous loss of a high molecular weight plasmid and the biocide of *Bacillus thuringiensis* var. *israelensis*. *Biochem. Biophys. Res. Commun.* 110: 477–482.

Kim, K. H.; Ohba, M.; and Aizawa, K. 1984. Purification of the toxic protein from *Bacillus thuringiensis* serotype 10 isolate demonstrating a preferential larvicidal activity to the mosquito. *J. Invertebr. Pathol.* 44: 214–219.

Krywienczyk, J.; Dulmage, H. T.; and Fast, P. G. 1978. Occurrence of two serologically distinct groups within *Bacillus thuringiensis* serotype 3a,b var. *kurstaki*. *J. Invertebr. Pathol.* 31: 372–375.

Pfannenstiel, M. A.; Ross, E. J.; Kramer, V. C.; and Nickerson, K. W. 1984. Toxicity and composition of protease-inhibited *Bacillus thuringiensis* var. *israelensis* crystals. *FEMS Microbiol. Lett.* 21: 39–42.

Sekar, V. 1986. Biochemical and immunological characterization of the cloned crystal toxin of *Bacillus thuringiensis* var. *israelensis*. *Biochem. Biophys. Res. Commun.* 137 (2): 748–751.

Sekar, V., and Carlton, B. C. 1985. Molecular cloning of the delta-endotoxin gene of *Bacillus thuringiensis* var. *israelensis*. *Gene* 33: 151–158.

Sekar, V.; Thompson, D. V.; Maroney, M. J.; Bookland, R. G.; and Adang, M. J. 1987. Molecular cloning and characterization of the insecticidal crystal protein gene of *Bacillus thuringiensis* var. *tenebrionis*. *Proc. Natl. Acad. Sci.* 84: 7036–7040.

Sen, K.; Honda, G.; Koyama, N.; Nishida, M.; Niki, A.; Sakai, H.; Himeno, M.; and Komano, T. 1988. Cloning and nucleotide sequences of the two 130-kDa insecticidal protein genes of *Bacillus thuringiensis* var. *israelensis*. *Agric. Biol. Chem.* 52: 873–878.

Tandeau de Marsac, N.; de la Torre, F.; and Szulmajster, J. 1987. Expression of the larvicidal gene of *Bacillus sphaericus* 1593M in the cyanobacterium *Anacystis nidulans* R2. *Mol. Gen. Genet.* 214: 42–47.

Thorne, L.; Garduno, F.; Thompson, T.; Decker, D.; Zounes, M.; Wild, M.; Walfield, A.; and Pollock, T. J. 1986a. Comparison of a gene that codes for a mosquitocidal toxin from *Bacillus thuringiensis* subsp. *israelensis* to a Lepidoptera-specific toxin gene from *B. thuringiensis* subsp. *kurstaki*. *Abstr. Ann. Mtg. ASM,* 149, H-130.

———. 1986b. Structural similarity between the Lepidoptera- and Diptera-specific insecticidal endotoxin genes of *Bacillus thuringiensis* subsp. *kurstaki* and *israelensis*. *J. Bacteriol.* 166 (3): 801–811.

Tungpradubkul, S.; Settasatien, C.; and Panyim, S. 1988. The complete nucleotide sequence of a 130-kDa mosquito-larvicidal delta-endotoxin gene of *Bacillus thuringiensis* var. *israelensis*. *Nucleic Acids Res.* 16: 1637–1638.

Visser, B.; van Workum, M.; Dullemans, A.; and Waalwijk, C. 1986. The mosquitocidal activity of *Bacillus thuringiensis* var. *israelensis* is associated with M_r 230,000 and 130,000 crystal proteins. *FEMS Microbiol. Lett.* 30: 211–214.

Waalwijk, C.; Dullemans, A. M.; van Workum, M. E. S.; and Visser, B. 1985. Molecular cloning and the nucleotide sequence of the M_r 28,000 crystal protein gene of *Bacillus thuringiensis* subsp. *israelensis*. *Nucleic Acids Res.* 13 (22): 8207–8217.

Walfield, A. M.; Garduno, F.; Thorne, L.; Zounes, M.; Decker, D. J.; Wild, M. A.; and Pollock, T. J. 1986. Cloning of a gene that codes for a mosquitocidal toxin from *Bacillus thuringiensis* var. *israelensis*. In *Bacillus molecular genetics and biotechnology applications,* ed. A. T. Ganesan and J. A. Hoch, 321–333. Orlando, Fla.: Academic Press.

Ward, E. S., and Ellar, D. J. 1983. Assignment of the δ-endotoxin gene of *Bacillus thuringiensis* var. *israelensis* to a specific plasmid by curing analysis. *FEBS Lett.* 158: 45–49.

———. 1987. Nucleotide sequence of a *Bacillus thuringiensis* var. *israelensis* gene encoding a 130 kDa delta-endotoxin. *Nucleic Acids Res.* 15: 7195.

Ward, E. S.; Ellar, D. J.; and Todd, J. A. 1984. Cloning and expression in *Escherichia coli* of the insecticidal δ-endotoxin gene of *Bacillus thuringiensis* var. *israelensis*. *FEBS Lett.* 175: 377–382.

Wu, D., and Chang, F. N. 1985. Synergism in mosquitocidal activity of 26 and 65 kDa proteins from *Bacillus thuringiensis* subsp. *israelensis* crystal. *FEBS Lett.* 190: 232–236.

Yamamoto, T.; Iizuka, T.; and Aronson, J. N. 1983. Mosquitocidal protein of *Bacillus thuringiensis* subsp. *israelensis:* Identification and partial isolation of the protein. *Curr. Microbiol.* 9: 279–284.

Yamamoto, T., and McLaughlin, R. E. 1981. Isolation of a protein from the parasporal crystal of *Bacillus thuringiensis* var. *kurstaki* toxic to the mosquito larva, *Aedes taeniorhynchus*. *Biochem. Biophys. Res. Commun.* 103 (2): 414–421.

Yamamoto, T.; Watkinson, I. A.; Kim, L.; Sage, M. V.; Stratton, R.; Akanda, N.; Li, Y.; Ma, D. P.; and Poe, B. A. 1988. Nucleotide sequence of the gene coding for a 130-kDa mosquitocidal protein of *Bacillus thuringiensis* var. *israelensis*. *Gene* 66: 107–120.

7

Transfer of the *Bacillus thuringiensis israelensis* Mosquiticidal Toxin Gene into Mosquito Larval Food Sources

KATHLEEN C. RAYMOND
HIROETSU WABIKO
ROBERT M. FAUST
LEE A. BULLA, JR.

7.1 INTRODUCTION

Mosquitoes and black flies are serious pests to humans and animals. Mosquitoes transmit diseases such as filariasis, elephantiasis, malaria, and yellow fever, all of which are still threats in tropical areas (Gillett 1971). Black flies act as vectors of filarial worms and blood protozoans among domestic and wild vertebrates and transmit parasites to humans (Crosskey 1981). Synthetic chemical pesticides have been effectively used to control these pests. However, the disadvantages of chemicals are that they persist for a long time in the environment; are hazardous to humans; and due to chemical longevity, genetically based resistance to the chemicals occurs. Consequently, the development of biological controls as an alternative means of pest control is desired.

Bacillus thuringiensis subsp. *israelensis* (*B.t.i.*) produces proteinaceous parasporal crystals during sporulation. These crystals are unlike those of the lepidopteran-specific *B.t.* subspecies in that they are composed of multiple proteins ranging in size from 26 to 135 kilodaltons (kDa), rather than a major polypeptide, that of the toxin. The crystals are lethal to larval and adult mosquitoes (Goldberg and Margalit 1977; Klowden, Held, and Bulla 1983) and to black fly larvae (Undeen and Nagel 1978). The larvicidal activity is high, rapid, and specific for mosquitoes and black flies (Goldberg and Margalit 1977; Undeen and Nagel, 1978). No effect is observed on other aquatic insects, fish, or frog larvae (Garcia and Goldberg 1978). For this reason, the World Health Organization (WHO) recommended use of *B.t.i.* as an agent for biological control of mosquitoes (Arata et al. 1978). This recommendation was approved by the Environmental Protection Agency (EPA) in the United

States. Since then, commercial sources of *B.t.i.* have been successfully used to control mosquitoes and black flies.

Much research is required to develop a long-lasting and viable product for field application. Direct application of *B.t.i.* in field conditions to control mosquitoes has complications. For example, Margalit et al. (1983) determined that 4 applications, at 10-day intervals, of *B.t.i.* powder suspended in water decreased larval mosquito populations, but about 48 hours after each application, mosquito larvae hatched from new eggs. To further control the larval population in the absence of larval predators, spraying with *B.t.i.* at 8- to 10-day intervals was necessary. In this chapter, we will discuss possible strategies for transferring the *B.t.i.* toxin gene into alternate hosts that are food sources for mosquitoes and black flies. Release of a genetically modified *B.t.i.* biocontrol agent for mosquitoes and black flies is feasible, and could be safer and more economical than chemical or fermentation processes currently used.

7.2 ASSIGNMENT OF TOXIC ACTIVITY

The parasporal crystals of *B.t.i.* are composed of several distinct proteins ranging in size from 26 to 135 kDa (Tyrell et al. 1981; Pfannenstiel et al. 1984; Lee, Eckblad, and Bulla 1985; Ellar et al. 1985). The major components in these crystals, however, have an apparent molecular weight of approximately 28 and 68 kDa. There has been considerable debate as to which protein possesses dipteran insecticidal activity. This debate centers around the 26/28- and 65/68-kDa proteins. The cytolytic activity of the 28-kDa protein is well documented (Hurley et al. 1985; Thomas and Ellar 1983a, 1983b), but there is controversy over which protein bestows larvicidal activity. Recently, it has been suggested that neither protein alone is larvicidal, but that synergism between the two proteins is necessary to produce larvicidal activity (Wu and Chang 1985; Ibarra and Federici 1986).

Thomas and Ellar (1983a, 1983b) showed that solubilized *B.t.i.* crystals caused rapid lysis of insect and mammalian cells, then, through further experimental analysis, assigned the mosquiticidal activity to the 26-kDa protein. They hypothesized that toxicity is manifested by interaction of the protein with the plasma membrane, causing a detergentlike rearrangement of the lipids, and thereby leading to disruption of the membrane and cytolysis (Ellar et al. 1985).

Armstrong, Rohrmann, and Beaudreau (1985) made an initial assumption that the toxin would be resistant to protease digestion, which is a common phenomenon among species of *Bacillus*, and purified a protease-resistant protein from *B.t.i.*. The crystals were solubilized with alkali, and the proteins were digested with trypsin and proteinase K. Using a combination of gel filtration and ion exchange chromatography, two forms of a 25-kDa protein, differing in size by two amino acids, were purified. This protein, in either form, lysed rabbit and human erythrocytes, lysed cultured mosquito cells, was insecticidal to mosquito larvae at 50 μg/ml, and was lethal to mice. When antibodies to the purified protein were tested against *B.t.i.* proteins, only those proteins from toxic *B.t.i.* strains cross-reacted with the antibodies. They concluded that this protein was derived from the 28-kDa peptide, which they therefore believe to be the *B.t.i.* toxin.

In contrast, Lee, Eckblad, and Bulla (1985) assigned the mosquiticidal activity to the 65-kDa protein. Parasporal inclusion bodies from *B.t.i.* were separated according to size through a 10–40% sucrose gradient. Based on protein content, the population designated as small dots was found to be most toxic to mosquito larvae. Analysis of the protein content of the small dots revealed that they were composed of the 38-kDa and 65-kDa proteins, but did not contain any detectable 28-kDa protein. Assignment of the toxic activity to the 65-kDa protein was made when further analysis showed a strong correlation between production of the 65-kDa protein and larvicidal activity during sporulation.

Gel filtration chromatography of alkali-solubilized *B.t.i.* crystals permitted separation of the 28- and 65-kDa proteins (Hurley et al. 1985). The 28-kDa protein was not mosquiticidal, but caused hemolysis of rat red blood cells. The 65-kDa protein, however, was found to be toxic to mosquito larvae (*Aedes aegypti*) with an LC_{50} of 180 ng/ml, and had no hemolytic activity. Purification of the toxic protein (65 kDa) from the unfractionated crystal protein mixture resulted in a seven-fold increase in specific activity, which correlated with their observation that the 65-kDa protein represented approximately one-seventh of the total proteins in the crystal preparations they used.

Sriram, Kamdar, and Jayaraman (1985) used the antibiotic netropsin to separate the sporulation and crystal formation events in *B.t.i.*. At low concentrations (1–3 μg/ml), the antibiotic inhibited sporulation, but had no effect on crystal production or larvicidal activity (LC_{50} = 50 ng/ml). However, at higher concentrations of netropsin (3–7 μg/ml), crystal inclusions were not formed, and there was a concomitant 10-fold reduction in the larvicidal activity (LC_{50} = 500 ng/ml). SDS-polyacrylamide gel electrophoretic analysis of alkali-solubilized proteins from cells treated with varying amounts of netropsin revealed that the acrystalliferous cells contained only the low molecular weight crystal peptides, and that the 26-kDa protein was present in all of the cells analyzed. From this observation, they concluded that the larger molecular weight proteins are responsible for crystal production and that the 28-kDa protein is the larvicidal component in the cells.

A synergism between the 26- and 65-kDa proteins has also been suggested (Wu and Chang 1985). *B.t.i.* crystals were alkali-solubilized overnight at 4° C and then chromatographed, resulting in separation of three major protein components of 26, 65, and 130 kDa. The 26-kDa protein was inactive against mosquito larvae at concentrations of 6.4 μg/ml. The 65-kDa protein also was inactive in their assay system at low concentrations, and only slightly active at higher concentrations. They attributed what little activity it had at concentrations of 1.6 μg/ml to the 2% level of contamination with 26-kDa protein. Maximum larvicidal activity against *Ae. aegypti* was observed only when both of the proteins were present simultaneously.

A similar conclusion was reached by Ibarra and Federici (1986). An inclusion body from *B.t.i.*, predominantly composed of the 65-kDa protein, was isolated and used in a larvicidal assay. This inclusion body was less toxic to mosquito larvae (*Ae. aegypti*) than were the native *B.t.i.* crystals in their assay system. They also found that the toxicity of this inclusion body was directly correlated to the extent of contamination with the 28-kDa protein.

Held, Huang, and Kawanishi (1986) removed the 28-kDa cytolytic protein from solubilized *B.t.i.* crystals by affinity chromatography using a monoclonal antibody directed against the 28-kDa protein. Bioassays were performed on third-instar *Ae. aegypti* larvae to determine LC_{50} values. Total solubilized crystals had an LC_{50} value of 0.64 μg/ml. The crystal protein fraction depleted of the 28-kDa protein was nonhemolytic and retained nearly full toxicity to mosquito larvae (LC_{50} = 0.75 μg/ml). The purified 28-kDa protein was hemolytic and relatively nontoxic to mosquito larvae (LC_{50} = 21.7 μg/ml). They conclude that the 65-kDa protein is the predominant protein in the flow-through fraction, which is devoid of the 28-kDa cytolytic protein. Therefore, the 65-kDa protein is the mosquiticidal toxin.

We believe that the 65-kDa protein is solely responsible for larvicidal activity (Lee, Eckblad, and Bulla 1985; Hurley et al. 1985). However, only by cloning each of the genes will a definitive assignment of the toxin activity to either the 28-kDa protein or the 65-kDa protein be possible.

7.3 CLONING OF THE MOSQUITO TOXIN GENE

B. thuringiensis subsp. *israelensis*, serotype H14, was analyzed by agarose-gel electrophoresis. It contains eight plasmids ranging in size from 3.3 to 135 megadaltons (MDa) and one linear piece of DNA of approximately 10 MDa (Gonzalez and Carlton 1984). Plasmid curing studies implicated the 75-MDa plasmid in crystal production, which is synonymous with larvicidal activity. Plasmid transfer experiments involving acrystalliferous (Cry^-) *B.t.i.* strains further implicated the 75-MDa plasmid as the one encoding the larvicidal activity. Other plasmids were transferred into the Cry^- strain, but only transfer of the 75-MDa plasmid converted the trancipient strain to crystal and toxin production.

Ward, Ellar, and Todd (1984) cloned *Hind* III fragments from this 75-MDa plasmid into the vector pUC13. Two clones, pIP173 and pIP174, both containing the 9.7-kilobase pair (kbp) *Hin*d III fragment, produced a 26-kDa protein in an *Escherichia coli* in vitro transcription-translation system. The protein produced by these clones was precipitable by antibody prepared against the 26-kDa protein from native *B.t.i.* crystals. *E. coli* cells harboring pIP174 were toxic to *Ae. aegypti* larvae, and protein extracts caused cytolysis of *Aedes albopictus* cells. The amount of protein required for cells to be lethal to *Ae. aegypti* larvae suggests that this clone does not encode the protoxin, but rather the cytolytic activity. Moreover, that cytolytic activity is observed substantiates this notion.

Using virtually the same approach, Waalwijk et al. (1985) also cloned a 9.7-kbp *Hin*d III fragment into pBR322. This clone, p425, produced a 28-kDa protein in an in vitro transcription-translation system. Amino acid composition of this protein agreed well with that determined for the 28-kDa protein from the native *B.t.i.* crystals. Insect toxicity was not examined, however. This clone most likely is identical to that previously obtained by Ward, Ellar, and Todd (1984); therefore, we believe that it contains the gene encoding the cytolytic activity and not the protoxin activity.

Because expression of the protoxin gene is sporulation specific in native *B.t.i.*, Sekar and Carlton (1985) employed a *Bacillus* cloning system, rather than an *E. coli* one, to enhance production of the cloned gene product. DNA from the 75-MDa plasmid of *B.t.i.* was partially digested with *Xba*I and ligated into the *Bacillus cereus* cloning vector, pBC16, carrying a tetracycline resistance marker. Polyethyleneglycol-induced protoplasts of *Bacillus megaterium* VTI660 were transformed with the resultant DNA, and tetracycline-resistant transformants were selected. The transformants were allowed to sporulate and lyse; the lysates were used directly for bioassays against *Ae. aegypti* larvae. A toxic clone, VB131, was produced that contained a 6.3-kbp insert composed of three *Xba*I fragments (2.7, 1.8, and 1.8 kbp). This clone produced phase-refractile bodies during sporulation that were toxic to *Ae. aegypti* larvae. Alkali-solubilized inclusion bodies from this clone produced a precipitin band in a double immunodiffusion assay using antiserum prepared against solubilized *B.t.i.* crystals. No polyacrylamide gel analysis was provided, and, therefore, it is not known what polypeptide species was produced from this clone.

Recently, Thorne et al. (1986) described the cloning of a fragment from the 75-MDa plasmid of *B.t.i.* that encodes larvicidal activity. The restriction map of this insert bears no resemblance to either of the clones previously described (Waalwijk et al. 1985; Sekar and Carlton 1985). Furthermore, the clone has no hemolytic activity on human red blood cells (Alan Walfield, pers. comm.). Thorne et al. (1986) have sequenced the region of DNA encoding the gene product responsible for larvicidal activity. Two large open reading frames (ORFs) were identified in the 3,750 nucleotides sequenced. The first ORF could code for a protein of 72 kDa, and the second ORF, which extends beyond the sequenced region, could encode a protein of at least 26 kDa. When gene expression was carried out in vitro with an *E. coli* transcription-translation system, polypeptides up to 72 kDa were generated; however, when expressed in *Bacillus subtilis*, a 58-kDa protein accumulates. The 5′-flanking sequence containing the promoter and ribosome binding site is virtually identical to that of *B. thuringiensis* subsp. *kurstaki* (toxic to lepidopteran species), which these investigators (Thorne et al. 1986) also have cloned and sequenced. The coding regions of the two different subspecies exhibited little DNA homology, although the deduced amino acid sequences were strikingly similar.

Both Waalwijk et al. (1985) and Thorne et al. (1986) have sequenced the DNA responsible for encoding what they believe to be the mosquito toxin, which is known to be expressed only during sporulation in *B.t.i.*. Interestingly, the promoter region of the gene encoded in the clone of Thorne et al. (1986) appears to be sporulation-specific, as evidenced by its increased production in *B. subtilis* during sporulation and a DNA sequence homologous to the promoter of the cloned toxin gene from *B.t. kurstaki*. The sequence of the promoter encoding the 28-kDa protein, which Waalwijk et al. (1985) claim to be larvicidal, however, is different from that described for other *Bacillus* genes expressed during sporulation.

From the deduced amino acid sequences of the two genes (Waalwijk et al. 1985; Thorne et al. 1986), the hydropathy and secondary structure of the molecules were predicted. Figure 7.1 compares the predicted hydropathy of these two molecules to that predicted for the protoxin that we have identified from the lepidopteran-killing *B. thuringiensis* subsp. *thuringiensis* (Wabiko, Raymond, and Bulla 1986). The predicted hydropathy of the 28-kDa protein (fig. 7.1a) bears no resemblance to the predicted hydropathy of either the 68-kDa protein from *B.t.i.* or the 130-kDa protoxin from *B.t. thuringiensis* (fig. 7.1b and 7.1c, respectively). Curiously, the 68-kDa and 130-kDa proteins are strikingly similar. Likewise, a comparison of the predicted secondary structure of the three molecules (fig. 7.2) lends additional evidence to the conclusion that the 68-kDa and 130-kDa proteins are quite similar (fig. 7.2b and 7.2c, respectively). These comparisons, showing a predicted similarity between the lepidopteran- and dipteran-killing toxins, along with the fact that the 68-kDa protein is not cytolytic to human red blood cells (Alan Walfield, pers. comm.), indicate that the 68-kDa protein is the *B.t.i.* toxin, as previous work indicated (Lee, Eckblad, and Bulla 1985; Hurley et al. 1985).

Since the first writing of this chapter, the debate over which protein in the *B.t.i.* crystal is larvicidal has expanded to include the 130-kDa protein. Visser et al. (1986) solubilized *B.t.i.* crystals and separated three proteins of 230, 130, and 28 kDa by sucrose gradient centrifugation. Mosquiticidal activity coincided with the 230- and 130-kDa peaks, whereas hemolytic activity was found only in the peak corresponding to the 28-kDa protein. The protein with the greatest mosquiticidal activity was the 130-kDa protein. They also

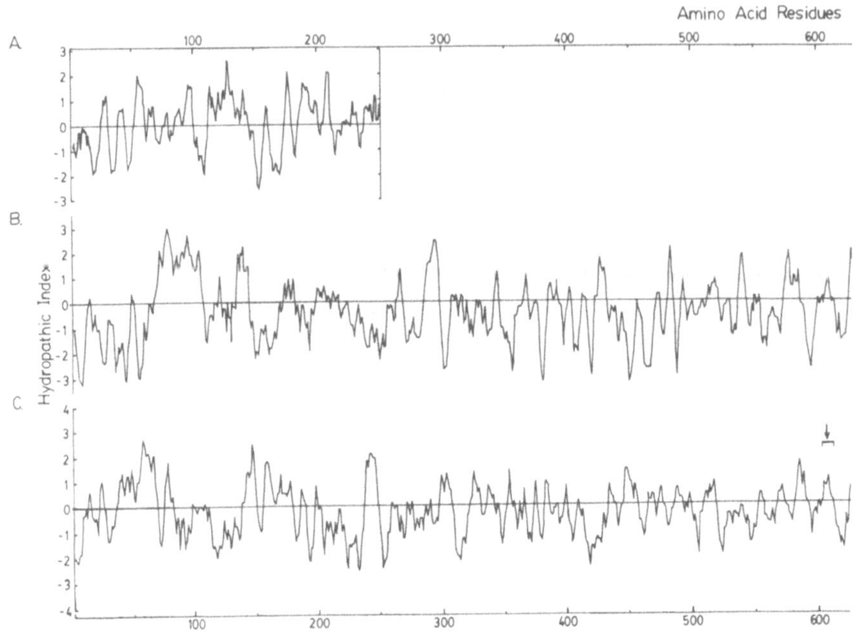

FIGURE 7.1 The predicted hydropathy of the 28-kDa and 68-kDa proteins from *B. t. i.* and the 135-kDa protoxin from *B. thuringiensis* subsp. *thuringiensis.* The hydropathy of each protein was predicted according to Kyte and Doolittle (1982). Windowing average at residue *i* was calculated across six residues, from i−3 through and including i+3. Positive values indicate hydrophobic regions, and negative values indicate hydrophilic regions. Panels *A* and *B* represent the predicted hydropathy of the

observed that solubilization of the proteins in NaOH pH 12.0, versus pH 9.5, favored production of a 65-kDa protein with a concomitant reduction in the amounts of the 230- and 130-kDa proteins. Unfortunately, they never did any bioassays with the 65-kDa protein, nor did they determine if the 230- or 130-kDa proteins are protoxins containing a 65-kDa toxic moiety. Their data does, however, strongly suggest that *B. t. i.* makes a 130-kDa protoxin and a 65-kDa toxin in a similar fashion as do the lepidopteran-specific *B. t.* subspecies; however, this was not investigated.

Three groups, using different restriction enzymes and cloning vectors, have cloned the same gene from *B. t. i.* that encodes a 130-kDa protein (Bourgouin, Klier, and Rapoport 1986; Angsuthanasombat et al. 1987; Ward and Ellar 1988). In all cases, the gene was cloned from the 72-MDa plasmid of *B. t. i.*. The 130-kDa gene product is mosquiticidal and has no hemolytic activity. A 68-kDa protein is also produced, again indicating that the 68-kDa protein could be the toxic portion generated from the 130-kDa protoxin. This possibility was not addressed by the investigators.

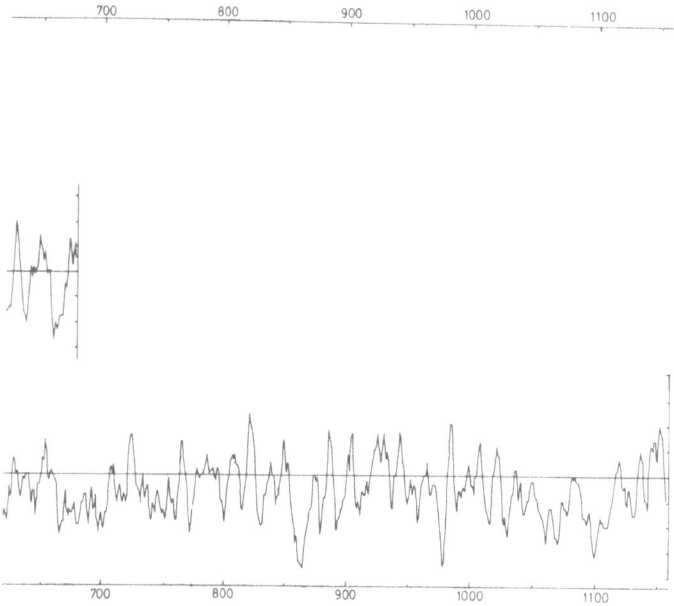

FIGURE 7.1 *(Continued)*
B. t. i. 28-kDa and 68-kDa proteins, respectively. Panel *C* represents the predicted hydropathy of the *B. thuringiensis* subsp. *thuringiensis* protoxin. The bracket and arrow above amino acid residues 604–613 in panel *C* represent the region that delimits the toxic portion of the subsp. *thuringiensis* molecule (amino acid residues 1–613; Wabiko, Raymond, and Bulla 1986).

Very recently, Bourgouin et al. (1988) cloned a gene from *B. t. i.* that produces a 125-kDa protein that is mosquiticidal. The insecticidal activity of the 125- and 130-kDa proteins were compared. The 125-kDa protein kills larvae of *Ae. aegypti, Culex pipiens,* and *Anopholes stephensi;* whereas the 130-kDa protein has no effect on *Cx. pipiens* larvae, but is effective against the other two. Restriction endonuclease mapping of the two genes indicates that the 3′-halves are similar, but the 5′-halves are not. Also, the gene encoding the 125-kDa protein is flanked by inverted repeat sequences. This is the first report of repeat sequences in *B. t. i.*

The predicted similarities between the *B. t. i.* 68-kDa protein and *B. t. thuringiensis* 130-kDa protein (fig. 7.2) suggest that there is a conserved structure/function relationship, or common mechanism of toxicity, between the two *B. t.* subspecies. With the recent cloning of a *B. t. i.* gene producing a 130-kDa protein (Bourgouin et al. 1988), which may in fact be a protoxin containing a 68-kDa toxin, the similarities between lepidopteran- and dipteran-specific *B. t.* subspecies are even more striking. More experimentation,

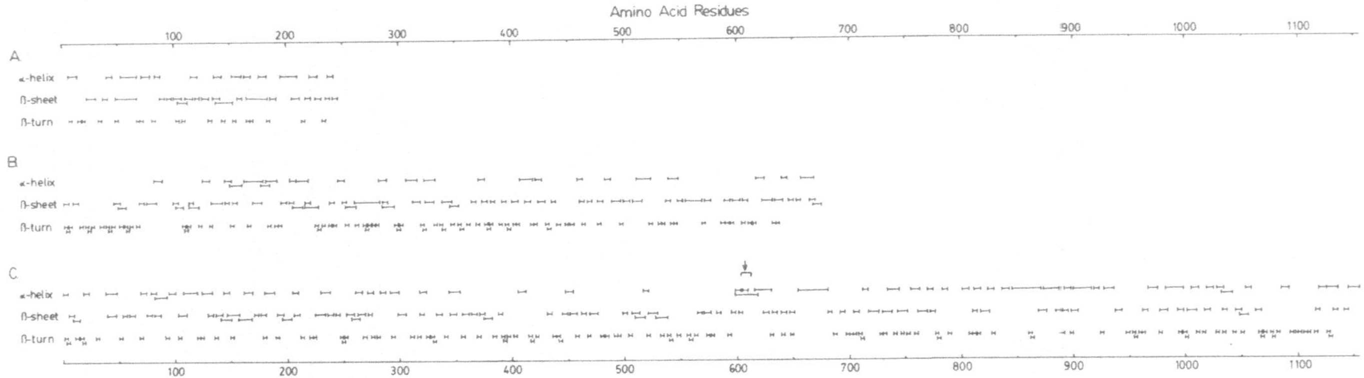

FIGURE 7.2 The predicted secondary structure of the 28-kDa and 68-kDa proteins from *B.t.i.* and the 135-kDa protoxin from *B. thuringiensis* subsp. *thuringiensis.* The secondary structure of each protein was predicted according to Chou and Fasman (1978). Panels *A* and *B* represent the predicted secondary structure of the *B.t.i.* 28-kDa and 68-kDa proteins, respectively. Panel *C* represents the predicted secondary structure of the *B. thuringiensis* subsp. *thuringiensis* protoxin. As in fig. 7.1, the region that delimits the toxic portion of the subsp. *thuringiensis* molecule is indicated (Wabiko, Raymond, and Bulla 1986).

however, is necessary to further understand the mechanism of toxic action and to determine what is responsible for the different spectrum of insect host range among the different *B.t.* subspecies.

7.4 MOSQUITO AND BLACK FLY LARVAL FOOD SOURCES

Mosquito larvae live in either stagnant or running water. They also are found at the edges of clear, running streams, especially ones with well-developed algal populations. Chloride is essential for larvicidal viability. The larvae of a few species (*Aedes australis* and *Aedes detritas*) actually live in pools containing seawater (Gillett 1971).

Mosquito larvae are filter feeders; large particles are excluded, whereas small particles enter the mouth and are further segregated before reaching the pharyngeal lumen (Gillett 1971). They feed on particulate matter ranging in size from that of microscopic bacteria to that of clearly visible particles (Horsfall 1955). Depending upon the mosquito species, feeding tendencies differ. *Anopheles* species feed close to the surface of water, whereas *Culex* species feed slightly below the surface as well as on the very bottom (Gillett 1971).

Larval black fly habitat, however, is restricted to running water (Jamnback 1981). Two modes of feeding are used by black flies: browsing and filter feeding. Particulate matter ingested by black fly larvae ranges in size from 0.5–300 μm in length and 0.5–120 μm in width (Chance 1970).

The gram-negative bacterium *E. coli*, present in domestic sewage, ranges from 2–5 μm in length during exponential growth (Cullum and Vicente 1978), which is within the size range of particulate matter capable of passing through the internal organs of larvae. The gram-positive bacterium *B. subtilis* is a normal inhabitant of soil. Both bacteria are present in storage waters, such as lakes and ponds that receive drainage (Fredeen 1964). During periods of snowmelt and runoff when black fly larvae are active, these particular bacteria are abundant.

Fredeen (1964) has tested various bacteria found in natural black fly larval habitats to determine their potential as larval food sources. He fed *Simulidae* larvae the alga *Chylamydamonas*, the gram-positive bacterium *B. subtilis*, and the gram-negative bacteria *Aerobacter aerogenes* and *E. coli*. Under laboratory conditions, *Simulium venustum*, *Simulium verecundum*, and *Simulium vittatum* larvae developed from first-instar larvae to pupae when they were provided only bacteria as a food source. Some of the pupae developed to adults, indicating that bacteria alone is an adequate larval food source. However, more reached adulthood when the alga was added as a food source.

Anderson and Dicke (1960) dissected black fly larvae to determine the gut microflora. They found that all species of larvae analyzed contained diatoms and algae. One in particular was the blue-green alga *Cyanobacter*. This alga is a photosynthetic prokaryote and is widely distributed not only in fresh water, but also in marine habitats (Fogg 1973; Whitton 1973; Stanier and Cohen-Bazire 1977). The size of *Cyanobacter* species, as well as other blue-green algae, render them a potential food source for both black fly and mosquito larvae.

7.5 INTRODUCTION OF THE MOSQUITO TOXIN GENE INTO LARVAL FOOD SOURCES

In our opinion, the ultimate insecticide for mosquito and black fly control involves incorporating the gene encoding entomicidal activity into these insects' natural food source. The blue-green alga is a logical candidate because genetic manipulation of this organism is possible and because it is a larval food source. Introduction of the dipteran toxin gene from *B. t. i.* into this alga would provide a form of pest control having persistence in nature and would alleviate multiple applications otherwise necessary with direct application of *B. t. i.* and chemical insecticides.

The cyanobacteria (blue-green algae) are photosynthetic procaryotes that carry out oxygenic photosynthesis similar to higher plants. Many species of cyanobacteria contain endogenous plasmids (Lau and Doolittle 1979; Lau, Sapienza, and Doolittle 1980; Simon 1978; van den Hondel et al. 1979). As yet, no function has been determined for these plasmids, which range in size from 1.8 MDa to 74 MDa. As many as five plasmids are present in various strains of *A. nidulans* and other species of *Anacystis*.

A number of laboratories have used these endogenous plasmids to construct hybrid plasmids capable of transformation and replication in *E. coli* and *A. nidulans*. These plasmids contain selective markers for antibiotic resistance and have at least one unique restriction site for cloning (van den Hondel et al. 1980; Gendel et al. 1983; Kuhlemeier et al. 1983). The major problem with these hybrid plasmids is that they are relatively large, thereby limiting the size of a potential cloned insert.

Similar hybrid plasmids were constructed that are capable of transforming the cyanobacterium *Agmenellum quadruplicatum* (PR-6). Buzby, Porter, and Stevens (1983) used the smallest plasmid endogenous to PR-6 (3.0 MDa) and combined it with various *E. coli* plasmids containing selective markers. To increase transformation efficiency they found that it was necessary to eliminate all *Ava*I restriction sites in the plasmid so that the exogenous DNA would not be restricted by the PR-6 restriction system (*Ava*I is an isoschizomer of the *Aqu*I restriction endonuclease of PR-6). Recently, Buzby, Porter, and Stevens (1985) observed expression of the *E. coli* lacZ gene from a plasmid vector in PR-6. Expression of the gene product in PR-6 was comparable to the level observed in *E. coli.* These investigators are now studying gene fusions with various PR-6 promoters in hope of finding foreign gene products formed in PR-6 under the control of the lac promoter. Genetic engineering of this kind would be of potential use when considering such systems for introduction of the toxin gene into cyanobacteria.

Wolk et al. (1984) constructed a plasmid capable of transforming the cyanobacter *Anabaena.* Construction of the plasmid was accomplished by combining DNA from a plasmid endogenous to the cyanobacter *Nostoc* to DNA from pBR322, a plasmid of *E. coli.* To increase transformation efficiency, it was necessary to delete that portion of the plasmid that could be restricted in *Anabaena.* The resultant plasmids were capable of transference between *E. coli* and *Anabaena* by a conjugationlike process.

Most studies have indicated that transformation of cyanobacteria with *E. coli* plasmids is not possible unless combined with DNA from plasmids endogenous to cyanobacteria. However, recently, Dzelzkalns and Bogorad (1986) stably transformed the cyanobacterium *Synechocystis* sp. PCC 6803 (6803) using a pretreatment involving ultraviolet irradiation. In *E. coli* and other prokaryotes, irradiation induces a repair process known as the SOS response. Apparently, irradiating the 6803 cells alleviates the host-controlled restriction system and low-frequency recombination occurs, allowing foreign DNA to integrate into the host chromosome.

Daniell, Sarojini, and McFadden (1986) report the transformation of *A. nidulans* 6301 by the *E. coli* plasmid pBR322. Transformed cells were selected by ampicillin resistance, and β-lactamase activity was detected in these cells. Transformation was performed with both intact cells and permeaplasts.

We believe that, in time, cyanobacter transformation techniques will be greatly improved as more knowledge is gained about the system. Indeed, the recent finding that cyanobacteria can be transformed with heterologous DNA without the presence of endogenous DNA in the vector is noteworthy (Dzelzkalns and Bogarad 1986; Daniell, Sarojini, and McFadden 1986). This fact now makes it much more feasible to introduce a foreign gene into cyanobacteria, alleviating recloning into shuttle vectors and thereby saving much time and effort.

Since the initial writing of this chapter, the gene encoding mosquito larvicidal activity from *Bacillus sphaericus* 1593M has been cloned and expressed in the cyanobacterium *Anacystis nidulans* R2 (Tandeau de Marsac, de la Torre, and Szulmajster 1987). The gene was cloned into the vector pHV33 (Primrose and Ehrlich 1981) for expression in *E. coli* and *B. subtilis,* and into the vector pUC303 (Kuhlemeier et al. 1983) for expression in *A. nidulans* R2. Bioassays against second-instar larvae of *Cx. pipiens* were performed to determine larvicidal activity expressed as 100% mortality per μg of protein per ml in 48 hours. The results indicated that expression of the larvicidal activity in *A. nidulans* R2 is comparable to that in *E. coli* (1 μg/ml).

These results bring us a step closer to developing a cyanobacterial-based biopesticide; however, the larvicidal activity is much reduced as compared to *B. sphaericus* 1593M [$1(10^{-3})$ µg/ml]. This may be resolved by putting the expression of the *B. sphaericus* gene under the control of a cyanobacterial promoter.

Because transformation vectors and procedures are available for various bacteria and cyanobacteria as well, several of which are known to be larval food sources, there are a number of potential systems that can be used in developing biological agents to control mosquitoes and black flies. A biopesticide would be much more economical and safer for the environment than chemical pesticides now widely in use (Faust and Bulla 1982).

7.6 SAFETY ASPECTS

B.t.i. itself has no effect on aquatic organisms such as water mites, shrimps, and oysters (Davidson 1982). The only affected species are the dipteran insects: mosquitoes and black flies. No effects from the *B.t.i.* toxin have been observed in the food chain after exposure to the bacterium, although alkali-solubilized parasporal crystals of *B.t.i.* are hemolytic to erythrocytes from many species, including humans. This general cytolytic or hemolytic activity has been assigned to the 28-kDa protein in the native parasporal crystals (Thomas and Ellar 1983a, 1983b; Hurley et al. 1985). Genetically engineering an organism to encode only the larvicidal activity (68 kDa), which is not cytolytic (Hurley et al. 1985; Alan Walfield, pers. comm.), would alleviate any safety problems concerning cytolytic activity. Introducing the *B.t.i.* entomicidal toxin gene into a larval food source capable of replicating naturally also would preclude repeated applications of the control agent. Furthermore, it would eliminate the costly fermentation process and chemical synthesis of pesticides now in use. If a biocontrol agent as described herein were developed and used, the animal and human population, as well as the environment, would be exposed to fewer chemical irritants and pollutants.

Systems are now available to engineer biopesticides. However, before they can be tested in the field, applications will have to be submitted to the EPA. As it stands now, applications are handled on a case-by-case basis. A potentially safe insecticide may be developed soon, but it will be some time before it can be field tested and before it can be approved for widespread use.

Acknowledgments

This chapter is based in part on the presentation given by Robert M. Faust at the fiftieth annual meeting of the American Mosquito Control

Association and the seventy-second annual meeting of the New Jersey Mosquito Control Association, Atlantic City, New Jersey, March 18–21, 1985. Contribution No. 1502 of the Wyoming Agricultural Experiment Station, Laramie, Wyo. 82071.

References

Anderson, J. R., and Dicke, R. J. 1960. Ecology of the immature stages of some Wisconsin black flies (Diptera: Simuliidae) *Ann Entomol. Soc. Am. J.* 3: 386–404.

Angsuthanasombat, C.; Chungjatupornchai, W.; Kertbundit, S.; Luxananil, P.; Settasatian, C.; Wilairat, P.; and Panyim, S. 1987. Cloning and expression of 130-kd mosquito-larvicidal delta-endotoxin gene of *Bacillus thuringiensis* var. *israelensis* in *Escherichia coli. Mol. Gen. Genet.* 208: 384–389.

Arata, A. A.; Chapman, H. C.; Cupello, J. M.; Davidson, E. W.; Laird, M.; Margalit, J.; and Roberts, D. W. 1978. Status of Biocontrol in Medical Entomology. *Nature* 276: 669–670.

Armstrong, J. L.; Rohrmann, G. F.; and Beaudreau, G. S. 1985. Delta endotoxin of *Bacillus thuringiensis* subsp. *israelensis. J. Bacteriol.* 161 (1): 39–46.

Bourgouin, C.; Delecluse, A.; Ribier, J.; Klier, A.; and Rapoport, G. 1988. A *Bacillus thuringiensis* subsp. *israelensis* gene encoding a 125-kilodalton larvicidal polypeptide is associated with inverted repeat sequences. *J. Bacteriol.* 170: 3575–3583.

Bourgouin, C.; Klier, A.; and Rapoport, G. 1986. Characterization of the genes encoding the haemolytic toxin and the mosquitocidal delta-endotoxin of *Bacillus thuringiensis israelensis. Mol. Gen. Genet.* 205: 390–397.

Buzby, J. S.; Porter, R. D.; and Stevens, S. E., Jr. 1983. Plasmid transformation in *Agmenellum quadruplicatum* PR-6: Construction of biphasic plasmids and characterization of their transformation properties. *J. Bacteriol.* 154 (3): 1446–1450.

———. 1985. Expression of the *Escherichia coli lacz* gene on a plasmid vector in a cyanobacterium. *Science* 230: 805–807.

Chance, M. M. 1970. The functional morphology of the mouthparts of black fly larvae (Diptera: Simuliidae). *Quaest. Entomol.* 6: 245–286.

Chou, P. Y., and Fasman, G. D. 1978. Prediction of the secondary structure of proteins from their amino acid sequence. *Adv. Enzymol.* 47: 45–148.

Crosskey, R. W. 1981. Simuliid taxonomy—the contemporary scene. In *Black flies,* ed. M. Laird, 3–18. London: Academic Press.

Cullum, J., and Vicente, M. 1978. Cell growth and length distribution in *Escherichia coli. J. Bacteriol.* 134: 330–337.

Daniell, H.; Sarojini, G.; and McFadden, B. A. 1986. Transformation of the cyanobacterium *Anacystis nidulans* 6301 with the *Escherichia coli* plasmid pBR322. *Proc. Natl. Acad. Sci.* 83: 2546–2550.

Davidson, E. W. 1982. Bacteria and the control of arthropod vectors of human and animal disease. In *Microbial control and viral pesticides,* ed. E. Kurstak, 289–315. New York: Marcel Decker.

Dzelzkalns, V. A., and Bogorad, L. 1986. Stable transformation of the Cyanobacterium *Synechocystis* sp. PCC 6803 induced by UV irradiation. *J. Bacteriol.* 165: 964–971.

Ellar, D. J.; Thomas, W. E.; Knowles, B. H.; Ward, S.; Todd, J.; Drobniewski, F.; Lewis, J.; Sawyer, T.; Last, D.; and Nicholls, C. 1985. Biochemistry, genetics, and mode of action of *Bacillus thuringiensis* δ-endotoxins. In *Molecular biology of microbial differentiation,* ed. J. A. Hoch and P. Setlow, 230–240. Washington, D.C.: American Society for Microbiology.

Faust, R. M., and Bulla, L. A., Jr. 1982. Bacteria and their toxins as insecticides. In *Microbial and viral pesticides,* ed. E. Kurstak, 75–208. New York: Marcel Dekker.

Fogg, B. E. 1973. Physiology and ecology of marine blue-green algae. In *The biology of blue-green algae,* ed. N. G. Carr and B. A. Witton, 368–414. Berkeley and Los Angeles: University of California Press.

Fredeen, F. J. H. 1964. Bacteria as food for black fly larvae (Diptera:Simuliidae) in laboratory cultures and in natural streams. *Can. J. Zool.* 42: 527–548.

Garcia, R., and Goldberg, L. J. 1978. *Univ. Calif. Mosq. Control Res., Ann. Rep. 1977,* 29.

Gendel, S.; Straus, N.; Pulleyblank, D.; and Williams, J. 1983. Shuttle cloning vectors for the cyanobacterium *Anacystis nidulans J. Bacteriol.* 156: 148–154.

Gillett, J. D. 1971. *The mosquito: Its life, activities, and impact on human affairs.* Bungay, Great Britain: Richard Clay (Chaucer Press).

Goldberg, L. J., and Margalit, J. 1977. A bacterial spore demonstrating rapid larvicidal activity against *Anopheles sergentii, Uranotaenia unquiculata, Culex univitattus, Aedes aegypti,* and *Culex pipiens. Mosq. News* 37: 355–358.

Gonzalez, J. M., Jr., and Carlton, B. C. 1984. A large transmissible plasmid is required for crystal toxin production in *Bacillus thuringiensis* variety *israelensis. Plasmid* 11: 28–38.

Held, G. A.; Huang, Y.-S.; and Kawanishi, C. Y. 1986. Effect of removal of the cytolytic factor of *Bacillus thuringiensis* subsp. *israelensis* on mosquito toxicity. *Biochem. Biophys. Res. Commun.* 141: 937–941.

Horsfall, W. R. 1955. *Mosquitoes: Their bionomics and relation to disease.* New York: Ronald Press.

Hurley, J. M.; Lee, S. G.; Andrews, R. E., Jr.; Klowden, M. J.; and Bulla, L. A., Jr. 1985. Separation of the cytolytic and mosquiticidal proteins of *Bacillus thuringiensis* subsp. *israelensis. Biochem. Biophys. Res. Commun.* 126: 961–965.

Ibarra, J. E., and Federici, B. A. 1986. Isolation of a relatively nontoxic 65-kilodalton protein inclusion from the parasporal body of *Bacillus thuringiensis* subsp. *israelensis. J. Bacteriol.* 165 (2): 527–533.

Jamnback. 1981. The origins of black fly control programmes. In *Black flies,* ed. M. Laird, 71–73. London: Academic Press.

Kim, K. H.; Ohba, M.; and Aizawa, K. 1984. Purification of the toxic protein from *Bacillus thuringiensis* serotype 10 isolate demonstrating a preferential larvicidal activity to the mosquito. *J. Invertebr. Pathol.* 44: 214–219.

Klowden, M. J.; Held, G. A.; Bulla, L. A., Jr. 1983. Toxicity of *Bacillus thuringiensis* subsp. *israelensis* to adult *Aedes aegypti* mosquitoes. *Appl. Environ. Microbiol.* 46: 312–315.

Kuhlemeier, C. J.; Thomas, A. A. M.; van der Ende, A.; van Leen, R. W.; Borrias, W. E.; van den Hondel, C. A.; and van Arkel, G. A. 1983. A host-vector system for gene cloning in the cyanobacterium *Anacystis nidulans* R2. *Plasmid* 10: 156–163.

Kyte, J., and Doolittle, R. F. 1982. A simple method for displaying the hydropathic character of a protein. *J. Mol. Biol.* 157: 105–132.

Lau, R., and Doolittle, W. 1979. Covalently closed circular DNAs in closely related unicellular cyanobacteria. *J. Bacteriol.* 137: 648–652.

Lau, R. H.; Sapienza, C.; and Doolittle, W. F. 1980. Cyanobacterial plasmids: Their widespread occurrence and the existence of regions of homology between plasmids in the same and different species. *Mol. Gen. Genet.* 178: 203–211.

Lee, S. G.; Eckblad, W.; and Bulla, L. A., Jr. 1985. Diversity of protein inclusion bodies and identification of mosquitocidal protein in *Bacillus thuringiensis* subsp. *israelensis. Biochem. Biophys. Res. Commun.* 126 (2): 953–960.

Margalit, J.; Zomer, E.; Erel, Z.; and Barak, Z. 1983. Development and application of *Bacillus thuringiensis* var. *israelensis* serotype H14 as an effective biological control agent against mosquitoes in Israel. *Biotechnology* 1: 74–76.

Pfannenstiel, M. A.; Ross, E. J.; Kramer, V. C.; and Nickerson, K. W. 1984. Toxicity and composition of protease-inhibited *Bacillus thuringiensis* var. *israelensis* crystals. *FEMS Microbiol. Lett.* 21: 39–42.

Primrose, S. B., and Ehrlich, S. D. 1981. Isolation and plasmid deletion mutants and study of their instability. *Plasmid* 6: 193–201.

Sekar, V., and Carlton, B. C. 1985. Molecular cloning of the delta-endotoxin gene of *Bacillus thuringiensis* var. *israelensis. Gene* 33: 151–158.

Simon, R. D. 1978. Survey of extrachromosomal DNA found in the filamentous cyanobacteria. *J. Bacteriol.* 136: 414–418.

Sriram, R.; Kamdar, H.; and Jayaraman, K. 1985. Identification of the peptides of the crystals of *Bacillus thuringiensis* var. *israelensis* involved in the mosquito larvicidal activity. *Biochem. Biophys. Res. Commun.* 132 (1): 19–27.

Stanier, R. Y., and Cohen-Bazire, G. C. 1977. Phototrophic prokaryotes: The cyanobacteria. *Ann. Rev. Microbiol.* 31: 225–274.

Tandeau de Marsac, N.; de la Torre, F.; and Szulmajster, J. 1987. Expression of the larvicidal gene of *Bacillus sphaericus* 1593M in the cyanobacterium *Anacystis nidulans* R2. *Mol. Gen. Genet.* 214: 42–47.

Thomas, W. E., and Ellar, D. J. 1983a. *Bacillus thuringiensis* var. *israelensis* crystal δ-endotoxin: Effects on insect and mammalian cells in vitro and in vivo. *J. Cell. Sci.* 60: 181–197.

———. 1983b. Mechanism of action of *Bacillus thuringiensis* var. *israelensis* insecticidal δ-endotoxin. *FEBS Lett.* 154: 362–368.

Thorne, L.; Garduno, F.; Thompson, T.; Decker, D.; Zounes, M.; Wild, M.; Walfield, A. M.; and Pollock, T. J. 1986. Structural similarity between the lepidoptera- and diptera-specific insecticidal endotoxin genes of *Bacillus thuringiensis* subsp. *kurstaki* and *israelensis*. *J. Bacteriol.* 166 (3): 801–811.

Tyrell, D. J.; Bulla, L. A., Jr.; Andrews, R. E., Jr.; Kramer, K. J.; Davidson, L. I.; and Nordin, P. 1981. Comparative biochemistry of entomocidal parasporal crystals of selected *Bacillus thuringiensis* strains. *J. Bacteriol.* 145 (2): 1052–1062.

Undeen, A. H., and Nagel, W. L. 1978. The effect of *Bacillus thuringiensis* ONR-60A strain (Goldberg) on *Simulium* larvae in the laboratory. *Mosq. News* 38: 524–527.

van den Hondel, C.; Keegstra, W.; Borrias, W.; and van Arkel, G. 1979. Homology of plasmids in strains of unicellular cyanobacteria. *Plasmid* 2: 323–333.

van den Hondel, C. A.; Verbeek, S.; van der Ende, A.; Weisbeek, P. J.; Borrias, W. E.; and van Arkel, G. A. 1980. Introduction of transposon Tn 901 into a plasmid of *Anacystis nidulans*: Preparation for cloning cyanobacteria. *Proc. Natl. Acad. Sci.* 77: 1570–1574.

Visser, B.; van Workum, M.; Dullemans, A.; and Waalwijk, C. 1986. The mosquitocidal activity of *Bacillus thuringiensis* var. *israelensis* is associated with M_r 230,000 and 130,000 crystal proteins. *FEMS Microbiol. Lett.* 30: 211–214.

Waalwijk, C.; Dullemans, A. M.; van Workum, M. E. S.; and Visser, B. 1985. Molecular cloning and the nucleotide sequence of the M_r 28,000 crystal protein gene of *Bacillus thuringiensis* subsp. *israelensis*. *Nucleic Acids Res.* 13 (22): 8207–8217.

Wabiko, H.; Raymond, K. C.; and Bulla, L. A., Jr. 1986. *Bacillus thuringiensis* entomocidal protoxin gene sequence and gene product analysis. *DNA* 5: 305–314.

Ward, E. S., and Ellar, D. J. 1988. Cloning and expression of two homologous genes of *Bacillus thuringiensis* subsp. *israelensis* which encode 130-kilodalton mosquitocidal proteins. *J. Bacteriol.* 170: 727–735.

Ward, E. S.; Ellar, D. J.; and Todd, J. A. 1984. Cloning and expression in *Escherichia coli* of the insecticidal δ-endotoxin gene of *Bacillus thuringiensis* var. *israelensis*. *FEBS Lett.* 175: 377–382.

Whitton, B. A. 1973. Freshwater plankton. *The biology of blue-green algae*, ed. N. G. Carr and B. A. Whitton, 353–367. Berkeley and Los Angeles: University of California Press.

Wolk, C. P.; Vonshak, A.; Kehoe, P.; and Elhai, J. 1984. Construction of shuttle vectors capable of conjugative transfer from *Escherichia coli* to nitrogen-fixing filamentous cyanobacteria. *Proc. Natl. Acad. Sci.* 81: 1561–1565.

Wu, D., and Chang, F. N. 1985. Synergism in mosquiticidal activity of 26 and 65 kDa proteins from *Bacillus thuringiensis* subsp. *israelensis* crystal. *FEBS Lett.* 190: 232–236.

8

Potential for Improved Formulations of *Bacillus thuringiensis israelensis* through Standardization and Fermentation Development

HOWARD T. DULMAGE
JOSE A. CORREA
GABRIEL GALLEGOS-MORALES

8.1 INTRODUCTION

Before we discuss the standardization and fermentation of *Bacillus thuringiensis* (*B. t.*), it is important to define certain precepts about *B. t.* that are generally accepted today and are germane to this chapter.

First, the species *B. thuringiensis* can be divided into over 25 subspecies, primarily but not solely based on the serology of the flagellae that are present in the motile stages of the development of this bacterium.

Second, isolates of different subspecies of *B. t.* can have different spectra of insecticidal activity, as illustrated in table 8.1. Thus, the two subspecies of *B. t.* that we will consider in this chapter differ markedly in the type of insect species that they kill. Subspecies ***kurstaki*** (HD-1) is active against many lepidopterous pests, has only very weak activity against mosquitoes, and has no activity against aquatic black flies. Subspecies *israelensis* is very active against mosquitoes and black flies, but has little or no activity against lepidopterous insects (de Barjac 1978a, 1978b, 1978c).

Third, the insecticidal activities of *B. t.* isolates are associated with crystalline parasporal bodies that appear in the cells at the time the *Bacillus* sporulates. These toxins are grouped under the general title δ-endotoxins. The toxins are all high molecular weight proteins. They are similar to each other in appearance, but differ in molecular weight and in the type of insects they kill. As has been shown by Dulmage and Cooperators (1981), a single isolate may produce more than one toxin, and the spectra of these toxins may differ widely. In some cases, the physical characteristics of the toxin may also differ, as in the small, amorphous P2 toxin described by Yamamoto and McLaughlin (1981).

Finally, there is no chemical assay for the δ-endotoxins, so (as will be discussed later in this chapter) we cannot express the quantity of endotoxin

TABLE 8.1.
Comparison of Spectra of Activities of Six Subspecies of *Bacillus thuringiensis*

Subspecies	Crystal type	Typical potency vs. species				
		Trichoplusia ni	*Heliothis virescens*	*Hyphantria cunea*	*Bombyx mori*	Mosquito
thuringiensis	*thu*	4,000	1,000	30,000	1,200	na
alesti	*ale*	na	na	51,000	30,000	na
kurstaki	*k-1*	18,000	7,600	6,000–100,000	16,000–40,000	+
galleriae	*g-9*	5,000	600	48,000	7,000	±
alizawai	*aiz*	11,000	900	80,000	46,000	na
israelensis	—	na	na	na	na	+++

NOTE: na = not active.

in a formulation in terms of weight of toxin. We can only measure the potency of the material containing the toxin and express activity in terms of an insecticidal unit derived by comparing the activity of the sample with that of a standard formulation. Potencies are then expressed in International Units (IU) in the case of formulations of subspecies *kurstaki* or in International Toxicity Units (ITU) in the case of formulations of subspecies *israelensis*.

8.2 HISTORY

8.2.1 Background

Winston Churchill once said that those who ignore history are doomed to repeat it. In a similar vein, Edward Steinhaus, the father of modern insect pathology, stressed the importance of understanding the thoughts and philosophies that guided our predecessors in their research on microbial insecticides as we develop our own programs on these entomopathogens. This is especially important in understanding the problems associated with the development of practical and effective formulations of *B. t.* If we are not aware of the mistakes of the past and their causes, we can make the same mistakes again.

To understand the problems that slowed the early development of this group of bioinsecticides, one must begin by examining the early concepts of this species and its use.

8.2.2 Discovery and Early History

Ishiwata was the first to isolate a culture of *B. t.*, which he found in dying larvae in a silkworm factory in Japan in 1901. He saw that the causative agent for the disease in the larvae was a bacterium, and we can recognize the organism as a *B. t.* from the descriptions in his publications. However, his descriptions were incomplete, and the name that he gave to his isolate was improper, so Ishiwata could not be credited with the initial discovery. In 1911, Berliner discovered a similar bacterium in the province of Thuringia in Germany, which was killing the Mediterranean flour moth, *Anagasta kuehniella*, in bins of stored grain. He isolated the causative organism, described it correctly, and proposed the epithet *Bacillus thuringiensis* for the isolate. Ironically, in the same year, Aoki and Chigasaki, students of Ishiwata, published a valid description of Ishiwata's isolate and proposed the name *Bacillus sotto*. However, Berliner's name had priority, and the name *B. thuringiensis* became the accepted one. The name *sotto* is retained as the name of a subspecies.

It is interesting to keep in mind the sources of these two isolates of *B. t.*

Ishiwata's microbe was found to cause a serious disease in a colony of valuable domestic insects; Berliner's microbe was found to help rid a valuable crop of a serious insect pest. These discoveries have colored research goals since then. Eastern silk-producing countries have usually considered *B.t.* a threat, while western countries have usually considered *B.t.* a potentially valuable ally in crop protection. The development of modern *B.t.* formulations with high activities against a wide variety of insect pests has fortunately blurred that distinction.

8.2.3 Standardizing and Measuring *B.t.* Products

8.2.3.1 Early Concepts. Most of the early studies of *B.t.* were predicated on the assumption that *B.t.* was a disease of insects—that its action was due to an infection caused by the *Bacillus.* There was evidence that this was true: if insects were exposed to *B.t.,* they died; if the dead larvae were examined, spores of the *Bacillus* could be seen; if these spores were subcultured, the subcultures could infect insects exposed to them. Apparently, Koch's postulates were satisfied.

What was not recognized by these early workers was the role played by the δ-endotoxin in poisoning the insects. Since they did not know the interactions between the insect and the *B.t.* toxins, they proceeded on the assumption that the viable cells and spores were responsible for the activities of *B.t.* powders and formulations. If this were true, then potencies of *B.t.* formulations could be expressed as spores per gram of formulation, with application rates being specified as the number of spores per hectare. At the time, this seemed to be the most logical approach to standardization. However, in view of the erratic performances of *B.t.* formulations in the field, there was, in retrospect, considerable warning that something was wrong with our standardization procedures. It is well to stop here and review these procedures in the light of our present knowledge.

8.2.3.2 The Spore Count. The easy way to measure the quantity of viable *Bacilli* is through a spore count in which varying amounts of pasteurized samples of the *Bacillus* preparation are mixed with nutrient or tryptose-phosphate agar, then poured into petri dishes and incubated for 24 to 36 hours, at which time the number of colonies on the plates are counted. Each spore of the *Bacillus* is presumed to give rise to one colony on the plate (a highly questionable assumption). However, the spore count does give a reasonably good estimate of the quantity of *B.t.* cells in the original formulation. Unfortunately, this method is wrong for two reasons (aside from the inaccuracies faced in any spore counts): (1) the spore is at best only a secondary factor in killing insects—a toxin produced by the *B.t.* during the fermentation (the δ-endotoxin described earlier), rather than the *Bacillus* itself, is responsible for most of the activity against insects; and (2) even if it were true that *B. t.* functioned by infecting the target insects, the ability of the *Bacillus* to infect the insect could vary, and a spore count could not measure such differences in the virulence of the *Bacilli.*

Nevertheless, our predecessors, in all good faith, relied on the spore count; indeed, the regulatory agencies in the United States required that commercial formulations be standardized by spore count and that the label contain a statement expressing potency as spores of *B. t.* per gram of formulations, an essentially meaningless exercise. As we shall discuss in the next section, the only way to standardize or evaluate *B. t.* formulations or measure the potencies of *B. t.* samples is through bioassay.

8.3 BIOASSAYS AND THE INTERNATIONAL UNIT

8.3.1 Philosophical Differences between Bioassays of Chemical and Microbial Insecticides

Scientists familiar with the assay of chemical insecticides enter a new world when they turn to assaying microbial insecticides. In the case of a chemical insecticide, the assayer already knows the quantity and purity of the insecticide. The assay is an accessory to the production process; it is only confirmatory. In the case of a microbial agent such as *B. t.,* the assay must measure all stages of the production process: the quality of the fermentation, what sort of losses of activity occur during the recovery of the active material from the fermenter beer, the quality of the product recovered, and the characteristics of the final formulation. All this must be known accurately, and this knowledge can be achieved only through bioassay. The bioassay is not peripheral or accessory to the *B. t.* process, it is central to it. Thus, the reliability and accuracy that we must demand from the assay of a microbial insecticide must be much greater than that demanded from the assay of a chemical.

8.3.2 Bioassays and the LC_{50}

A bioassay essentially measures the interaction between a test insect and the toxin being assayed. The most dramatic response of an insect to a microbial insecticide, and the one that is easiest to observe, is death; and the most accurate expression of the killing power of a *B. t.* sample is the LC_{50}, the concentration of sample in the diet that will theoretically kill 50% of the larvae. This concentration is determined by exposing groups of larvae to differing concentrations of the sample in their diet, incubating them for a set period of time, determining the percentage killed at each concentration to which they have been exposed, and then using regression analysis to determine the LC_{50}.

Determining the LC_{50} is not difficult. What is difficult is achieving the necessary accuracy in the assay. This requires large numbers of healthy, uniform insects and considerable care and attention throughout the entire process. The uniformity of the insect colonies used, the number of dilutions of the sample tested, the number of larvae used at each dilution, the mechanics of how the test larvae are exposed to the toxin, and the slopes of the regression curves involved in the analyses all influence the accuracy. Deficiencies in technique or insect can be overcome, but only by increasing the number of insects required in the assay—a costly and wasteful procedure.

8.3.3 Choice of Insect Species for Bioassay

The principal use of *B.t.* in the early days was for the control of *Trichoplusia ni,* so it was logical to select *T. ni* as the assay insect. In addition to being a major target insect, it was also easy to rear. The insect was noncannibalistic and, depending on the desires of the assayer, could be used as neonate or as third-instar larvae. All this led to the adoption of *T. ni* as the assay insect for the "official" U.S. assay proposed by Dulmage et al. (1971) after the first bioassay recommendations on *Pieris brassicae* made by Burgerjon (1957) and Bonnefoi, Burgerjon, and Grison (1958). Other insect species can be used, and it is desirable to have assays against more than one kind of insect, as will be discussed later in this chapter. In the laboratories at Brownsville, most assays have been carried out with both *T. ni* and *Heliothis virescens* (Dulmage, Martinez, and Pena 1976). However, *H. virescens* is cannibalistic, and neonate larvae are recommended for use in this assay. Neonate larvae have proved very satisfactory, and most assays at Brownsville, regardless of species of insect, have been carried out using neonate larvae. A fuller discussion of the bioassays at Brownsville is given by Dulmage (1973b) and Dulmage et al. (1971).

Habits of insects often dictate procedures. In the cooperative program described by Dulmage and Cooperators (1981), many of the bioassays were conducted against lepidopterous insects fed on the artificial agar-based diet referred to above. These included, in addition to the *T. ni, H. virescens, Spodoptera exigua,* and *Spodoptera frugiperda* tested at Brownsville, *Bombyx mori, Hyphantria cunea,* and *Spodoptera litura* (Aizawa, Fujiyoshi, and Ohba), *Ostrinia nubilalis* (Lewis), and *Lymantria dispar* (Dubois). Some of the Lepidoptera studied in the program had to be assayed on other types of diets, for example, *Galleria mellonella* on a granular-type-medium (Burges and Jarrett) and *Ephestia cautella* and *Plodia interpunctella* on a grain-based diet (McGaughey and Dicke). (Names in parentheses are those of the scientists participating in assaying those particular insects in the program.) In a similar program, Salama and Foda (1984) and Salama, Foda, and El-Sharaby (1981) assayed *Hyphantria armigera* and *Spodoptera littoralis* on

TABLE 8.2.
Bioassay Standards

Name of Standard	Potency (ITU/mg)
IPS-78[a]	1,000
IPS-80	10,000
IPS-82[b]	15,000
HD-968-S-1983[c]	4,740

[a] Primary standard on which all other activities are based.
[b] Present World Health Organization (WHO) standard.
[c] Secondary standard for use in the United States (potency determined in assays against IPS-78).

an agar-based diet. Also, Mechalas and Anderson (1964) and Mechalas and Dunn (1964) assayed *Estigmene acrea* by dipping bouquets of alfalfa leaves into a series of different suspensions of the test samples. The accuracy of all these assays seemed similar, no matter which procedure was used.

Finally, several methods of assaying mosquito activities have been devised. De Barjac and Coz (1979) and de Barjac and Larget-Thiery (1979, 1984) developed assay procedures suitable to assay formulations of several mosquito species, including *Anopheles sergentii, Anopheles stephensi, Culex pipiens, Culex univittatus,* and *Uranotaenia unguiculata.* Also, as we shall discuss later in this chapter, de Barjac (1983) has presented a detailed protocol for bioassays of commercial formulations of *B.t.i.* against *Aedes aegypti.* Similarly, McLaughlin et al. (1984) and Dulmage et al. (1985) have described a protocol for use in bioassaying commercial formulations in the United States. In other studies, Jafri et al. (1982) have shown that the susceptibility of larvae of *An. stephensi, Aedes aegypti,* and *Aedes albopictus* varied between different geographical locations and even on different colonies from the same source of insects. Mosquito larvae, of course, cannot be grown on the agar diet used for the assay of lepidopterous insects but must be grown in water (dechlorinated tap water, for example). Mosquito assays will be discussed at length later in this chapter.

8.3.4 The International Unit

The LC_{50} is a valuable tool, but insects vary in their vigor and their response to a toxin from day to day, and the LC_{50} likewise will vary. These fluctuations limit the value and accuracy of the LC_{50}. The solution has been to introduce a standard formulation into the assay, determining its LC_{50} in an assay run in parallel, at the same time, and against the same population of insects used to determine the LC_{50}s of the samples being tested (Dulmage 1973a). One can then report that the sample is twice as active as the standard or one-third as active, or whatever the case might be. This is awkward, and it is much better to assign a potency in units to the standard and then define the activity of the samples in units compared to the standard (Burgerjon and Dulmage 1977). The units are assigned to the standard by agreement of the scientific community or are derived by the use of assays with an agreed-upon primary standard. In the case of *B.t.*, the potencies of the standards are expressed in IUs or in ITUs. It has been customary to express potencies of formulations designed for lepidopterous insects as IU and to express potencies of the mosquiticidal δ-endotoxins as ITU. Four standards have been or are being used in the assay of samples derived from *B.t.i.* fermentations. These are listed in table 8.2.

The calculation of the potencies of the test samples in ITU uses the following formula:

$$\text{Potency sample (ITU/mg)} = \frac{LC_{50}\ \text{Standard}}{LC_{50}\ \text{Sample}} \times \text{Potency Standard, ITU/mg}$$

When IPS-82 is used as the standard, the formula becomes

$$\text{Potency sample (ITU/mg)} = \frac{LC_{50}\ \text{Standard}}{LC_{50}\ \text{Sample}} \times 15{,}000\ \text{ITU/mg}$$

8.3.5 The de Barjac Protocol for *B.t.i.*: Its Design and Principles

The ITU is designed to compensate for the day-to-day fluctuations in insect response to *B.t.i.* so that the potencies determined for each sample are reproducible. And, as we shall discuss in this section, the ITU achieves this purpose. However, this does not allow anarchy in the protocols used in the assay. It would be well if everyone responsible for bioassaying *B.t.i.* would use exactly the same procedure, but from a practical standpoint this is impossible to achieve because of the differences in design and supplies of different laboratories. Nevertheless, it is possible and desirable to devise a standard protocol, which everyone should follow as closely as possible, allowing only minor differences in technique.

Several methods of bioassaying *B.t.i.* have been proposed, differing primarily in the insect used and/or its age or instar. While all the assay procedures may perform satisfactorily, the most widely used are two very similar protocols. One was devised by a group organized by WHO under the leadership of de Barjac (1983) and Rishikesh and Quelennec (1983); the other was devised by a group organized in the United States (McLaughlin et al. 1984) in a cooperative effort between the USDA Agricultural Research Service, the University of California, and representatives of the major companies producing *B.t.i.* The differences between the two protocols represent minor adjustments in the protocol to fit the assay better to U.S. needs. The principles upon which the protocol is based are similar to those used for bioassays of formulations of subspecies *kurstaki,* an assay that has been used successfully for years against lepidopterous larvae. While the procedures had to be modified considerably to fit the habits of a different type of insect, the principles remain the same as those that apply to most bioassays of microbial entomocidal poisons. The procedures specified in the de Barjac protocol are summarized below. The bioassay is based on the comparison of mosquito larval mortalities produced by a *B.t.i.* preparation and by the IPS-82 *B.t.i.* standard titrating 15,000 ITU/mg against *Ae. aegypti.*

8.3.5.1 Preparation of Stock Suspension of the Standard. A stock suspension of the IPS-82 standard is prepared by homogenizing 50 mg of the standard in 10 ml water, shaking the suspension with 6-mm glass beads in a 20-ml penicillin flask. After 10 minutes of homogenization, the suspension is used to prepare a stock solution by adding 0.1 ml of the homogenate to 9.9 ml deionized water. It is important that all water to which mosquito larvae will be exposed should be deionized and chlorine-free. This stock suspension is thoroughly mixed just before use by agitation on a Vortex agitator for a few seconds.

Subsequent dilutions of this stock suspension are directly prepared in plastic cups that have been filled with 188 ml of deionized water. First, 25 early fourth-instar (L4) larvae of *Ae. aegypti* in 2 ml of water are added to each cup by means of a Pasteur pipette. Then 120, 90, 60, 30, 24, and 15 μl of the stock suspension are added to the cups in order to obtain concentrations of IPS-82 of 0.04, 0.03, 0.02, 0.01, 0.008, and 0.005 mg/100 ml, respectively. Four similar cups containing only deionized water are used as controls. No food is added to any of the cups.

8.3.5.2 Preparation of Suspension of the Test Samples. For bioassays of preparations of unknown activity, an initial homogenate is made and comparable dilutions prepared as for the IPS-82 standard. The range of dilutions should exceed that of the standard in order to obtain a reliable regression curve. Since the unknown sample may differ considerably in potency from IPS-82, time may be saved by first making a range-finding bioassay, with widely spaced concentrations of the unknown. The data from this assay can then be used to fix the dilutions needed for an accurate bioassay.

8.3.5.3 Specifications for Larvae Used in Assay. Early L4 larvae are chosen for the assay because they are more representative of the total susceptibility of the target populations and also allow for greater convenience in handling. It is very important to use a homogenous population of early fourth-instar larvae, which, in de Barjac's standardized way of rearing, are usually 5 days old and 4.5 mm long.

Each bioassay series should involve at least 400 larvae for the IPS-82 standard, 500–1,000 larvae for unknown preparations, and 100 larvae for the control.

8.3.5.4 Reading the Assay. The assay is read after an exposure period of exactly 24 hours, at which time the numbers of live larvae are counted. (The results are based on live larvae because larvae tend to be cannibalistic at lower concentrations.) If more than 10% of the control larvae are dead, discard the assay.

Mortalities are determined at 24 hours, as discussed above. However, it can be useful to make a second reading at 48 hours. Because of the very rapid killing action of *B. t. i.*, there is usually no difference between the two readings. Changes in the 48-hour reading will detect possible intervention of factors other than *B. t. i.* components.

8.3.5.5 Evaluations of Assays and Their Reproducibility. Dulmage (1983) summarized the criteria for evaluating individual assays as follows:

1. There must be < 10% dead in the control larvae
2. Dilutions must be selected so that at least five concentrations of each sample and seven concentrations of the standard are valid, with no more than 90% or no less than 10% mortality
3. Slopes of the regression curves must be reasonable, similar to those obtained previously against the test insect, and, within the errors of the day's samples, parallel to each other
4. If computer analyses are available, the 95% confidence limits around the LC_{50} should be such that the maximum limit/minimum limit is < 2.0
5. Similarly, the 95% confidence limits around the ITUs should be determined, and the maximum limit/minimum limit should be < 2.0

Assays should be replicated on three separate days, the ITUs determined on each of those days should be averaged, and the standard deviation should be determined. These data should be used to determine the coefficient of variation (CV) between the replicated assays (standard deviation/average). This forms the key to the trust that can be placed in the assay. Experience has taught us that reliable assays will have a CV < 0.20 (possibly < 0.15 for assays with *Ae. aegypti*). CVs higher than this indicate that something is wrong with the sample being tested, with the insect colony being used, or with the techniques of the assayers.

8.4 ACTIVITY RATIOS

8.4.1 Definition of Activity Ratios

Dulmage and Cooperators (1981) have shown the overwhelming diversity in spectra of insecticidal activity within and between different powders of *B. t.* Krywienczyk, Dulmage, and Fast (1978) and Krywienczyk et

TABLE 8.3.
Hypothetical Results of Bioassays on Powders Containing One of Two Delta-endotoxins with Different Spectra of Activity

	Powder A	Powder B	Powder C	Powder D
Assays against *Trichoplusia ni*:				
LC_{50} standard, μg/ml	11.2	11.2	11.2	11.2
LC_{50} sample, μg/ml	20.0	4.0	4.0	16.0
Potency of sample, IU/mg	10,000	50,000	50,000	12,500
Assays against *Heliothis virescens*:				
LC_{50} standard, μg/ml	2.8	2.8	2.8	2.8
LC_{50} sample, μg/ml	5.0	1.0	2.0	2.0
Potency of sample, IU/mg	10,000	50,000	25,000	25,000
Tn/Hv ratio[a]	1.0	1.0	2.0	0.5

SOURCE: Adapted from Dulmage, 1979. Activity ratios are discussed at length by Dulmage and Cooperators, 1981.

NOTE: Powder A and Powder B are homologous to themselves and to the standard, and Powder C and Powder D are neither homologous to the standard nor to each other.

[a] $\text{Tn/Hv ratio} = \frac{\text{Potency vs. } T.\ ni \text{ (IU/mg)}}{\text{Potency vs. } H.\ virescens \text{ (IU/mg)}}$

al. (1981) have shown through serological analyses that more than one type of crystal can be present in a single powder. One way to identify these different toxins is through the use of an activity ratio. This ratio takes into account that the IUs determined in these assays are determined through a comparison with a standard formulation. If the test sample and the standard are homologous (i.e., contain the same toxin), the potencies determined will be the same, regardless of which insect is used in the bioassay, so long as the insect is susceptible to the toxin. If the test sample and the standard are not homologous, then the potencies determined against different insect species may differ. This difference can be expressed as a ratio by dividing the potency determined against one insect species by the potency determined against a second insect species. Table 8.3 shows the calculation of an activity ratio based on assays against *T. ni* and *H. virescens*. The insects used in the ratio are indicated by the initials of their genus and species names, thus the Tn/Hv ratio in the example in table 8.3.

8.4.2 Use and Significance of Activity Ratios

The numerical value of the activity ratio is independent of potency. It indicates the character rather than the quantity of a toxin in a powder. It will detect differences between toxins produced by different subspecies of *B. t.* or any changes in the toxin produced by the same organism but under different fermentation conditions. The reproducibility of the activity

TABLE 8.4.
Influence of Medium on the Production of Delta-endotoxin by *Bacillus thuringiensis* Strain HD-1 in 14-liter Fermentors

	Medium Ingredients (g/liter)[a]			Yield (kIU/ml)[b]		
Experiment no.	Cottonseed flour	Glucose	Corn steep	*Trichoplusia ni*	*Heliothis virescens*	Tn/Hv ratio[c]
1	20	15	—	414	298	1.39
	30	40	25	834	365	2.28
	30	20	25	893	472	1.89
2	20	15	10	711	278	2.56
	30	40	10	797	381	2.09
	30	40	10*	515	246	2.09
3	30	40	50*	1730	—	—

[a]All media, except those designated by *. Media contained yeast extract, 2.0; peptone, 2.0; $MgSO_4.7H_2O$, 0.3; $FeSO_4.7H_2O$, 0.02; $ZnSO_4.7H_2O$, 0.02; $CaCO_3$, 1.0.
[b]Measured against *T. ni* and *H. virescens*, using HD-1-S-1971 as a referenced standard. HD-1-S-1971 contains 18,000 IU/mg; kIU = international units × 10^3.
[c]IU/ml measured against *T. ni* divided by IU/ml measured against *H. virescens*. Reproducibility of Tn/Hv ratio: n = 6; Avg = 2.050; SD = 0.394; CV = 0.192.

ratio should be less than that expected from assays against a single insect species because a second variable has been introduced—activity versus a second insect species. Experience with many activity ratios indicates that a CV of 0.33 or less should be acceptable (Dulmage, unpub. data).

8.4.2.1 Tn/Hv Ratios of subspecies *kurstaki*. Table 8.4 illustrates the value of the activity ratio. The table shows the potencies of powders of subspecies *kurstaki* (HD-1) measured in assays versus *T. ni* and *H. virescens*. The powders have been produced in several different media and also differ from each other in potency. However, the CVs of the various powders averaged 0.192, well within experimental error, indicating that the toxin present in each powder was the same despite differences in the quantities of the powders produced or the media in which it was produced.

8.4.2.2 Cq/Aa Ratios of subspecies *israelensis*. The activity ratios studied in the preceding sections demonstrate that the activity ratio is a useful tool to compare the lepidopterous activities produced by the various isolates of *B.t.* Less work has been done evaluating activity ratios of powders of subspecies *israelensis*. In one experiment, powders of five isolates of subspecies *israelensis* were tested against *Culex quinquefasciatus* and *Ae. aegypti*, and Cq/Aa activity ratios were determined. Results are shown in table 8.5 and are similar to results seen when testing lepidopterous larvae. If the powder tested and the standard IPS-78 contain homologous toxins, then the Cq/Aa ratios should be 1.00. The average ratios of the toxins produced by the five isolates were 1.261, close enough to 1.00 to be within

TABLE 8.5.
Comparison of Formulations of Five Isolates of *Bacillus thuringiensis* subsp. *israelensis* Bioassayed with Two Species of Mosquitoes

		Potency—IU/mg vs. Species		
Culture no.	Run no.[a]	Aedes aegypti	Culex quinquefasciatus	Cq/Aa ratio[b]
HD-500	R-179	206	—	—
	R-966-A	680	644	0.947
HD-519	R-933-B	179	—	—
	R-966-B	152	181	1.19
HD-522	R-967-A	1940	3460	1.784
HD-563	R-967	2170	3500	1.61
	FR-201-AB-AP	3250	—	—
HD-567	R-967-C	1970	1530	0.777
	FR-201-CD-AP	2910	—	—

[a]Cultures with a run number starting with *R* were grown in 500-ml flasks on a rotary shaker. Cultures with a run number starting with *FR* were grown in submerged culture in 14-liter fermentors.
[b]Reproducibility of Cq/A ratios: avg = 1.26, SD = 0.422, CV = 0.339 (SD = standard deviation; CV = coefficient of variation). Ratios with a CV less than 0.33 indicate that the ratios of the different materials are identical.

experimental error. However, some caution must be used in interpreting the results. The Cq/Aa activity ratios of three of the five powders were very close to 1.00. However, the ratios of the other two powders averaged 1.70, and the CV of the five powders was 0.34; this value is close to, but slightly above, the limits of acceptability estimated for activity ratios (Dulmage and McLaughlin, unpub. data).

8.4.2.3 Reproducibility of Activity Ratios in Fermentation Studies. During our fermentation studies, powders of five isolates of *B. thuringiensis* were produced under a wide variety of conditions (including such variables as type of fermentation equipment, media, age at harvest, and temperatures of incubation). Activity ratios of each powder were determined and compared to learn if changes in fermentations of the individual isolates changed the activity ratios observed. The number of formulations tested and the number of times each powder was assayed varied. Results are shown in table 8.6. As the table shows, it was possible to distinguish powders produced by each isolate by the Tn/Hv activity ratio associated with them. However, the activity ratios of powders produced by the same isolate were closely similar, with CVs ranging from 0.23 to 0.29, all within experimental error. It should be noted that the activity ratios of 5/62 powders of HD-129 differed widely from the ratios of the other 57. There was no logic to which powder was different, and the differences seemed to be due to error in the

TABLE 8.6.
Reproducibility of Activity Ratios of Formulations from Fermentations of Four Isolates *Bacillus thuringiensis* in 14-liter Fermentors

Culture	Number of Formulations Tested[a]	Average Tn/Hv Ratio[b]	Coefficient of Variation
HD-1	60	2.40	0.25
HD-129	59	5.63	0.32
HD-241	26	0.63	0.24
HD-244	26	1.54	0.23
HD-263	136	0.44	0.29

[a] Includes formulations from different media or fermentation conditions and from different times of harvest.
[b] International unit (IU)/ml measured against *Trichoplusia ni* divided by IU/ml measured against *Heliothis virescens*. IU determined against HD-1-S-1971 as a standard. HD-1-S-1971 contains 18,000 IU/mg.

Heliothis assay, rather than a change in HD-129 toxin, so the abnormal powders were omitted from consideration. At least for the five isolates in this study—and we hope for all isolates—it seems that the toxin produced by an individual isolate will always be the same, regardless of changes in fermentation conditions or procedures (Dulmage, unpub. data).

There is some evidence that quite drastic modifications in the fermentation can be made without changing the toxin produced. Nishiitsutsuji-uwo, Wakisaka, and Eda (1975) and Wakisaka et al. (1982) treated cells of subspecies *alesti, aizawai,* and *kurstaki* with mutagens to derive asporogenous mutants that would still produce δ-endotoxin. They reported that the toxins produced were the same as the parent isolates. Also, Blokhina et al. (1985) studied the continuous fermentation of an isolate of subspecies *galleriae.* The continuous process was marked by the appearance of S-form and R-form variants. The R-forms produced virulent phages when they were grown under chemostat conditions or in the presence of vancomycin. The S-form did not produce phage. When the S-forms were grown in the chemostat, they did not revert to R-forms. However, in spite of the differences between the R and S forms, the biological activity seemed unchanged.

The apparent reproducibility of the *B.t.* fermentations is very encouraging—and necessary if *B.t.* fermentations are going to be practical. A company producing *B.t.* cannot afford a fermentation that is not reproducible. Producers must be able to rely upon the culture they are growing to produce the same toxin from fermentation to fermentation. Otherwise, the costs of production and quality control would be so high as to prevent the development of an economical and reliable product.

8.5 POTENTIAL FOR IMPROVEMENTS IN PRODUCTION OF THE *B.t.i.* TOXIN

8.5.1 Fermentation

The fermentations of the different isolates of *B. t.*, regardless of subspecies, have some general characteristics in common. They all use sugar (usually glucose, molasses, or starch), producing acid during the fermentation. In general, they have similar requirements for proteins or protein hydrolysates, can use NH_4^+ salts, and respond similarly to minerals. However, the individual isolates are unique entities, and a particular medium that may support good growth or toxin production by one isolate may be less satisfactory for another (Dulmage 1970, 1971; Dulmage and de Barjac 1973; Salama et al. 1983a, 1983b; Salama, Foda, and Dulmage 1983; and Smith 1982). Also, as discussed at length in Dulmage and Cooperators (1981), different isolates of *B. t.* may produce toxins with different spectra of activities.

8.5.1.1 Strain Selection. Strain selection has been one of the classical means of improving a fermentation. The data in table 8.5 illustrate this. The potencies of powders of the five isolates of subspecies *israelensis* vary from < 200 to > 3,000 ITU/mg. Obviously, the production capabilities of the five isolates vary widely, and one must choose the isolate to use in the production of H14 toxin carefully. Such large changes are usually seen only in the early stages of development of a fermentation, but strain selection is a valuable tool even during advanced studies where incremental changes are usually much smaller (20–30%) (Dulmage 1979).

Strain selection is especially valuable in the case of *B. thuringiensis*, where gains can also be reaped by the discovery of *B. thuringiensis* toxins with high activity against a particular target insect. This has been shown for lepidopterous insects by Dulmage and Cooperators (1981) and for subspecies *israelensis* by Smith (1982) as discussed below.

8.5.1.2 Aeration. Adequate aeration is very important to the *B. t.* fermentation. Dulmage and McLaughlin (unpub. data; table 8.5) grew two isolates of subspecies *israelensis* (HD-563 and HD-567) in 500-ml Erlenmeyer flasks and in 14-liter fermentors. Aeration was much more efficient in the 14-liter fermentors, and assays with *Ae. aegypti* showed that yields increased 50% with the improved aeration. Foda, Salama, and Selim (1985) grew subspecies *entomocidus* under conditions of high and low aeration in the presence of high or low glucose concentrations. Considerable acid was produced in both glucose media. Consequently, the culture failed to grow in either medium unless the pH was controlled, either by daily adjustments of the pH of the beer, or by buffering the medium with $Ca(OH)_2$. When the pH was controlled, the culture grew in the presence of both concentrations of glucose. However, the culture failed to sporulate when grown under conditions of low aeration, although it sporulated and lysed normally when grown under high aeration. Moraes, Santana, and Hokka (1980) examined the influence of oxygen concentration on the production of δ-endotoxin. They reported that respiration increased when the oxygen concentration in the medium was increased to 2.5–10% of the saturation value but decreased if the oxygen concentration was increased further.

Zamola and Kajfez (1977) developed and patented a novel intermittent use of aeration in a *B.t.* fermentation. The authors found that interrupting aeration after 12–14 hours of fermentation prevented a lysis of cells. The beer was held without aeration for 3–6 hours and then returned to normal aeration. The period with no aeration prevented sporulation but allowed cell growth. The return of aeration allowed sporulation to proceed. As a result, when the beer was harvested after 35–40 hours, sporulation and lysis were normal and were essentially completed, but increased yields of insecticidal toxin were obtained.

In another study, Goldberg et al. (1980) described a continuous fermentation and reported that a high growth rate and a high cell concentration resulted in a high spore yield. The aeration rate in their fermentors was 3–4 v/v/m. Yields in their fermentors were not increased when the sugar level was raised above their standard 15 g/l, indicating that, in their media, glucose was not a limiting factor. Goldberg also carried out his fermentation studies on three different media and found that yields of toxin varied between the media, indicating that further research on media would be valuable. In another study on continuous fermentation, Freiman and Chupin (1973) obtained good yields in their fermentations. However, to get the production of spores, flow through the second-stage fermentors had to be stopped, and the fermentation had to continue to final spore formation. In this work, too, it seemed that there was a correlation between aeration, carbohydrate, and spore-toxin production.

Changes in the fermentation media can also lead to higher yields of δ-endotoxins. Smith (1982) tested three isolates of subspecies *israelensis* on six different media containing such nitrogen sources as $(NH_4)SO_4$, cottonseed meal, soybean flour, and casamino acids. Glucose was the primary carbon source studied, but sucrose, glycerol, and soy oil were also tested. The three isolates grew with different vigor on the various media, but all isolates did grow well on at least one medium, and no one medium was best for all isolates. When one considers the variety of nitrogen and carbon sources used successfully in this study and the equally wide variety of yields obtained in the various media, the outlook for developing improved strains and fermentation conditions looks very promising.

8.5.1.3 Selection of Nutrients. As stated above, the general requirements of isolates for nutrients can vary; what is a good combination of nutrients for one isolate may not necessarily be good for another. Therefore,

TABLE 8.7.
Examples of Potential Inexpensive Media Ingredients Available in Developing Countries

Liquids
- Coconut milk[a] (waste product)
- Crude sugar,[a] e.g., jaggery
- Whey[a] (waste product)
- Molasses[a]
- Corn steep liquid[a]

Inorganic Nitrogen
- $(NH_4)_2SO_4$

Materials of Plant Origin
- Legumes and other seeds, chick peas, peanuts, lima beans,[a] cowpeas,[a] soya beans,[a] bambara beans,[a] kidney beans,[a] cotton seed meal,[a] peanut cake, soy peptone, cottonseed, hydrolysate, horse beans, lentils
- Cereals, corn, guinea corn millets, wet mash from breweries, wheat flour, wheat bran
- Carbohydrate, dextrin, maltose, sucrose, glucose
- Plant extracts, potato tubers, sweet potato roots, minced citrus peels, ground seed of dates, carrots
- Tubers, cassava, yams, sweet potatoes
- Yeast powder,[a] fodder yeast

Materials of Animal (nonmammalian) Origin
- Fishmeal[a]

Materials of Mammalian Origin
- Blood,[a] chicken slaughterhouse residue

[a]Tried and found useful. The others either have not been tried or have given poor results but may be useful in combination with other ingredients.

it is impossible to recommend a fermentation medium that will be best for all isolates. Certain factors must be kept in mind, however, when designing a fermentation.

First, if the fermentation is to be economic, the cost of nutrients must be minimized. Therefore, it is best to use local nutrients, either inexpensive grains or waste products. For example, my laboratory is in the U.S. cotton belt. Cottonseed flour is relatively inexpensive. It is a good nutrient; therefore, I use it in many of my media. For a laboratory in an area where cotton is not grown, transportation costs for shipping cottonseed flour might be prohibitive. Table 8.7 lists some ingredients that have been proposed or used for the production of *B. t.*

Second, carbohydrates are important, as I have discussed in a previous section. However, *B. t.* subspecies produce acid from carbohydrates, and care must be taken to maintain a proper balance of acid-forming carbohydrate and alkali-forming proteins or polypeptides. If too much carbohydrate is added to a fermentation medium, the pH of the beer can drop below pH 5.5–5.7. Most isolates of *B. t.* cannot continue to grow when the pH is this low, and the fermentation will stop. This is a more frequent occurrence than is generally recognized, and occasionally there are papers in the literature reporting inhibitory effects of the use of a crude nutrient, where the author has failed to recognize that the inhibition is due to a failure to maintain a proper carbohydrate balance. The pH can be controlled by regular adjustment with alkali or, sometimes, by increasing the protein concentration in the medium.

The role of carbohydrate in a *B.t.* fermentation is more complex than just changing pH. Glucose has a two-fold effect on the fermentation. Glucose stimulates cell growth and can increase cell yields. However, glucose can also inhibit spore formation. A careful balance is needed between these two properties of carbohydrate metabolism. The picture may be further complicated by a more direct effect of carbohydrate metabolism on crystal formation. Scherrer, Lüthy, and Trumpi (1973) showed that when the glucose levels in their basic medium were increased, the crystals of the δ-endotoxin were increased in size and potency.

Third, the formation of the crystal and the production of the δ-endotoxin are intertwined with the metabolism of the cell and the whole factor of sporulation. Much needs to be learned. Indirect studies may contribute greatly, as, for example, the work of Garcia-Patrone (1985), who found that the addition of the antibiotic bacitracin increased the size of the parasporal crystals and the spores in *B. thuringiensis.* This is interesting since this antibiotic is produced by another species of *Bacillus.*

Other factors may play an important role in the production of δ-endotoxin. Wakisaka, Masaki, and Nishimoto (1982) demonstrated the stimulatory effect of K^+ ions on the production of endotoxin by *B. thuringiensis.* This effect was seen on media containing several different K^+ salts but was missing when the K^+ was replaced by Na^+, indicating that it was the potassium that was responsible for the increased production observed. Foda, Salama, and Selim (1985) reported a similar response to K^+ in their media. K^+ plays many important parts in the metabolism of cells, so, in a sense, it cannot be too surprising to find this relationship with K^+. Also, there is some interaction between the production and use of poly-hydroxybutyric acid and the activity of the K^+ ion. It will be very interesting, and possibly very important, to find what role K^+ plays in endotoxin formation.

Finally in contrast to most fermentations, the particle size of the nutrients used in *B.t.* fermentations must be controlled so that the particle size of any solids used in the medium are < 200 mesh. This comes about because in order to insure that the cost of the *B.t.* formulations are competitive with other methods of control, each step of the fermentation and recovery processes must be designed to be as inexpensive as possible. Thus the purification of the δ-endotoxin from the beer is minimal; if we strip the process to its bare essentials, we can see that the only real purification of the toxin is in the first step, when the beer is centrifuged. In this step, we rid ourselves of a large amount of water and dissolved solubles, but we retain, in addition to the toxin, a large quantity of unused nutrients or waste products, all contained in a thick cream.

TABLE 8.8.
Recovery Process for Spore-crystal Complex of *Bacillus thuringiensis*: Flow Sheet for Pilot Plant and Laboratory-scale Recovery

	Whole beer—pH 8.4–8.7
	Adjust to pH 7.0 with HCl
Supernate (discard)	Centrifuge
	Residue
	Suspend in 1/10–1/20 vol 4–6% lactose (vol. based on original beer)
	Add slowly while stirring 4–5 vol acetone
	Stir 30 minutes
	Let stand 10–30 minutes
Filtrate (discard)	Filter with suction
	Residue
	Stir with small-volume acetone
Filtrate (discard)	Filter with suction
	Residue
	Stir with small-volume acetone
Filtrate (discard)	Filter with suction
	Residue
	Dry overnight

SOURCE: After Dulmage, Correa, and Martinez, 1970.

This cream is essentially the final product. The only additional treatment given to this cream is to add various agents to make it more suitable for use in the field. We add to the cream, we do not purify it. Thus, if there are any coarse particles in the fermentation beer, they will be carried over to the final product. If this formulation is a flowable or wettable powder, the particles that it contains must be small enough to pass through a 200-mesh screen in order to avoid plugging the pipes and nozzles of the spray apparatus. This limits the choice of nutrients used in the fermentation. For example, soybean meal and soybean flour both support the growth of *B. thuringiensis* about equally. However, soybean meal, which is almost entirely composed of coarse particles, is unsatisfactory as a fermentation ingredient because the residual flakes in the medium are too coarse. Soybean flour, on the other hand, which is almost entirely > 200 mesh, is highly satisfactory. This is discussed further in Dulmage (1983).

TABLE 8.9.
Recovery Process for Spore-crystal Complex of *Bacillus thuringiensis:* Flow Sheet for Pilot Plant and Production-scale Recovery

Spray Drying

Whole beer—pH 8.4–9.0

Centrifuge
(tubular bowl continuous flow centrifuge)

Residue (weigh cream) — Supernate (discard)

Prepare slurry
based on weight of cream

Add 10% lactose
1% wetting agent
1% dispersant

Mix thoroughly

Adjust total solids
to approximately 20%

Spray dry
(centrifugal atomization feed)

Inlet temperature: 125–150° C

Outlet temperature: 75–100° C

Formulation

SOURCE: After Dulmage, 1983.

8.5.2 Recovery

8.5.2.1 Spore-Crystal Formulations. The reasons for the choice of recovery procedures to separate the *B.t.* toxin from the beer are given in the previous section. All recovery begins with centrifuging the beer. This centrifugation can be batch (used in laboratory-scale recovery); spinning tube, continuous-flow centrifugation; or multidisc continuous centrifugation (Zamola et al. 1981). Whatever method is chosen, the centrifugation yields a thick cream containing the activity. If the toxin is to be marketed as a flowable, the cream is blended with water or oils and stabilizers, wetting agents, or other additives the producer selects to make a stable suspension or emulsion. In the laboratory and for many field applications, a wettable powder may be preferred. Wettable powders are easily prepared in the laboratory by using the acetone-lactose coprecipitation technique (Dulmage, Correa, and Martinez 1970). On a larger scale, spray drying, with the desired additives being incorporated into the cream before the drying, is a common procedure. Flowsheets of processes of recovering wettable powders are given in tables 8.8 and 8.9 (Dulmage 1983).

8.5.2.2 Viable Spore-free Formulations. As discussed earlier in this chapter, most scientists originally believed that the activity of *B.t.* against susceptible insects was due to the ability of the spores of *B.t.* to multiply and cause a septicemia in susceptible insects. Most scientists now believe, after Angus's first announcement (1956), that a toxin, the crystalline δ-endotoxin produced by the *Bacillus*, is responsible for its activity. Nevertheless, some regulatory agencies and some scientists are concerned that releasing *B.t.* spores might induce an epizootic among desirable insects and advise that caution about releasing viable spores is needed. I think that the experiences of the past 15 years give adequate reassurance that the spore is harmless, both in the laboratory and in nature. However, in some areas a spore-free formulation is considerably more acceptable than a formulation containing spores and crystals. Therefore, some effort has been put into developing spore-free *B.t.* This can be done by following the pathways of researchers (Nishiitsutsuji-uwo, Wakisaka, and Eda 1975; Wakisaka, Masaki, and Nishimoto 1982; Wakisaka et al. 1982) who have developed sporeless mutants of *B.t.* that retain the production capacity and type of the normal parent. Some theoretical aspects to this approach might be significant for future research. Each cell of *B.t.* has enough energy to produce a normal amount of both toxin and spore proteins. When the bacterium is freed of the ability to produce spore proteins, it could be postulated that it would have more energy with which to make δ-endotoxin, so that potential yields of toxin would be much higher.

Alternatively, one can prepare a mixture of spores and crystals in the normal manner and then kill the spores. Under this procedure, there still would be no effort to separate crystals and nonviable spores. This has been done successfully by destroying the heat resistance of the *B.t.* with such chemicals as H_2O_2, and then pasteurizing the mixture. The spores will die, but the toxin will be unharmed. A Japanese patent has been issued on the technique (Shimizu et al. 1973; Utsumi 1973). The company holding the patent (Toa Gosei Chemical Co.) has told me that they see no difference in field effectiveness between spore-killed and normal, viable spore formulations.

8.6 SUMMARY

8.6.1 Bioassays and the Production of *B.t.*

Improvements in fermentations, once the initial stages of discovery are past, come in relatively small increments compared with the 20–30% gains discovered earlier. Thus, if we are to improve fermentations significantly, we must be able to measure changes of this size. The accuracy of the bioassays we have discussed in this chapter, if done carefully, will detect these changes. However, if the assay is allowed to become less accurate (by reducing the number of larvae in the assay, the number of dosage levels used, etc.), the lesser, but important improvements that we need in research on fermentation and recovery will be masked by the increased error and not be found.

8.6.2 Bioassays and the Future of *B.t.*

Bioassays, as we have discussed in this chapter, are tedious, expensive, time-consuming, and absolutely necessary. It is tempting to look for easy alternatives, to accept inaccurate methods in the interest of time and staff. Although it is tempting, it is vital to the future of *B.t.* that this temptation be resisted. All of our work to date has indicated that both *B.t.i.* and *B. sphaericus* are very promising microbial insecticides. But no matter how effective these microbes can be under field conditions, they will not be accepted unless they work reliably. And the word *reliable* is extremely important to the future of these two agents. The users—farmers or field applicators—are not particularly interested in using microbial control; they are interested in protecting crops or controlling mosquitoes. They are accustomed to reliable control with applications of chemical agents and will accept no less from biologicals. If we deliver this reliability in our commercial formulations, the future of *B.t.* is bright. The culture is easy to ferment, is not at all fastidious, and offers many opportunities for improving both the yield and the potencies of the toxins produced by the different subspecies of *B.t.* The bioassay gives us the tools to follow improvements in production, and the variations we have seen already give us many paths to explore in our drive to improve the value of all these subspecies, but particularly subspecies *israelensis.*

References

Angus, T. A. 1956. Association of toxicity with protein-crystalline inclusions of *Bacillus sotto* Ishiwata. *Can. J. Microbiol.* 2: 122–131.

Barjac, H. de. 1978a. A new candidate for biological control of mosquitoes: *Bacillus thuringiensis* var. *israelensis* (in French). *Entomophaga* 23: 309–319.

———. 1978b. Toxicity of *Bacillus thuringiensis* var. *israelensis* for larvae of *Aedes aegypti* and *Anopheles stephensi* (in French). *C. R. Acad. Sci.* (Paris) 286D: 1175–1178.

———. 1978c. Une nouvelle variété de *Bacillus thuringiensis* très toxique pour les moustiques: *B. thuringiensis* var. *israelensis* sérotype 14. *C. R. Acad. Sci.* (Paris) 286D: 797–800.

———. 1983. Bioassay procedure for samples of *Bacillus thuringiensis israelensis* using IPS-82 standard. WHO Report TDR/VED/SWG (5) (81.3).

Barjac, H. de, and Coz, J. 1979. Sensibilité comparee de six espèces différentes de moustiques a *Bacillus thuringiensis* var. *israelensis. Bull. de l'Org. Mond. Sante* (Paris) 57 (1): 139–141.

Barjac, H. de, and Larget-Thiery, I. 1979. Proposals for the adoption of a standardized bioassay method for the evaluation of insecticidal formulations derived from serotype H.14 of *Bacillus thuringiensis.* WHO/VBC/79.744.

———. 1984. Characteristics of IPS 82 as standard for biological assay of *Bacillus thuringiensis* H-14 preparations. WHO/VBC/84.892.

Blokhina, T. P.; Rautenstein, Y. I.; Sakharova, Z. V.; Rabotnova, I. L.; and Zavoyskaya, T. A. 1985. Biological peculiarities of *Bacillus thuringiensis* subsp. *galleriae* variants formed during continuous cultivation. *Mikrobiologiya* (USSR) 54 (4): 683–684.

Bonnefoi, A.; Burgerjon, A.; and Grison, P. 1958. Titrage biologique des preparations de spores de *Bacillus thuringiensis* Berliner. *C. R. Acad. Sci.* (Paris) 247: 1418–1420.

Burgerjon, A. 1957. L'utilisation des chenilles de *Pieris brassicae* L. comme insecte-test de laboratoire dans un service de controle de preparations pathogenes insecticides. *Entomophaga* 2: 129–135.

Burgerjon, A., and Dulmage, M. 1977. Industrial and international standardization of Microbial Pesticides: I. *Bacillus thuringiensis. Entomophaga* 22: 121–129.

Dulmage, H. T. 1970. Production of the spore-endotoxin complex by variants of *Bacillus thuringiensis* in two fermentation media. *J. Invertebr. Pathol.* 16 (3): 385–389.

———. 1971. Production of δ-endotoxin by eighteen isolates of *Bacillus thuringiensis,* serotype 3, in 3 fermentation media. *J. Invertebr. Pathol.* 18 (3): 353–358.

———. 1973a. Assay and standardization of microbial insecticides. *Ann. N.Y. Acad. Sci.* 217: 187–199.

———. 1973b. *B. thuringiensis* U.S. assay standard: Report on the adoption of a primary U.S. reference standard for assay of formulations containing the δ-endotoxin of *Bacillus thuringiensis. Bull. Entomol. Soc. Am.* 19 (4): 200–202.

———. 1979. Genetic manipulation of pathogens: Selection of different strains. In *Genetics in relation to insect management,* ed. M. A. Hoy and J. J. McKelvey, Jr., 116–127. Rockefeller Foundation Working Papers.

———. 1983. Guidelines for production of *Bacillus thuringiensis* H-14, ed. M. Vandekar and H. T. Dulmage. Proc. of Consultation, Geneva, Switzerland, 25–28 October 1982. UNDP/WORLD BANK/WHO Special Programme for Research and Training in Tropical Diseases.

Dulmage, H. T., and Barjac, H. de. 1973. HD-187: A new isolate of *Bacillus thuringiensis* that produces high yields of δ-endotoxin. *J. Invertebr. Pathol.* 22 (2): 273–277.

Dulmage, H. T.; Boening, O. P.; Rehnborg, C. S.; and Hansen, G. D. 1971. A proposed standardized bioassay for formulations of *Bacillus thuringiensis* based on the international unit. *J. Invertebr. Pathol.* 18 (2): 240–245.

Dulmage, H. T., and Cooperators. 1981. Insecticidal activity of isolates of *Bacillus thuringiensis* and their potential for pest control. In *Microbial control of pests and plant diseases, 1970–1980,* ed. H. D. Burges, 191–220. London: Academic Press.

Dulmage, H. T.; Correa, J. A.; and Martinez, A. J. 1970. Coprecipitation with lactose as a means of recovering the spore-crystal complex of *Bacillus thuringiensis. J. Invertebr. Pathol.* 15 (1): 15–20.

Dulmage, H. T.; McLaughlin, R. E.; Lacey, L. A.; Couch, T. L.; Alls, R. T.; and Rose, R. I. 1985. HD-968-S-1983: A proposed U.S. standard for bioassays of preparations of *Bacillus thuringiensis* subsp. *israelensis*-H-14. *Bull. Entomol. Soc. Am.* 31 (2): 31–34.

Dulmage, H. T.; Martinez, A. J.; and Pena, T. 1976. Bioassay of *Bacillus thuringiensis* (Berliner) δ-endotoxin using the tobacco budworm. U.S. Dept. Agric. Tech. Bull no. 1528.

Foda, M. S.; Salama, H. S.; and Selim, M. 1985. Factors affecting the growth physiology of *Bacillus thuringiensis. Appl. Microbiol. Biotechnol.* 22: 50–52.

Freiman, V. B., and Chupin, A. A. 1973. Aspects of continuous cultivation of spore-forming microbes from the group *Bacillus thuringiensis. Biotechnol. Bioeng.* 4: 259–265.

Garcia-Patrone, M. 1985. Bacitracin increases size of parasporal crystals and spores in *Bacillus thuringiensis. Mol. Cell. Biochem.* 68: 131–137.

Goldberg, I.; Sneh, B.; Battat, E.; and Klein, D. 1980. Optimization of a medium for a high yield production of spore-crystal preparation of *Bacillus thuringiensis* effective against the Egyptian cotton leaf worm, *Spodoptera littoralis Boisd. Biotechn. Lett.* 2 (10): 419–426.

Jafri, R. H.; Aslamkhan, M.; Nasreen, S.; and Asif, M. 1982. Variation among geographic strains of *Anopheles stephensi, Aedes aegypti,* and *Aedes albopictus* larvae in their susceptibility to *Bacillus thuringiensis* H-14 endotoxin. *Biologia* 28 (1): 101–116.

Krywienczyk, J.; Dulmage, H. T.; and Fast, P. G. 1978. Occurrence of two serologically distinct groups within *Bacillus thuringiensis* serotype 3ab var. *kurstaki. J. Invertebr. Pathol.* 31: 372–375.

Krywienczyk, J.; Dulmage, H. T.; Hall, I. M.; Beegle, C. C.; Arakawa, K. Y.; and Fast, P. G. 1981. Occurrence of *kurstaki k-1* crystal activity in *Bacillus thuringiensis* serotype I. *J. Invertebr. Pathol.* 37: 62–65.

McLaughlin, R. E.; Dulmage, H. T.; Alls, R.; Couch, T. L.; Dame, D. A.; Hall, I. M.; Rose, R. I.; and Versoi, P. L. 1984. U.S. standard bioassay for the potency assessment of *Bacillus thuringiensis* serotype H-14 against mosquito larvae. *Bull. Entomol. Soc. Am.* 30: 26–29.

Mechalas, B. J., and Anderson, N. B. 1964. Bioassay of *Bacillus thuringiensis* Berliner-based microbial insecticides: II. Standardization. *J. Insect Pathol.* 6 (2): 218–224.

Mechalas, B. J., and Dunn, P. H. 1964. Bioassay of *Bacillus thuringiensis* Berliner-based microbial insecticides: I. Bioassay procedures. *J. Insect Pathol.* 6: 214–217.

Moraes, I. O.; Santana, M. H. A.; and Hokka, C. O. 1980. The influence of oxygen concentration on microbial insecticide production. In *Advances in biotechnology.* Vol. 2, *Fuels, chemicals, foods, and waste treatment,* ed. M-Y. Murray, 75–79. Proc. Sixth Intnl. Fermentation Sym., London, Canada, 20–25 July 1980.

Nishiitsutsuji-Uwo, J.; Wakisaka, K.; and Eda, M. 1975. Sporeless mutants of *Bacillus thuringiensis. J. Invertebr. Pathol.* 25: 355–361.

Rishikesh, N., and Quelennec, G. 1983. Introduction to a standardized method for the evaluation of the potency of *Bacillus thuringiensis* serotype H-14-based products. *WHO Bull.* 61 (1): 93–97.

Salama, H. S., and Foda, M. S. 1984. Studies on the susceptibility of some cotton pests to various strains of *Bacillus thuringiensis. Z. Pflanzenkr. Pflanzenschutz* 91 (1): 65–70.

Salama, H. S.; Foda, M. S.; and Dulmage, H. T. 1983. Novel fermentation media for production of δ-endotoxins from *Bacillus thuringiensis. J. Invertebr. Pathol.* 41: 8–19.

Salama, H. S.; Foda, M. S.; and El-Sharaby, A. 1981. Potency of spore-δ-endotoxin complexes of *Bacillus thuringiensis* against some cotton pests. *Z. ang. Ent.* 91: 388–398.

Salama, H. S.; Foda, M. S.; El-Sharaby, A.; and Selim, M. H. 1983a. A novel approach for whey recycling in production of bacterial insecticides. *Entomophaga* 28 (2): 151–160.

Salama, H. S.; Foda, M. S.; Selim, M. H.; and El-Sharaby, A. M. 1983b. Utilization of fodder yeast and agro-industrial by-products in production of spores and biologically-active endotoxins from *Bacillus thuringiensis. Zbl. Mikrobiol.* 138: 553–563.

Scherrer, P.; Lüthy, P.; and Trumpi, B. 1973. Production of δ-endotoxin by *Bacillus thuringiensis* as a function of glucose concentrations. *Appl. Micriobiol. Biotechnol.* 25 (4): 644–646.

Shimizu, S.; Ohmori, I.; Ito, H.; and Suzuki, S. 1973. Insecticides based on a toxin from *Bacillus thuringiensis* sporangia. Japanese Patent Kokai 73,22,620. 23 March 1973.

Smith, R. A. 1982. Effect of strain and medium variation on mosquito toxin production by *Bacillus thuringiensis* var. *israelensis. Can. J. Microbiol.* 28: 1089–1092.

Utsumi, S. 1973. Sporicidal activity with treatment of some chemicals for *Bacillus thuringiensis. Kyota Kogei Sen'i Daigaku Sen'igakubu Gakujutsu Hokoku* 7: 68–73.

Wakisaka, Y.; Masaki, E.; Koizumi, K.; Nishimoto, Y.; Endo, Y.; Nishimura, M. S.; and Nishiitsutsuji-Uwo, J. 1982. Asporogenous *Bacillus thuringiensis* mutant producing high yields of δ-endotoxin. *Appl. Environ. Microbiol.* 43 (6): 1498–1500.

Wakisaka, Y.; Masaki, E.; and Nishimoto, Y. 1982. Formation of crystalline δ-endotoxin or poly-β-hydroxybutyric acid granules by asporogenous mutants of *Bacillus thuringiensis. Appl. Environ. Microbiol.* 43 (6): 1473–1480.

Yamamoto, T., and McLaughlin, R. E. 1981. Isolation of a protein from the parasporal crystal of *Bacillus thuringiensis* var. *kurstaki* toxic to the mosquito larvae, *Aedes taeniorhynchus. Biochem. Biophys. Res. Commun.* 103 (2): 414–421.

Zamola, B., and Kajfez, F. 1977. Biosynthesis of a microbial insecticide. Swiss Patent no. 587,014. 15 June 1977.

Zamola, B.; Rendic, S.; Kajfes, F.; and Tamburasev, G. 1970. Investigation of aeration effect on the sporulation of *Bacillus thuringiensis. Mikrobiologija* 7 (1): 117–122.

Zamola, B.; Valles, P.; Meli, G.; Miccoli, P.; and Kajfez, F. 1981. Use of the centrifugal separation technique in manufacturing a bioinsecticide based on *Bacillus thuringiensis. Biotechnol. Bioeng.* 23 (5): 1079–1086.

9

Activity, Field Efficacy, and Use of *Bacillus thuringiensis israelensis* against Mosquitoes

MIR S. MULLA

9.1 INTRODUCTION

In recent years, the use of synthetic organophosphate insecticides for the control of mosquito larvae has been on the decline in California (Mulla 1976) and elsewhere. This drastic decline is primarily due to the development of resistance in mosquito larvae, mammalian toxicity, environmental considerations, and the use of biological control agents such as larvivorous fish and *Bacillus thuringiensis* subsp. *israelensis* (*B.t.i.*). In the past decade, research on the development and use of biological control agents in vector control programs has been greatly stimulated and expanded thanks to the efforts of the United Nations Development Programme/World Bank/World Health Organization Special Program for Research and Training in Tropical Diseases (TDR). This agency has been involved in providing extramural funds for basic and applied research and coordinating research activities in the area of biological control of vectors. The Scientific Working Group and the Steering Committee on Biological Control of Vectors were established in 1976 to provide technical evaluation and to foster current and future direction for relevant research in this area. These bodies have been instrumental in promoting research on the development and wide use of new and currently available biological control agents for the management and suppression of disease vectors as well as pest species affecting the quality of life and well-being of millions of people around the world.

Biological control agents are defined here in the broad sense. They include biological agents and their products that can be employed alone or in combination with other strategies for the suppression or regulation of disease vectors. Microbial control agents such as *B.t.i.,* the subject of this chapter, fall into this category. It should also be pointed out that microbial agents, which do not recycle in the treated habitat, can be employed as selective agents in an integrated control technology of vectors. By selectively removing the target species during peak production periods, the regulating action of naturally occurring predators can be greatly enhanced, thus precluding the need for frequent treatments. This aspect of vector control technology, using *B. t. i.* and other microbial control agents, constitutes a basic approach to the control of disease vectors. This promising area for future research could lead to the development of cost-effective control methodologies. Its applicability in vector control programs will be emphasized in this chapter to stimulate further research on the wide use of *B. t. i.* and similar organisms in the control of mosquito vectors in some important but expansive aquatic habitats producing high populations of mosquitoes.

Over the past several years some comprehensive and excellent reviews and bibliographies of available literature on biological control of vectors, including microbial control agents, have been published. The reviews and works published by Chapman (1974), Davidson (1982), Lacey (1985), Lacey and Undeen (1986), Mulla (1985), Roberts and Strand (1977), and Roberts and Castillo (1980) include detailed references and information on microbial control agents, encompassing basic as well as applied investigations. The recent publications of Davidson (1982), Lacey (1985), Lacey and Undeen (1986), and Mulla (1985) have updated available information on bacterial agents, especially *B. t. i.* Therefore, extensive references to original publications on *B. t. i.* listed in these works will not be included here. Interested readers should refer to one or more of these comprehensive published works.

Several strains or varieties of *Bacillus thuringiensis* Berliner have been isolated and tested against mosquito larvae. Hall et al. (1977) studied 127 strains of this pathogen, but only 2 strains showed activity against *Aedes triseriatus* larvae. Tests against other species were limited and not promising. Aizawa and Ohba (1985) presented a detailed discussion of isolates other than *B. t. i.* However, the discovery of *B. thuringiensis* subsp. *israelensis* serotype H14 in 1976 (Goldberg and Margalit 1977; de Barjac 1978) availed an agent that showed a good level of activity against several species of mosquitoes. Additionally, this toxin-producing agent could be produced with the help of currently available fermentation technology, and the primary products were amenable to formulating manipulations, packaging, and shipment.

Development of *B. t. i.* for mosquito and *Simulium* control proceeded rapidly. Some formulations of this bacterial agent received registration from the U.S. Environmental Protection Agency in 1980. Since 1980, the use of *B. t. i.* formulations for mosquito control has greatly increased in California (CMVCA 1984, 1985, 1986). Total use of this agent in three years in billions of toxic units was as follows: 1983, 5,538; 1984, 18,500; and 1985, 83,560. This trend toward increasing use of *B. t. i.* is likely to continue for a few years in the future, although not at the same rate (fig. 9.1). From 1984 to 1985, use increased more than four-fold. Since many formulations with potencies of 200 to 3,000 ITU/mg were used, the total amount reported is given in potency units.

The discussion of the efficacy and use of *B. t. i.* in this chapter will include general trends in the activity and use patterns of this bacterial agent and

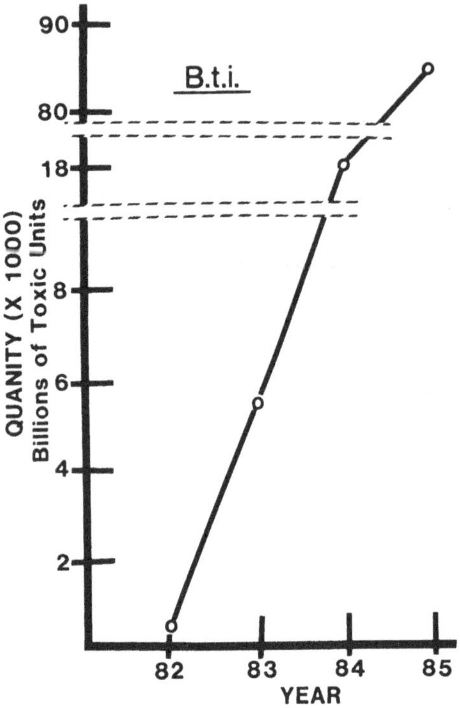

FIGURE 9.1 Use of various formulations of *B.t.i.* (in toxic units) as larvicides in mosquito control programs in California, United States.

suggestions as to how and where this selective biological control agent can be employed in vector control programs. Emphasis will be placed on the use of microbial control agents in an integrated management program against disease vectors, where the action of natural enemies coexisting with mosquito larvae is preserved or enhanced through the use of selective agents.

9.2 LABORATORY EVALUATION

Before field assessment and evaluation of the efficacy of control agents can be advanced, it is necessary to test primary or technical materials and formulated products in the laboratory against various larval instars of different species of mosquitoes under different test conditions. In these initial studies the relative activity, suspensibility, flowability, and other physical properties of the products are scrutinized for suitability in field applications. The data gathered in preliminary laboratory studies often provide a basis for undertaking further development of the product and its formulations. From laboratory data, approximate field dosages can be predicted, and the range of field efficacy can be determined with fewer tests.

9.2.1 Screening Procedures

In the initial screening procedure, technical or primary products are tested in glass containers or disposable treated cups. Contamination of glassware and containers used in bioassay tests is at times a vexing problem. The use of disposable supplies, such as cups, pipettes, and larval transfer containers, are advisable in situations where contamination from synthetic insecticides or microbial agents poses a problem.

Various test methods have been employed in assessing the potency of *B.t.i.* in the laboratory. Although different techniques are used in different laboratories, they all have a common denominator. The tests are run on various instars of larvae of different species reared in the laboratory. For species that have not been colonized, field-collected larvae have been successfully used in initial screening. In general, second to fourth instars have been employed. Since the principal toxin in *B.t.i.* is a particulate material and has no contact activity, the product has to be ingested in sufficient amounts to cause mortality. The *B.t.i.* toxin is fast acting, and mortality readings are taken 24 hours after initiation of exposure.

For testing technical (liquid or primary powder), FC (flowable concentrate), AS (aqueous suspension), or other formulations, the products are mixed and suspended in water; serial dilutions are also made in water. Stock suspensions and dilutions should be prepared in distilled water if available, otherwise tap or field water may be satisfactory. Aliquots of desired strength are added to 100–200 ml of water in cups or beakers containing 20–25 larvae of the desired instar. Since *B.t.i.* induces high level of mortality in 24 hours, food is generally not provided. The containers are replicated two to four times, and the tests are repeated on three or four different days to overcome biological variation in the test organisms.

After the desired exposure period (24 hours), mortality readings are taken. Mean mortalities for each concentration (mg/liter) are calculated, and dosage-response lines are plotted on log probit paper. The LC_{50} and LC_{90} (mg/liter) are ascertained from the dosage-response lines or obtained from computers using programs designed for probit analysis. The LC_{50} and LC_{90}, along with appropriate statistical data, yield needed information on the relative activity and potency of primary products and formulations. From these data approximate rates and dosages can be predicted for field testing.

Additional tests in larger volumes of water or simulated field conditions may be carried out in the laboratory. In such tests the influence of water pollution, depth of water, volume-to-surface ratio, and other factors influencing activity and efficacy can be studied. The role of formulations having floating or sinking properties, quick or slow release characteristics, and other parameters may be evaluated before actual field tests are implemented. By carrying out these simple tests, a great deal can be learned about the potency and use of a given product or formulation.

TABLE 9.1.
Relative Activity of *B.t.i.* Preparations and Formulations against Fourth-instar Larvae of *Culex quinquefasciatus*

Preparation	Manufacturer	ITU/mg	LC_{90} mg/liter
Bactimos®	Biochem	3,500	0.09
Vectobac® A	Abbott Labs	2,000	0.38
Vectobac® B	Abbott Labs	2,000	0.60
Teknar®	Sandoz	1,500	0.39
ABG-6108 II G[a]	Abbott Labs	250	0.52
ABG-6108 II G[b]	Abbott Labs	250	0.54
Bactisand®[c]	Summit Chemical	112	0.26

[a] Floating granules
[b] Sinking granules
[c] Fine, powdery granules

9.2.2 Preliminary Screening

As a new product is introduced for development, it is only available in small quantities (1–5 g) to a given researcher. Before extensive laboratory tests are conducted, the material is tested against some of the most common species of mosquitoes. The cosmopolitan species of mosquitoes found in many laboratories is *Culex pipiens quinquefasciatus*. In our laboratory, all preliminary tests are carried out against this species. Subsequently, pilot plant batches and formulated materials are also tested against this and other species of mosquitoes.

In the early 1980s, experimental and some commercial preparations of *B.t.i.* were tested in the laboratory. The various preparations or formulations showed different levels of activity against *Cx. quinquefasciatus* depending on the potency (ITU/mg) and type of preparation or formulation (table 9.1). For example, Bactimos®, which has the highest potency, showed the highest activity. It should be noted that there is no consistent relationship between potency (ITU/mg) and activity against larvae. Note that ABG-6108 II G (granules), with only 250 ITU/mg, manifested activity equal to that of Vectobac® B, with 2,000 ITU/mg. It is evident that formulation plays an important role in eliciting the real activity of the product.

Laboratory screenings to date indicate that most primary products show an LC_{90} (mg/liter) of 0.04 to 0.5 against most *Culex* and *Aedes* mosquitoes. *Anopheles* in general are more tolerant, and the LC_{90} can vary a great deal depending on the species (Ali, Sauerman, and Nayer 1984; Mulla 1985; Mulla and Darwazeh 1985; Mulla, Federici, and Darwazeh 1982; Mulla et al. 1982, 1985). Most *Anopheles* are susceptible in the range of 0.5 to 5.0 mg/liter. However, floating-type formulations and new preparations are more active than the earlier formulations.

TABLE 9.2.
Relative Activity of *B.t.i.* Formulations and Preparations against Fourth-instar Larvae of Mosquitoes (LC_{90} mg/liter)

Material	*Culex quinq.*	*Culex tarsalis*	*Aedes aegypti*	*Anopheles quad.*
Bactimos®[a]	0.090	0.062	0.170	1.30
Vectobac®[b]	0.380	—	—	—
Teknar®[c]	0.386	—	—	—
IPS-78[d]	0.655	0.380	0.480	1.40

[a] 3,500 ITU/mg
[b] 2,000 ITU/mg
[c] 1,500 ITU/mg
[d] 1,000 ITU/mg

9.2.3 Species Specificity

B.t.i. is quite specific, showing larvicidal activity against larvae of mosquitoes and *Simulium* species. Among mosquitoes, various species exhibit different levels of susceptibility to the same *B.t.i.* preparation. In general, *Culex* larvae are more susceptible; *Aedes* larvae are equally or slightly less susceptible, and *Anopheles* larvae are quite tolerant to primary products or currently available formulations (table 9.2). The two *Culex* species, *Cx. quinquefasciatus* and *Cx. tarsalis* showed similar level of susceptibility. *Aedes aegypti* was close to these two in susceptibility, while *Anopheles quadrimaculatus* was 2- to 15-fold less susceptible than *Cx. quinquefasciatus*. It should be noted that even among the members of the same genus, some species are more susceptible than others. For example, *Anopheles albimanus* is more susceptible than *Anopheles quadrimaculatus*. The activity of *B.t.i.* can be greatly increased against less susceptible species by means of formulation techniques.

Further studies with experimental preparations of *B.t.i.* against several species of mosquitoes documented the differential susceptibility to this material. Again *Culex* and *Aedes* species showed similar levels of susceptibility to the three experimental preparations (table 9.3). The LC_{90} in mg/liter of the various products was about 0.14–0.70 for *Aedes*, *Culex*, and *Psorophora* species. The values against *An. quadrimaculatus* were 1.40 to 2.30.

It should be pointed out that the experimental batches and preparations of *B.t.i.* produced in the early stages of its development had low potency, were difficult to suspend uniformly, and settled out from water rather rapidly. Subsequent developmental research yielded more potent primary products, and advanced formulation techniques have further increased the potency and efficacy of the technical or primary products. In general, manufacturers have been able to increase potency and efficacy by a factor of 2- to 10-fold for some of their products.

TABLE 9.3.
Activity (LC_{90} mg/liter) of *B.t.i.* Preparations against Fourth-instar Larvae of Mosquitoes

Species	Larval Source	IPS-78	R-153	Sandoz 402
Ae. aegypti	Lab	0.48	0.38	0.70
Ae. nigromaculis	Irrig. pastures	0.20	0.14	0.45
Ae. taeniorhynchus	Lab	0.15	0.22	0.45
Ps. columbiae	Irrig. pastures	0.20	0.32	0.42
An. quadrimaculatus	Lab	1.40	1.40	2.30
Cx. peus	Dairy lagoon	0.22	0.13	0.50
Cx. quinquefasciatus	Lab	0.43	0.35	0.60
Cx. tarsalis	Lab	—	0.29	0.58

9.2.4 Instar Susceptibility

Various laboratories have employed different larval instars in bioassay tests. The use of first-instar larvae is not advisable because they are hard to handle and can suffer higher levels of mortality from handling.

For almost all species tested, the younger-instar larvae are more susceptible than the older ones. As the larvae get older, their susceptibility declines significantly. In general, second instars are 1.5- to 5-fold more susceptible than the fourth instars (table 9.4). Late fourth instars that feed little or have ceased feeding are much less susceptible because of lack of ingestion of a lethal dose in a short period of time. Prepupae and pupae are refractory to *B.t.i.* due to the lack of feeding and ingestion of the toxin particles by these organisms.

Further expansion of the relationship between activity and larval age is

TABLE 9.4.
Species and Instar-related Activity of *B.t.i.* Experimental Preparations against Larvae of Two Species of Mosquitoes (LC_{90} in mg/liter)

		Cx. quinquefasciatus		*An. quadrimaculatus*	
Preparation	ITU/mg	2ds	4ths	2ds	4ths
IPS-78	1,000	0.120	0.43	1.0	1.4
R-153-78	1,000	0.068	0.35	0.34	1.4
ABG-6108	250	0.210	1.00	1.60	7.5
Sandoz 402	1,500	0.310	0.60	2.20	2.3

TABLE 9.5.
Species and Instar-related Activity of Strains *B.t.i.* against Larvae of Various Species of Mosquitoes in Laboratory

Strain or Preparation	LC$_{50}$–LC$_{90}$ mg/liter		
	Cx. tarsalis	*Ae. aegypti*	*An. quadrimaculatus*
		2d-instar larvae	
2013-9	0.02–0.04	0.01–0.03	0.05–0.14
Bactimos®[a]	0.005–0.01	0.02–0.03	0.03–0.09
IPS-78[b]	0.03–0.07	0.03–0.08	0.36–1.00
2013-1	0.24–0.48	0.24–0.55	0.56–1.30
2013-5	0.24–0.82	0.30–0.60	0.52–1.30
		4th-instar larvae	
2013-9	0.04–0.08	0.070–0.14	0.35–0.80
Bactimos®[a]	0.04–0.06	0.80–0.17	0.22–1.40
IPS-78[b]	0.17–0.38	0.25–0.48	0.85–1.40
2013-1	0.84–1.80	1.00–1.70	4.00–>10.00
2013-5	0.52–1.50	0.88–1.50	2.50–>10.00

[a] 3,000 ITU/mg
[b] 1,000 ITU/mg

presented in table 9.5. In *Cx. tarsalis* the second-instar larvae were 2–5 times more susceptible at the LC$_{90}$ concentration than the fourth instar. Similarly, in *Ae. aegypti* the factor was 2- to 6-fold. For *An. quadrimaculatus*, the second-instar larvae were approximately 2–15 times more susceptible than the fourth to the various materials. First-instar larvae, although not tested here, are believed to be 2–3 times more susceptible than the second-instar larvae.

Since fourth-instar larvae are less susceptible than the younger instars, it is advisable to use third or early fourth in screening tests. Dosages and concentrations that will induce mortality in the fourths will definitely cause mortality in the younger instars. In asynchronous species such as *Culex*, *Anopheles*, and some *Aedes*, all larval instars prevail in the breeding sources. To control these heterogeneous larval populations, administration of maximum dosages geared to kill the older larvae will be necessary. However, as pointed out earlier, late fourths and prepupae do not feed and therefore should not be used in laboratory tests. The use of late third and early fourth instars is recommended for obtaining data on the relative potency of primary products and formulations.

9.2.5 Biotic and Abiotic Factors Influencing Activity

The activity of *B.t.i.* is influenced by a variety of biological and environmental factors (Lacey 1985). Some of these factors are commonly encountered in the natural larval habitats of mosquitoes. A number of studies have been conducted under laboratory and simulated field conditions to determine the role of some of these factors affecting the activity and potency of *B. t. i.* preparations.

Among the many biological factors, species specificity, instar-related sensitivity, and feeding behavior influence activity of *B. t. i.* Species specificity and instar sensitivity have been amply documented and discussed above. Research on feeding behavior of larvae, although not extensive, has provided some evidence that elucidates the relationship between level of activity and the feeding behavior of larvae. Rashed and Mulla (1989) studied the roles of numerous factors that influence feeding behavior and ingestion rates of larvae feeding on particulate matter. For example, *Culex* and *Aedes* larvae feed actively through the whole column of a shallow body of water. Since the toxin particles in most preparations and formulations settle rapidly in a column of water, larvae of these two genera are prone to ingest lethal dosages in a short period of time. On the other hand, *Anopheles* larvae, which primarily browse and feed at the surface of water (Aly and Mulla 1986; Rashed and Mulla 1989), will not be able to ingest a lethal quantity of the particulate toxins in a short period of time. Particles that leave the surface layer and settle are not available to be ingested by *Anopheles* larvae. Therefore, formulations that confine the toxin particles to the surface film are deemed more desirable, and development of such formulations is being actively pursued. Some of these types of formulations have already been tested in the laboratory (Aly and Mulla 1986; Aly et al. 1987; Cheung and Hammock 1985; Margalit, Markus, and Pelah 1984), but many more laboratory and field studies are needed before floating formulations showing high level of activity against *Anopheles* larvae are developed for use in vector control programs. With the microlipid droplet encapsulation method, the δ-endotoxin of *B. t. i.* was made equally effective against *Anopheles freeborni* and *Culex* mosquitoes (Cheung and Hammock 1985). The nonencapsulate material had much lower activity against the former. Similarly, Aly et al. (1987) tested floating-type formulations, which significantly increased activity against *Anopheles* larvae.

Another important biological factor that influences larvicidal activity of *B. t. i.* toxin is the ratio of the amount of toxin administered to the number (density) of larvae. Both laboratory and field research have provided ample evidence for this relationship (Aly et al. 1988; Becker and Ludwig 1983; Mulla et al. 1982). A given dosage of *B. t. i.* that will control 95–100% of larvae prevailing at low density will not produce the same results when larval density per unit volume is increased markedly. Higher-density populations will require higher concentrations or dosages to produce mortalities equal to that of low-density populations. In practice, concentrations and dosages are calculated on the basis of volume or surface area of water to be treated and not larval density. In general, denser populations of larvae (50–100 larvae per dip) will require 1.5–2 times more material than the low-density populations

(5–20 larvae/dip) to yield equal mortalities. Other factors that result in denser larval populations, such as organic pollution and decaying vegetation, also require high dosages (as discussed below).

Of the environmental factors, organic pollution and presence of colloidal particles, including food particles, influence activity patterns of *B.t.i.* (Margalit and Bobroglio 1984; Margalit et al. 1985; Ramoska, Watts, and Rodrigues 1982). There seems to be a direct correlation between the extent and magnitude of organic pollution and the dosage of bacterial toxin required to yield a given level of mortality. Apparently, fewer toxin particles are ingested per unit time in the presence of organic and inorganic particles and floating materials than in the absence of extraneous materials. It is also an established fact that in the presence of high organic pollution, denser populations of mosquito larvae are produced. On both these counts (high pollution and high density) higher rates of application will be necessary to control mosquito larvae.

Other factors commonly encountered in nature are flowing water, depth of water, vegetative cover, and low water temperature. Flowing water will dilute the toxin material and will displace it from the application site. This feature, however, may be beneficial in situations where the target species (such as *Simulium* spp. and certain mosquitoes) breed downstream from the site of the treatment. Vegetative cover also decreases the efficacy of *B.t.i.* and other larvicides. Plant canopy intercepts liquid sprays, dusts, and part of the granular formulations. Some of the materials after penetration may lodge on plants and plant material, remaining out of the zone of feeding activity of mosquito larvae.

Water depth is an important consideration; the deeper the water the greater the dosage required per unit surface area. Since *B.t.i.* toxin settles out of the column of water, in deeper water the toxin particles will be mostly absent from the feeding zone of larvae. Water temperature is an important factor that needs to be taken into consideration. Although *B.t.i.* has been found to be active at low temperatures, its effectiveness may be reduced in cold water due to cessation or low rate of feeding of some larvae and decrease in metabolism.

9.2.6 Delayed Effects

Larvicidal materials, if administered at high enough concentrations and rates, will yield complete or almost complete mortality of the exposed populations. However, in practice, under diverse environmental conditions, it is not possible to achieve uniform coverage of the treated habitat and to facilitate exposure of all target organisms to equal but lethal concentrations of a larvicide. It is a foregone conclusion that in nature some organisms will experience exposure to lethal or above-lethal concentrations, while others will receive and be exposed to only sublethal dosages. A number of chemical larvicides and mosquito control agents have been shown to manifest delayed effects (beyond the treated stage) at sublethal dosages in the survivors. Effects such as delayed mortality, reduced survivorship of mature insects, reduction in the production of viable eggs, and reduction in fecundity and survivorship of F_1 individuals. Some of these types of effects, for example, have been documented for insect growth regulators (IGRs) in mosquitoes (Arias and Mulla 1975).

Studies on the delayed effects of *B.t.i.* in mosquitoes are meager. There are some indications that sublethal dosages of *B.t.i.* produce delayed effects beyond the stage treated. Hare and Nasci (1986) noted some delayed mortality in surviving larvae of *Ae. aegypti* exposed to LC_{50} concentration of *B.t.i.* However, they did not detect any other noticeable negative effects on larvae surviving sublethal concentrations. Further studies are warranted to document the occurrence of these subtle but important effects beyond the duration of the preimaginal stages treated.

9.3 FIELD EVALUATION AND EFFICACY TRIALS

Various preparations and formulations of *B.t.i.* have been field tested against different species of mosquitoes in a variety of habitats. Biotic and abiotic factors of the environment greatly influence initial activity and persistence of this microbial agent. Many studies have been carried out against culicine mosquitoes; studies against anophelines are still limited in scope.

Published reports on field evaluation and efficacy are scattered throughout the literature, but the recent review of Lacey (1985) has brought together most of the studies carried out to date under semifield and field conditions.

9.3.1 Spectrum of Field Activity

The range of activity of *B.t.i.* preparations and formulations varies a great deal depending on the species and type of breeding source treated. Even against the same species, range of effective dosages can vary by a factor of two to five in breeding sources possessing different biotic and abiotic characteristics.

9.3.1.1 Floodwater Mosquitoes. For *Aedes* and *Psorophora* species where larvae inhabit relatively shallow bodies of waters, *B.t.i.* concentrate formulations have provided excellent control (90–100%) of larvae at the rates of 0.10 to 2.0 kg/ha (Dame et al. 1981; Eldridge, Washino, and Hennenberger 1985; Hougard, Duval, and Escafre 1985; Lacey 1985;

Mulla et al. 1982, 1985; Schnetter et al. 1983). In some situations the required dosages may be below or above this range, but most species will be controlled for short periods using these dosages. Higher dosages are required in polluted and deeper bodies of water or where late-instar larvae are in preponderance. Vegetated breeding sources, as well as those where dilution of water is a factor, will require higher dosages.

The rates of dilute formulations (containing 1–5% of the primary products) will be higher. For the control of *Aedes* the rates of application are in the order of 5–10 kg/ha (Lacey 1985; Mulla et al. 1985; Sutherland et al. 1982). Lower or higher rates may be necessary depending on the formulation and habitat to be treated.

9.3.1.2 *Anopheles* Mosquitoes. Studies on the efficacy of *B.t.i.* against *Anopheles* species are indeed limited in scope. Only a few species and situations have been investigated. In general, most *Anopheles* species studied to date were controlled at 0.5 to 6 kg/ha of the concentrate formulations (Dame et al. 1981; Hougard, Darriet, and Bakayoko 1983; Lacey 1985; Lacey and Inman 1985; Majori, Ali, and Sabatinelli 1987; Sandoski et al. 1985; Standaert 1981), although in some cases, rates of formulations were as high as 19.1 kg/ha. The rates of granular formulations administered were generally in the range of 3 to 6 kg/ha. From published reports it appears that *Anopheles* in general could be effectively controlled at the rates of 0.5 to 4 kg/ha of the concentrate and 4 to 10 kg/ha of the dilute (granules and others) formulations.

9.3.1.3 *Culex* Mosquitoes. Investigations on the field efficacy of *B.t.i.* preparations and formulations have been carried out against many species of *Culex* mosquitoes of both pest and disease significance. In general, most *Culex* species are quite susceptible in biotopes having low levels of organic pollutants. Literature pertaining to the various studies on *B.t.i.* against *Culex* mosquitoes has been recently compiled and reviewed by Lacey (1985) and Mulla (1985).

Culex mosquitoes as a group are quite susceptible to the *B.t.i.* toxins. Primary products and concentrate formulations have been shown to be quite efficacious. *Culex* in clear-water situations have been readily controlled with *B.t.i.* in the dosage range of 0.10 to 0.56 kg/ha of the various preparations or formulations (Garcia et al. 1983; Majori and Ali 1984; Mulla 1985; Mulla, Federici, and Darwazeh 1982). In some situations, however, the administered dosage may be below or above this range. *Culex* species breeding in brackish or polluted waters require somewhat higher dosages. *Culex pipiens*, *Culex peus*, and *Cx. quinquefasciatus* breeding in polluted and highly polluted water will require higher rates of application. These and other *Culex* species in polluted water have been controlled at the rates of 0.25 to 2.2 kg/ha of the various preparations and formulations (Eldridge and Callicrate 1982; Hougard, Darriet, and Bakayoko 1983; Lacey 1985; Lacey, Urbina, and Heitzman 1984; Majori and Ali 1984; Majori, Ali, and Sabatinelli 1987; Mulla

1985; Mulla, Federici, and Darwazeh 1982; Mulla and Darwazeh 1984; Mulla, Darwazeh, and Aly 1986; Sinégre et al. 1980). The higher range of dosages was required for those materials that were experimental preparations.

Other species of mosquitoes have shown varying degrees of susceptibility. *Culiseta inornata* and *Culiseta incidens* were not as susceptible as *Culex* species (Garcia et al. 1983; Mulla, Federici, and Darwazeh 1982). Similarly, one of the *Culex* species, *Cx. tritaeniorhynchus*, required high concentrations (10 ppm) for effective control in rice paddies (Yu et al. 1981). *Mansonia uniformis* also showed a low level of susceptibility to Teknar® (Foo and Yap 1983), and *Coquelletidia perturbans* was not effectively controlled in the field by the maximum recommended dosages of Bactimos® and Teknar® formulations of *B. t. i.* (Sjogren, Batzer, and Junemann 1986). Apparently these formulations are not suitable for the control of bottom-dwelling larvae. *Psorophora ciliata*, unlike *Psorophora columbiae*, was not controlled at 2.7 kg/ha of Teknar® (Sutherland et al. 1982). The latter species was found to be quite susceptible to most preparations and formulations of *B. t. i.* (Dame et al. 1981; Mulla et al. 1982; Sandoski et al. 1985; Stark and Meisch 1983). It is very likely that the current more potent and improved preparations and formulations will show higher activity against these species found to be less susceptible than the *Culex* species to the earlier formulations.

Granular formulations of *B. t. i.*, especially those developed in the last two to three years, have shown excellent activity against floodwater and *Culex* mosquitoes. These formulations have provided satisfactory control of these mosquitoes in the range of 2 to 10 kg/ha of the various formulations (Lacey 1985; Lacey and Inman 1985; Lacey, Urbina, and Heitzman 1984; Mulla 1985; Mulla and Darwazeh 1984, 1985). As the production and formulation techniques are improved, the efficacy and physical properties of formulations are also improved.

The currently available concentrate and granular formulations of *B. t. i.* are capable of providing satisfactory control of *Aedes* and *Culex* mosquitoes at reasonable costs. The concentrate formulations (e.g., Bactimos®, Teknar®, and Vectobac®) can yield 95% + control of these mosquitoes at the rates of 0.2 to 0.5 kg/ha. For most *Anopheles* this range of rates has to be increased to 0.5 to 2.0 kg/ha. The granular formulations with potency of 200–300 ITU/mg are deemed to provide satisfactory control of most *Aedes* and *Culex* species at the rates of 5 to 10 kg/ha of the formulation. For *Anopheles* the rate may be doubled. However, further studies are now warranted against *Anopheles* on the use of new and improved preparations and formulations.

9.3.2 Persistence and Recycling

In mosquito control programs, especially those using biological control agents, it is highly desirable to have materials and agents that will yield long-lasting control with one or few treatments. The reason for this demand is the high cost of application (including cost of administration, technical services, surveillance, and application equipment), which usually amounts to five to ten times the cost of material. Therefore, those agents that persist or recycle in the treated habitat and provide satisfactory control of larvae over prolonged periods are greatly desired.

Field research data gathered to date indicate that most commercial formulations (concentrate and dilute) of *B. t. i.* yield high levels of initial control, but effectiveness declines rapidly. It has been observed that in asynchronous species, young mosquito larvae appear within three to four days in habitats treated with standard larvicidal rates. This situation prevails even though *B. t. i.* sporulates and produces crystals in larval cadavers (Aly, Mulla, and Federici 1985) as well as in gut of mosquito larvae (Aly 1985). In detailed field studies where four biweekly treatments were made with Vectobac® powder and Teknar® flowable concentrate, no residual activity was indicated two weeks after each treatment (Mulla 1985). The larval populations of *Cx. tarsalis* (third and fourth instars) reached treatment threshold levels (three larvae/dip or more) within two weeks after each treatment.

Persistence and residual activity may be enhanced through formulation techniques. Some specially made slow-release formulations have been found to yield persistent release and control of larvae (Lacey, Urbina, and Heitzman 1984). However, no such formulations, with the exception of briquettes (used in specific situations), have been prepared commercially and tested under field conditions. The potential for slow-release formulations looks good, and this area of endeavor requires a great deal of interdisciplinary research.

Although *B. t. i.* preparations and formulations lack recycling and provide no persistent control, their usefulness has been established in an integrated approach toward the control of asynchronous species. In conjunction with naturally occurring predatory organisms, *B. t. i.* treatments could yield long-lasting control without resorting to frequent treatments (Mulla 1986). The use of *B. t. i.* in cost-effective integrated control of asynchronous mosquitoes is treated in detail in the following section.

9.4 IMPACT ON NONTARGET ORGANISMS

Public policy matters and environmental considerations dictate that proof of safety to nontarget organisms (NTO) and fish and wildlife be documented before any pest-control agent can be employed in actual pest-control programs. Microbial control agents are not exempt from this mandate. Therefore, extensive environmental and biological impact data are necessary before microbial agents such as *B. t. i.* can be cleared for public use.

A comprehensive review of the safety of *B. t. i.* and *Bacillus sphaericus* to

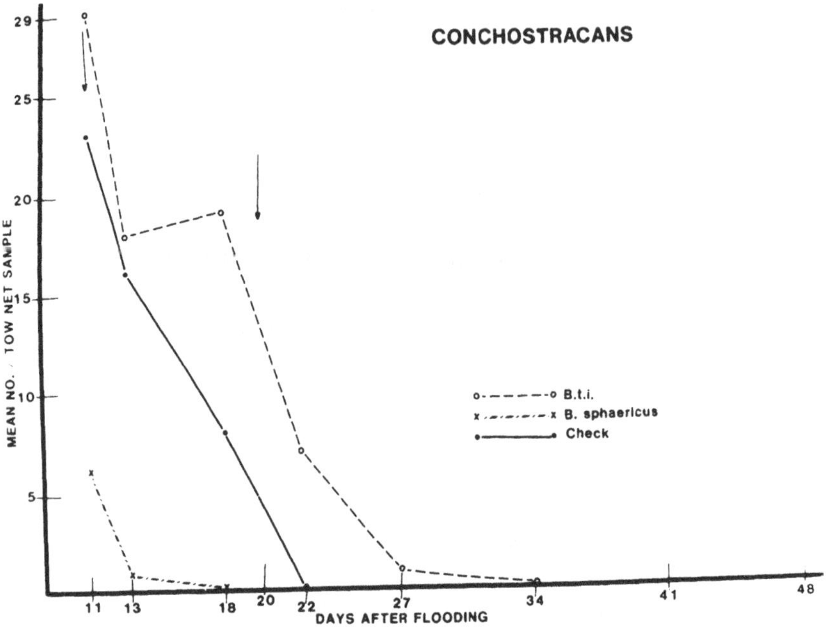

FIGURE 9.2 Lack of adverse impact of 2 microbial control agents on population trends of the conchostracan (*Eulimnadia texana*) in 35m² mesocosms treated twice for the control of mosquito larvae. Arrows indicate time of treatment: first treatment at full larvicidal dosage, second treatment at ½ the full rate.

NTO has been rendered by Lacey and Mulla (1990). For a more detailed analysis of available data, readers should see this work. A concise analysis of published data indicates that all laboratory and field data point to the safety of *B. t. i.* to NTO and fish and wildlife. Since this agent is specifically toxic to mosquito larvae and has to be ingested, it was postulated earlier that other filter-feeding organisms, such as simuliids, chironomid midge larvae, caldocerans, ostracods, and others, would be prone to the toxic action of *B. t. i.* toxin. With the exception of simuliids (which are pests and vectors of disease) and some chironomine pests, other filter-feeding organisms have not been affected at mosquito larvicidal rates. A filter-feeding crustacean, *Gammarus lacustris*, ingests *B. t. i.* but shows no toxic symptoms (Brazner and Anderson 1986). Larvae of some chironomid midges were found to have a low degree of susceptibility to *B. t. i.* preparations (Ali 1981), but the range of toxic concentrations were higher than those that might be encountered in practical mosquito control programs. Garcia, Des Rochers, and Toser (1980) reported on the safety of *B. t. i.* to some 40 species of NTO, including amphibians, fish, and macroinvertebrates. However, they showed some toxic effects on dixid, chironomid, and ceratopogonid larvae under laboratory conditions. Miura,

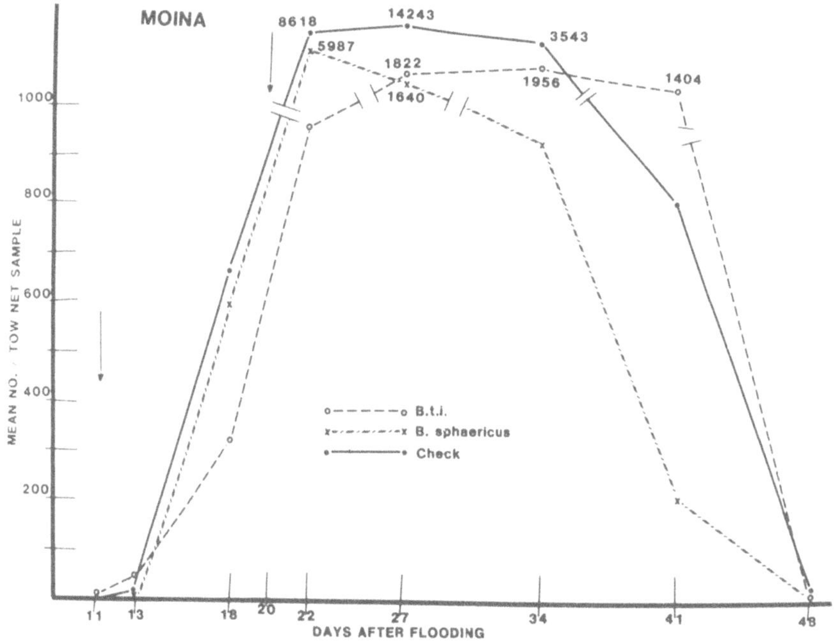

FIGURE 9.3 Lack of adverse impact of 2 microbial control agents on population trends of the cladoceran (*Moina rectirostris*) in 35m² mesocosms treated twice for the control of mosquito larvae. For additional details see fig. 9.2.

Takahashi, and Mulligan (1980) also alluded to low-level effects on chironomid larvae of larvicidal application of *B. t. i.* when organisms were held in laboratory. However, when field populations of chironomids were sampled from the treated plots, there was little or no marked effect on chironomid densities. In both of these studies (Garcia, Des Rochers, and Tozer 1980; Miura, Takahashi, and Mulligan 1980), no mortality or trends in the check populations are included in the data to show the natural population die off or fluctuations occurring due to factors other than the treatments. However, from all studies carried out to date it appears that *B. t. i.* is specific, producing toxic effects only in mosquito and simuliid larvae at practical larvicidal rates. All data gathered to date point to a good record of safety for a variety of macroinvertebrates cohabiting with mosquito larvae (Ali 1981; Lacey and Mulla 1990).

During the course of our field testing of *B. t. i.* formulations and preparations against mosquito larvae, detailed sampling of macroinvertebrates was carried out in replicated treated and untreated mesocosms. In all these studies we detected no significant impact attributable to treatments of this microbial agent at larvicidal rates on any of the macroinvertebrates sampled (Mulla, Federici, and Darwazeh 1982). No marked effects of the treatments (up to

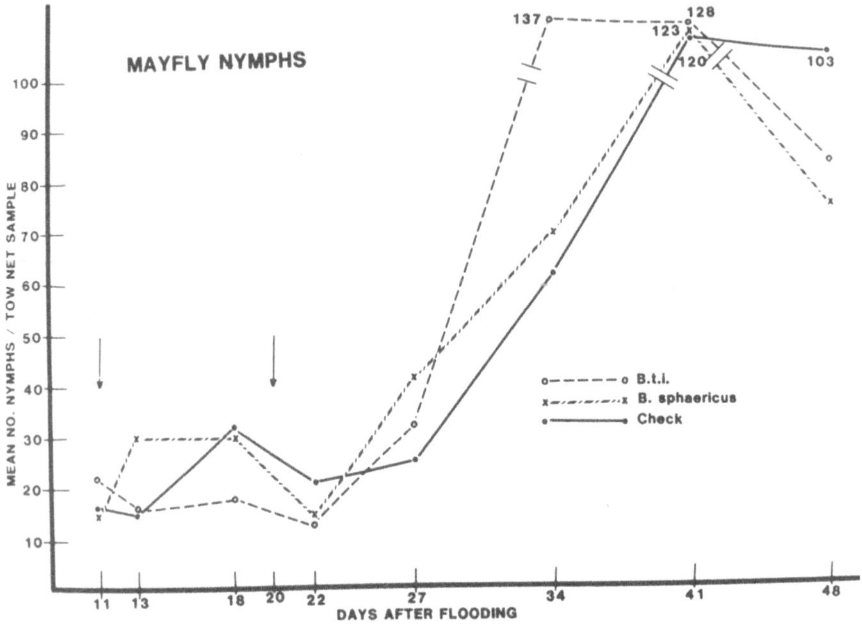

FIGURE 9.4 Lack of adverse impact of 2 microbial control agents on population trends of naiads of the mayfly (*Callibaetis pacificus*) in 35m^2 mesocosms treated twice for the control of mosquito larvae. For more details see fig. 9.2.

2.24 kg/ha) were noted on mayfly and dragonfly naiads, diving beetle larvae, or ostracods.

To further assess the impact of larvicidal rates of two microbial control agents (*B.t.i.* and *B. sphaericus*), detailed studies were carried out in 35 m^2 mesocosms at outdoor facilities (as reported in Mulla 1985) on natural populations of target and nontarget organisms. The treatment and check ponds were replicated four times each, and the organisms, including mosquito larvae, were sampled by dipping and tow-net techniques. Population trends of target organisms and NTOs were determined once a week during a 48-day study period. The ponds were treated twice with *B.t.i.* and *B. sphaericus* as required by the level of third- and fourth-instar populations of mosquito larvae. The first treatment was made at full dosage *B.t.i.* (Abbott Laboratories, Vectobac®) at 0.56 kg/ha and *B. sphaericus* at 0.22 kg/ha, and the second treatment was made at half these rates.

The two treatments at full larvicidal and half larvicidal rates had no noticeable effects on the conchostracan *Eulimnadia texana* (fig. 9.2). Populations of this organism followed similar trends in the check and treated mesocosms. The initial density of *E. texana* in *B. sphaericus*–treated ponds was quite low, and the species declined quickly. This decline was seemingly

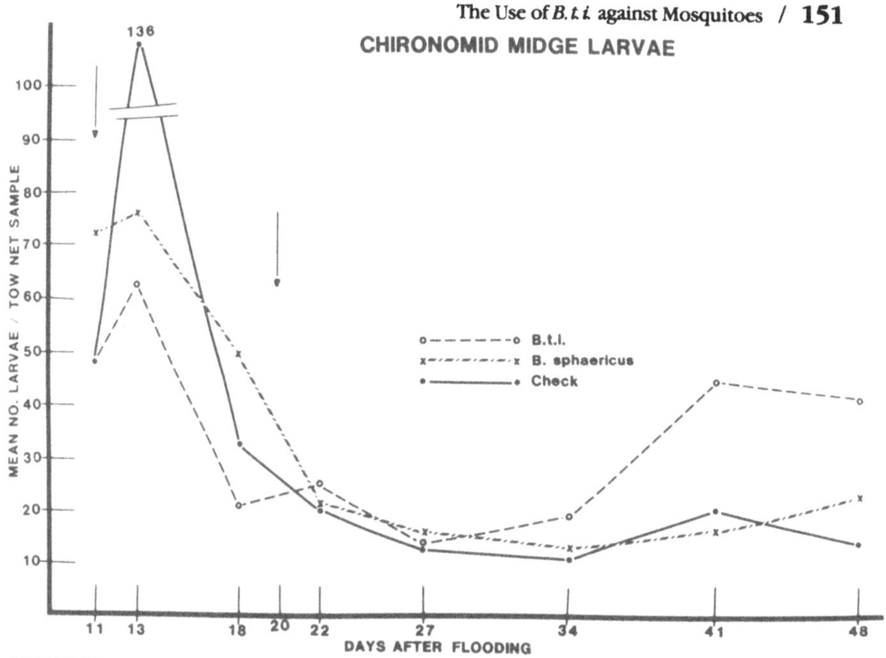

FIGURE 9.5 Lack of adverse impact of 2 microbial control agents on population trends of chironomid midges (several species, chironomines and tanypodines) breeding in 35m² mesocosms treated twice for the control of mosquito larvae. For more details see fig. 9.2.

not due to the treatment, but probably due to other, unknown factors. Natural decline in populations of this organism, which is adapted to flooding and drying regimens, occurred in all ponds (treated and untreated), reaching very low levels within 18–27 days after flooding.

The cladoceran *Moina rectirostris* was not affected by the two treatments of each of the microbial control agents (fig. 9.3). This organism started to appear within two weeks after flooding, peaked within three weeks, and maintained peak populations for another three weeks. Thereafter, its populations crashed, reaching zero level 48 days after flooding. Under hot, sunny desert conditions, this organism disappears within five to six weeks after flooding of temporary breeding sources. This organism feeds on planktonic organisms and particles, thus preventing or reducing water turbidity.

The mayfly (*Callibaetis pacificus*) is a dominant herbivore in mosquito larval habitats, feeding on filamentous algae and other organic detritis. Naiads of this aquatic insect were not adversely affected by treatments of the microbial control agents (fig. 9.4). The naiads started to appear about 10 days after flooding. These organisms matured and started to emerge within 3–4 weeks after flooding. The naiad population peaked 5–6 weeks after flooding and started to decline 7 weeks after flooding. From the population trends in treated and check ponds, there was no marked adverse impact due to the treatments.

Chironomid midge larvae, sampled by the dipper and tow-net methods, showed similar trends (fig. 9.5). In the check ponds, chironomid larval density was twice as high as in the treated ponds two days after treatment. However, after this period, populations in all the ponds (treated and untreated) declined sharply, prevailing in low numbers during the period of three to five weeks postflooding. There was slight to moderate rise in larval numbers six and seven weeks postflooding. The impact of *B.t.i.* on chironomid midge larvae is critically evaluated and reviewed by Lacey and Mulla (1990).

It should be pointed out that the midge larvae sampled here were nektonic, prevailing in the water column. We did not sample benthic larvae occurring at the surface or within the mud. It is possible that the benthic larvae may be more or less susceptible than the nektonic populations.

9.5 MICROBIAL LARVICIDES IN INTEGRATED CONTROL OF MOSQUITOES

In natural breeding sites, preimaginal mosquitoes provide a source of food for a variety of microinvertebrate and macroinvertebrate natural enemies. Immatures of most culicines (with the exception of some *Aedes*) and anophelines are regulated by a host of macroinvertebrates, which require little or no intervention. In permanently and semipermanently flooded sources of mosquitoes, a succession of a number of groups of natural enemies occurs following flooding of a given larval habitat by rain, irrigation, flooding, or disposal of wastewater. If flooding of the habitats continues, the natural enemy complex and immature mosquitoes reach an equilibrium level; the natural regulating agents suppress and maintain mosquito larvae at a low-density level. In general, most permanently flooded habitats (in equilibrium) hardly produce significant numbers of adult mosquitoes. It is during the first few weeks of the postflooding period when peak mosquito production occurs, before natural enemies catch up with the larval population.

Natural enemies, including predators, invade mosquito developmental sites and coexist with mosquito larvae in a variety of semipermanently flooded biotopes. Christie (1958) noted a variety of predaceous macroinvertebrates feeding on *Anopheles gambiae.* Service (1977) studied mortalities in *An. gambiae* complex in rice fields and temporary pools and found that natural predators play an important role in suppressing larval mosquitoes. Similar studies were conducted on the role of natural predators in suppressing larval populations of *Aedes cantans* (Service 1973).

During the past twenty years or so we have noted (Mulla, unpublished data) that high larval populations of *Cx. tarsalis* are produced following flooding of the larval habitat. The peak populations prevail for a period of

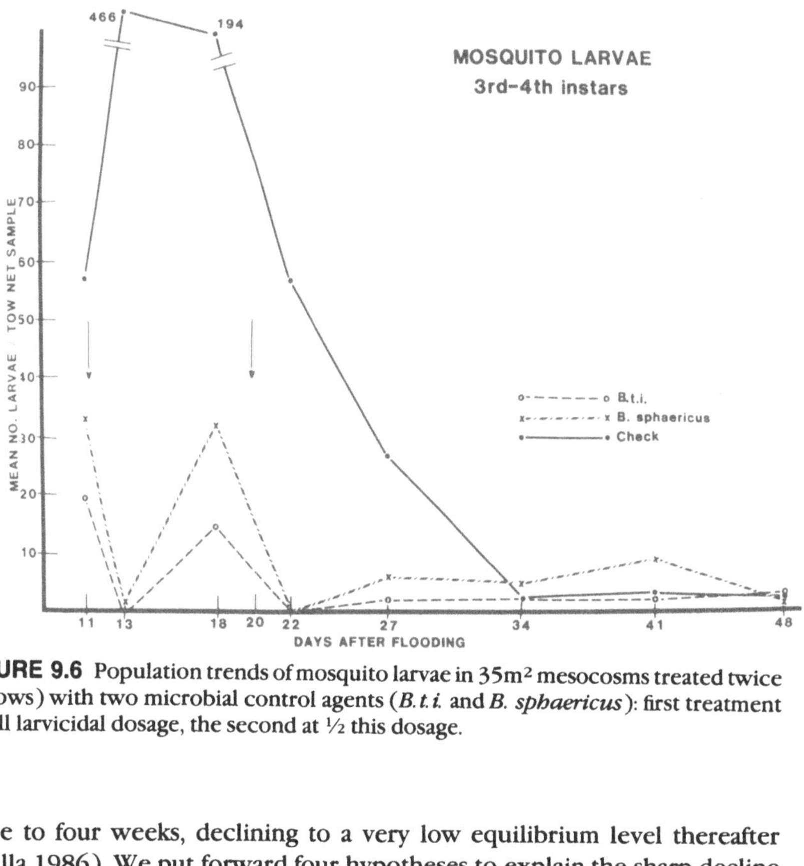

FIGURE 9.6 Population trends of mosquito larvae in 35m² mesocosms treated twice (arrows) with two microbial control agents (*B.t.i.* and *B. sphaericus*): first treatment at full larvicidal dosage, the second at ½ this dosage.

three to four weeks, declining to a very low equilibrium level thereafter (Mulla 1986). We put forward four hypotheses to explain the sharp decline in larval densities postflooding: (1) the intervention and succession of natural predators, (2) the unattractiveness of the aquatic habitat for oviposition, (3) a combination of the two, or (4) depletion of food supply.

Since the natural decline in larval populations has been noted consistently over the years, we further hypothesized that in semipermanently flooded biotopes, it will be necessary to make only one or two treatments with a selective larvicide (such as *B.t.i.* or *B. sphaericus*), thus precluding the need for weekly or biweekly treatments, a regimen used at the present time in many larvicidal programs. To provide evidence and to develop a model, we initiated a three-year study using replicated mesocosms (Mulla 1985). An arbitrary treatment threshold of three larvae/dip (third and fourth instars) was established for administering treatments. In the first year of the study, two full-dosage larvicidal treatments with *B.t.i.* and *B. sphaericus* were applied (*B.t.i.* Vectobac® AS at 0.56 kg/ha and *B. sphaericus* BS-2362 powder at 0.26 kg/ha). The second year, two treatments of the same larvicides were applied, the first treatment at full dosage and the second at one-half the rate. In the third year, the first treat-

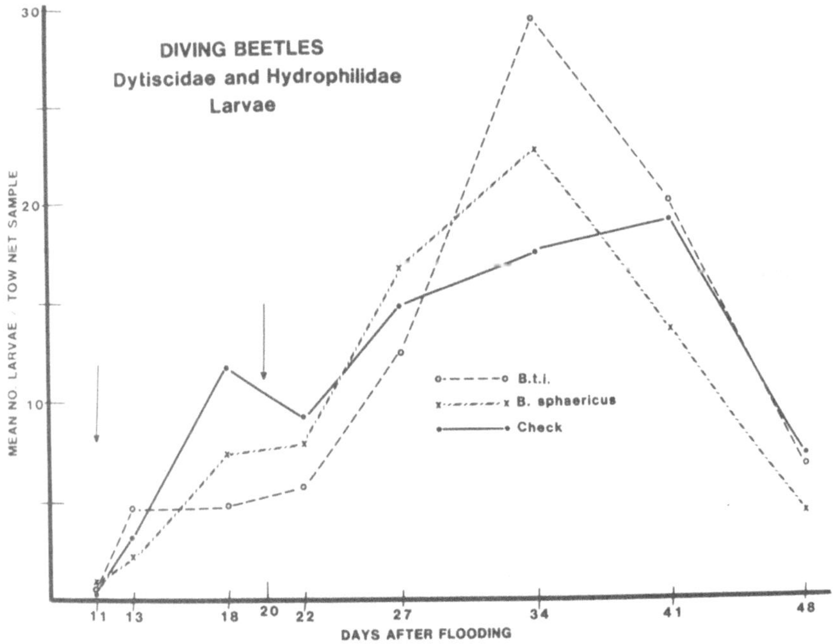

FIGURE 9.7 Population trends of predaceous diving beetle larvae (6–8 dominant species) in 35m² mesocosms treated twice (arrows) with 2 microbial control agents: first treatment at full larvicidal dosage, the second at ½ this dosage.

ment was applied at one-half the dosage, and the second treatment was applied at one-fourth the dosage.

Two sampling methods for mosquito larvae and macroinvertebrates were employed. The dipping and tow-net techniques provided similar trends of populations of some 20 species of organisms sampled. Also, the population trends and succession of *Cx. tarsalis* larvae and predaceous, herbivorous, and detritivorous macroinvertebrates were essentially the same in the three years of study.

To shed light on the interaction of *Cx. tarsalis* larvae and their natural enemies, data for some important key predators (adequately sampled by the methods noted above) are presented here for the second-year study, when full-dosage (see above) and half-dosage treatments were made. As observed before, we clearly documented the natural decline of third- and fourth-instar larvae of *Cx. tarsalis* in the replicated mesocosms (fig. 9.6). Peak larval populations prevailed in check ponds during the postflooding period of about one to four weeks, declining thereafter to a very low level of equilibrium in the untreated mesocosms. In the treated plots, the first treatment (*B.t.i.* and *B. sphaericus*) suppressed *Cx. tarsalis* larvae to the nil level; a slight increase a

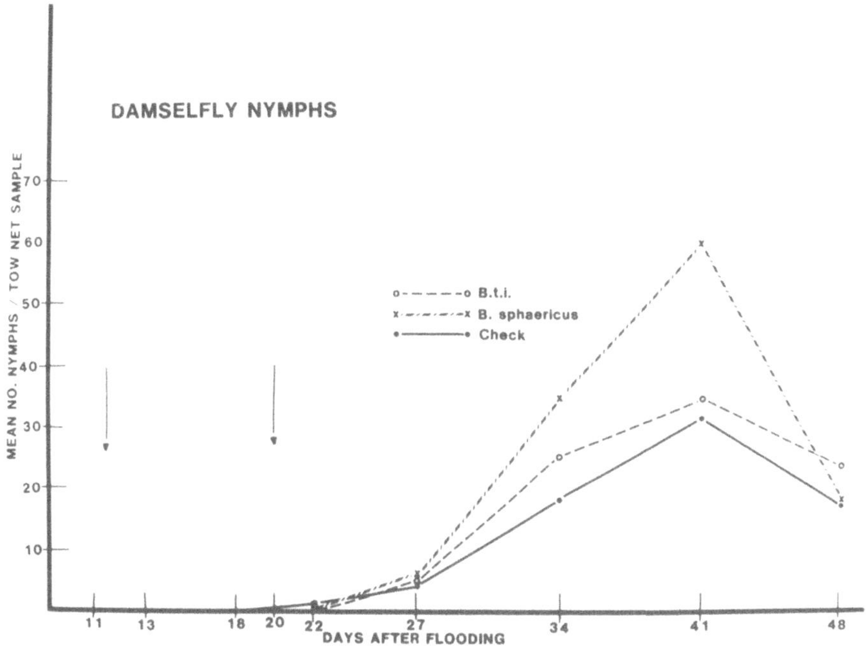

FIGURE 9.8 Population trends of damselfly nymphs (2 species) in 35m² mesocosms treated twice (arrows) with two microbial control agents: first treatment at full larvicidal dosage, the second at ½ this dosage.

week after the first treatment required a second treatment, which was applied at half the full dosage. The second treatment at reduced dosage suppressed larvae to the nil level, with no further marked increase in larval densities. The rational for reducing the dosage of the second treatment was based on observations made by many workers and ourselves that low-density larval populations require lower amounts of particulate toxins.

After noting the tremendous natural decline in larval populations, we proceeded to establish the relationship between key predators, their succession, and mosquito larvae. The first group of predaceous organisms that invaded and reproduced in the flooded sites were the diving beetles (*Dytiscidae* and *Hydrophilidae*). We encountered between 8 and 12 species of these predators, their larvae appearing within 10–15 days after flooding (fig. 9.7). The diving beetle larval population peaked 4–6 weeks postflooding, declining sharply 7 weeks postflooding. It should be noted that the peak beetle larval population coincides with the sharp decline in *Cx. tarsalis* larvae. It is also quite evident that the two treatments with each of the two microbial larvicides did not adversely affect larval populations of diving beetles.

Damselfly naiads (two to three species), like the diving beetle larvae, feed voraciously on mosquito larvae as well as on other organisms. Damselfly

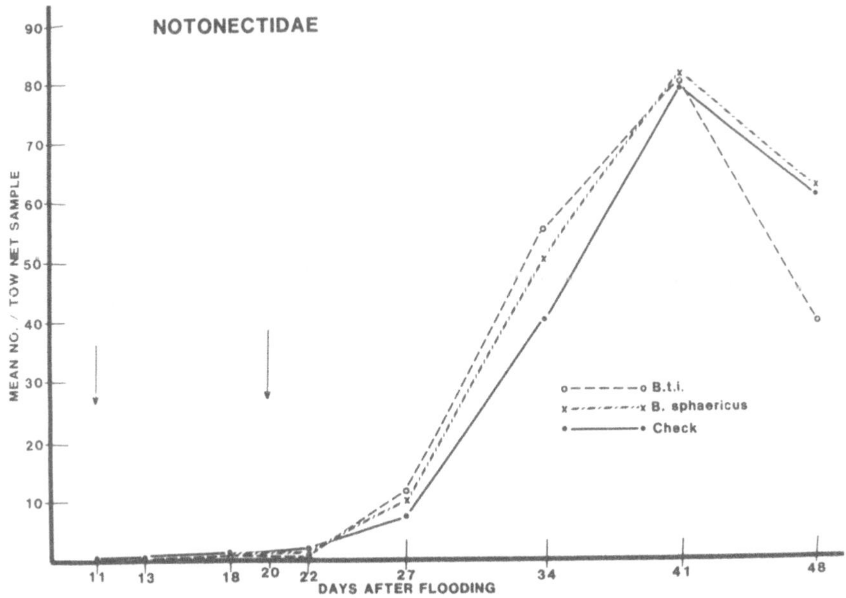

FIGURE 9.9 Population trends of notonectids (adults and immatures of 2 species, *Boanoa* and *Notonecta*) in 35m² mesocosms treated twice (arrows) with 2 microbial control agents: first treatment at full larvicidal dosage, the second at ½ this dosage.

naiads appeared later than beetles, about four weeks after flooding (fig. 9.8), reaching a peak five to six weeks after flooding and then declining to a moderate level. These predators provided an additional means of suppressing mosquito larvae, especially at a time when diving beetle larvae were on the decline. Again the two larvicidal treatments had no adverse impact on the population trends of these predators.

The intensity of predation was further enhanced by the appearance of notonectids (two to three species, belonging to *Boanoa* and *Notonecta*) four weeks after flooding (fig. 9.9). Their populations peaked six weeks after flooding, declining somewhat but still prevailing at high numbers seven weeks postflooding. The two treatments had no significant impact on the appearance and population trends of these predaceous organisms.

Although there are many other predaceous macroinvertebrate organisms cohabiting with mosquito larvae, in our test area the three groups discussed here were in preponderance. Naiads of several species of dragonflies also cohabit with mosquito larvae, and these naiads probably are important in regulating mosquito larvae. They appear late, about the same time as the notonectids. None of the treatments had any significant adverse impact on dragonfly naiads.

From the foregoing data it is quite clear that some or most asynchronous species of mosquitoes can be effectively controlled with fewer treatments and reduced cost using selective larvicides such as *B.t.i.* or *B. sphaericus.* These larvicides produce 98–100% control of peak larval populations at the outset, when the habitats are devoid of sufficient numbers of key natural predators. Once peak larval populations are eliminated, the appearance of abundant natural enemies suppresses new cohorts of mosquitoes, if any, thus precluding the administration of weekly or biweekly treatments with non-selective larvicides as is practiced now in many larvicidal programs. It is thus obvious that selective larvicides could be used cost effectively in an integrated control scheme of many species of pest and vector mosquitoes as shown in our studies and elsewhere (Yu et al. 1983). Further research is warranted on the application of this concept to mosquito control in some expansive developmental sites of mosquitoes such as rice paddies, duck marshes, and other semipermanently flooded breeding habitats of mosquitoes.

Acknowledgments

Many scientists provided information and advice regarding the completion of this chapter. Our sincere thanks to all those who contributed to and reviewed this chapter. Special thanks are due H. Axelrod, J. D. Chaney, H. A. Darwazeh, J. Rodcharoen, and W. A. Walton of the Department of Entomology, University of California, Riverside, California, who assisted in some of the studies reported in this chapter.

References

Aizawa, K., and Ohba, M. 1985. Screening of effective *Bacillus thuringiensis* isolates other than subspecies *israelensis* for mosquito and blackfly control. *Integrated mosquito control methodologies,* ed. M. Laird and J. W. Miles, 2: 199–212. San Diego: Academic Press.

Ali, A. 1981. *Bacillus thuringiensis* serovar *israelensis* against chironomids and some non-target aquatic invertebrates. *J. Invertebr. Pathol.* 38: 264–272.

Ali, A.; Sauerman, D. M.; and Nayar, J. K. 1984. Pathogenecity of industrial formulations of *Bacillus thuringiensis* serovar *israelensis* to larvae of some Culicine mosquitoes in the laboratory. *Fla. Entomol.* 67: 193–197.

Aly, C. 1985. Germination of *Bacillus thuringiensis* var. *israelensis* spores in the gut of *Aedes aegypti* larvae (Diptera:Culicidae). *J. Invertebr. Pathol.* 45: 1–8.

Aly, C., and Mulla, M. S. 1986. Orientation and ingestion rates of larval *Anopheles albimanus* in response to floating particles. *Entomol. Exp. Appl.* 42: 83–90.

Aly, C.; Mulla, M. S.; and Federici, B. A. 1985. Sporulation and crystal production by *Bacillus thuringiensis* var. *israelensis* in cadavers of mosquito larvae (Diptera:Culicidae). *J. Invertebr. Pathol.* 46: 251–258.

Aly, C.; Mulla, M. S.; Schnetter W.; and Xu, Bo-Zhao. 1987. Floating bait formulations increase effectiveness of *Bacillus thuringiensis* var. *israelensis* against *Anopheles larvae. J. Amer. Mosquito Control Assoc.* 3: 583–588.

Aly, C.; Mulla, M. S.; Xu, Bo-Zhao; and Schnetter, W. 1988. Rate of ingestion by mosquito larvae (Diptera:Culicidae) as a factor in the effectiveness of a bacterial stomach toxin. *J. Med. Entomol.* 25: 191–196.

Arias, J., and Mulla, M. S. 1975. Postemergence effects of two insect growth regulators on the mosquito *Culex tarsalis* (Diptera: Culicidae). *J. Med. Entomol.* 12: 317–322.

Barjac, H. de. 1978. Une nouvelle variété de *Bacillus thuringiensis* très toxique pour les moustiques: *B. thuringiensis* var. *israelensis* sérotype 14. *C. R. Acad. Sci.* (Paris) 286D: 797–800.

Becker, N., and Ludwig, H. W. 1983. Mosquito control in West Germany. *Bull. Soc. Vector Ecol.* 8: 85–93.

Brazner, J. C., and Anderson, R. L. 1986. Ingestion and adsorption of *Bacillus thuringiensis* subsp. *israelensis* by *Gammarus lacustris* in the laboratory. *Appl. Environ. Microbiol.* 52: 1386–1390.

Chapman, H. C. 1974. Biological control of mosquito larvae. *Ann. Rev. Entomol.* 19: 33–59.

Cheung, P.Y.K., and Hammock, B. D. 1985. Micro-lipid-droplet encapsulation of *Bacillus thuringiensis* subsp. *israelensis* δ-endotoxin for control of mosquito larvae. *Appl. Environ. Microbiol.* 50 (4): 984–988.

Christie, M. 1958. Predation on larvae of *Anopheles gambiae* Giles. *Trop. Med. Hyg.* 1958 (July): 168–176.

CMVCA. 1984. California Mosquito and Vector Control Association, Inc., Sacramento, California. *Year Book.*

______. 1985. California Mosquito and Vector Control Association, Inc., Sacramento, California. *Year Book.*

______. 1986. California Mosquito and Vector Control Association, Inc., Sacramento, California. *Year Book.*

Dame, D.; Savage, K.; Meisch, M.; and Oldacre, S. 1981. Assessment of industrial formulations of *Bacillus thuringiensis* var. *israelensis*. *Mosq. News* 41: 540–546.

Davidson, E. W. 1982. Bacteria and the control of arthropod vectors of human and animal diseases. In *Microbial control and viral pesticides*, ed. E. Kurstak, 289–315. New York: Marcel Dekker.

Eldridge, B. F., and Callicrate, J. 1982. Efficacy of *Bacillus thuringiensis* var. *israelensis* de Barjac for mosquito control in a western Oregon pond. *Mosq. News* 42: 102–105.

Eldridge, B. F.; Washino, R. K.; and Hennenberger, D. 1985. Control of snow pool mosquitoes with *Bacillus thuringiensis* ser. H-14 in mountain environments in California and Oregon. *J. Amer. Mosq. Control Assoc.* 1: 69–75.

Foo, A.E.S., and Yap, H. H. 1983. Field trials on the use of *Bacillus thuringiensis* serotype H-14 against *Mansonia* mosquitoes in Malaysia. *Mosq. News* 43: 306–310.

Garcia, R.; Des Rochers, B.; and Tozer, W. 1980. Studies on the the toxicity of *Bacillus thuringiensis* var. *israelensis* against organisms found in association with mosquito larvae. *Calif. Mosq. and Vector Control Assoc. Proc. & Papers* 48: 33–36.

Garcia, R.; Des Rochers, B.; Tozer, W.; and McNamara, J. 1983. Evaluation of *Bacillus thuringiensis* var. *israelensis* serotype H-14 for mosquito control. *Calif. Mosq. and Vector Control Assoc. Proc. & Papers* 51: 25–29.

Goldberg, L. J., and Margalit, J. 1977. A bacterial spore demonstrating rapid larvicidal activity against *Anopheles sergentii, Uranotaenia unquiculata, Culex univitattus, Aedes aegypti,* and *Culex pipiens*. *Mosq. News* 37: 355–358.

Hall, I. M.; Arakawa, K. Y.; Dulmage, H. T.; and Correa, J. A. 1977. The pathogenecity of strains of *Bacillus thuringiensis* to larvae of *Aedes* and to *Culex* mosquitoes. *Mosq. News* 37: 246–251.

Hare, S.G.F., and Nasci, R. S. 1986. Effects of sublethal exposure to *Bacillus thuringiensis* var. *israelensis* on larval development and adult size in *Aedes aegypti*. *J. Amer. Mosq. Control Assoc.* 2: 325–328.

Hougard, J. M.; Darriet, F.; and Bakayoko, S. 1983. Evaluation en milieu naturel de l'activité larvicide de *Bacillus thuringiensis* sérotype H-14 sur *Culex quinquefasciatus* Say, 1823 et *Anopheles gambiae* Giles, 1902 s.l. (Diptera: Culicidae) en Afrique de l'Ouest. *Cah. ORSTOM, sér. Ent. méd. et Parasitol.* 21 (2): 111–117.

Hougard, J. M.; Duval, J.; and Escaffre, H. 1985. Évaluation en milieu naturel de l'activité larvicide de une formulation de *Bacillus thuringiensis* H-14 sur *Aedes aegypti* (L.) dans un foyer épidemique de fieure jaune en Côte d'Ivoire. *Cah. ORSTOM, sér. Ent. méd. et Parasitol.* 23: 235–240.

Lacey, L. A. 1985. *Bacillus thuringiensis* serotype H-14. *Amer. Mosq. Control Assoc. Bull.* 6: 132–158.

Lacey, L. A., and Inman, A. 1985. Efficacy of granular formulations of *Bacillus thuringiensis* (H-14) for the control of *Anopheles* larvae in rice fields. *J. Amer. Mosq. Control Assoc.* 1: 38–42.

Lacey, L. A., and Mulla, M. S. 1990. Safety of *Bacillus thuringiensis* var. *israelensis* and *Bacillus sphaericus* to nontarget organisms in the aquatic environment. In *Safety of microbial insecticides*, ed. Laird, M., Lacey, L. A., and Davidson, E. W. CRC Press. In press.

Lacey, L. A., and Undeen, A. H. 1986. Microbial control of black flies and mosquitoes. *Ann. Rev. Entomol.* 31: 265–296.

Lacey, L. A.; Urbina, M. J.; and Heitzman, C. M. 1984. Sustained release formulations of *Bacillus sphaericus* and *Bacillus thuringiensis* (H-14) for control of container breeding *Culex quinquefasciatus*. *Mosq. News* 44: 26–32.

Majori, G., and Ali, A. 1984. 1984. Laboratory and field evaluation of industrial formulations of *Bacillus thuringiensis* serovar *israelensis* against some mosquito species of central Italy. *J. Invertebr. Pathol.* 43: 316–323.

Majori, G.; Ali, A.; and Sabatinelli, G. 1987. Laboratory and field efficacy of *Bacillus thuringiensis* var. *israelensis* and *B. sphaericus* against *Anopheles gambiae* s.l. and *Culex quinquefasciatus* in Ouagadougou, Burkina Faso. *J. Amer. Mosquito Control Assoc.* 3: 20–25.

Margalit, J., and Bobroglio, H. 1984. The effect of organic materials and solids in water on the persistence of *Bacillus thuringiensis* var. *israelensis*. *Z. Angew. Entomol.* 97: 516–520.

Margalit, J.; Markus, A.; and Pelah, Z. 1984. Effect of encapsulation on the persistence of *Bacillus thuringiensis* var. *israelensis* (H-14). *Appl. Microbiol. Biotechnol.* 19: 382–383.

Margalit, J.; Pascar-Gluzman, C.; Bobroglio, H.; Barak, Z.; and Lakhim-Tsror, L. 1985. Biocontrol of mosquitoes in Israel. In *Integrated mosquito control methodologies*, ed. M. Laird and J. W. Miles, 2: 361–374. San Diego: Academic Press.

Miura, T.; Takahashi, R. M.; and Mulligan, F. S. 1980. Effects of the bacterial mosquito larvicide *Bacillus thuringiensis* serotype H-14 on selected aquatic organisms. *Mosq. News* 40: 619–622.

Mulla, M. S. 1976. Evolution of chemical control strategies for mosquitoes: Conventional, novel, and natural chemicals. *Calif. Mosq. Control Assoc. Proc. & Papers* 44: 71–77.

———. 1985. Field evaluation and efficacy of bacterial agents and their formulations against mosquito larvae. In *Integrated mosquito control methodologies*, ed. M. Laird and J. W. Miles, 2: 227–250. San Diego: Academic Press.

———. 1986. Role of *B.t.i.* and *Bacillus sphaericus* in mosquito control programs. In *Fundamental and applied aspects of invertebrate pathology*, ed. R. A. Sampson, J. M. Vlak, and D. Peters, 494–496. Foundation Fourth International Colloquium of Invertebrate Pathology. Wageningen, The Netherlands.

Mulla, M. S., and Darwazeh, H. A. 1984. Larvicidal efficacy of various formulations of *Bacillus thuringiensis* serotype H-14 against mosquitoes. *Bull. Soc. Vector Ecol.* 9: 51–58.

———. 1985. Efficacy of formulations of *Bacillus thuringiensis* H-14 against mosquito larvae. *Bull. Soc. Vector Ecol.* 10: 14–19.

Mulla, M. S.; Darwazeh, H. A.; and Aly, C. 1986. Laboratory and field studies on new formulations of two microbial control agents against mosquitoes. *Bull. Soc. Vector Ecol.* 11: 255–263.

Mulla, M. S.; Darwazeh, H. A.; Ede, L.; Kennedy, B.; and Dulmage, H. T. 1985. Efficacy and field evaluation of *Bacillus thuringiensis* (H-14) and *B. sphaericus* against floodwater mosquitoes in California. *J. Amer. Mosq. Control Assoc.* 1: 310–315.

Mulla, M. S.; Federici, B. A.; and Darwazeh, H. A. 1982. Larvicidal efficacy of *Bacillus thuringiensis* ser. H-14 against stagnant-water mosquitoes and its effects on nontarget organisms. *Environ. Entomol.* 11: 788–795.

Mulla, M. S.; Federici, B. A.; Darwazeh, H. A.; and Ede, L. 1982. Field evaluation of the microbial insecticide *Bacillus thuringiensis* ser. H-14 against floodwater mosquitoes. *Appl. Environ. Microbiol.* 43: 1288–1293.

Ramoska, W. A.; Watts, S.; and Rodrigues, R. E. 1982. Influence of suspended particulates on the activity of *Bacillus thuringiensis* ser. H-14 against mosquito larvae. *J. Econ. Entomol.* 75: 1–4.

Rashed, S. S., and Mulla, M. S. 1989. Factors influencing ingestion of particulate materials by mosquito larvae (Diptera: Culicidae). *J. Med. Entomol.* 26: 210–216.

Roberts, D. W., and Castillo, J. M. 1980. Bibliography on pathogens of medically important arthropods. WHO Bull., suppl. 58: 1–197.

Roberts, D. W., and Strand, M. A. 1977. Pathogens of medically important arthropods. *WHO Bull.*, suppl. 1. 55: 1–419.

Sandoski, C. A.; Yates, M. W.; Olson, J. K.; and Meisch, M. V. 1985. Evaluation of Beecomist® applied *Bacillus thuringiensis* (H-14) against *Anopheles quadrimaculatus* larvae. *J. Amer. Mosq. Control Assoc.* 1: 316–319.

Schnetter, W.; Engler-Fritz, S.; Aly, C.; and Becker, N. 1983. Anwendung von *Bacillus thuringiensis* var. *israelensis* Präparaten gegen Stechmücken an Oberrhein. *Mitt. dtsch. Ges. allg. angew Ent.* 4: 18–25.

Service, M. W. 1973. Study of the natural predators of *Aedes cantans* (Meigen) using the precipitin tests. *J. Med. Entomol.* 10: 503–510.

______. 1977. Mortalities of the immature stages of species B of the *Anopheles gambiae* complex in Kenya: Comparison between rice fields and temporary pools, identification of predators, and effects of insecticidal spraying. *J. Med. Entomol.* 13: 535–545.

Sinégre, G.; Gaven, B.; Jullien, J. L.; and Crespo, O. 1980. Activité du sérotype H-14 de *Bacillus thuringiensis* vis-à-vis des principales espèces de moustiques anthropophiles du littoral Mediterranean Francaise. *Parasitologia* 22: 223–231.

Sjogren, R. D.; Batzer, D. P.; and Junemann, M. A. 1986. Evaluation of methoprene, temephos, and *Bacillus thuringiensis* var. *israelensis* against *Coquilletidia perturbans* larvae in Minnesota. *J. Amer. Mosq. Control Assoc.* 2: 276–279.

Standaert, J. Y. 1981. Persistence et l'efficacité de *Bacillus thuringiensis* H-14 sur les larves de *Anopheles stephensi. Z. Angew. Entomol.* 91: 292–300.

Stark, P. M., and Meisch, M. V. 1983. Efficacy of *Bacillus thuringiensis* serotype H-14 against *Psorophora columbiae* and *Anopheles quadrimaculatus* in Arkansas ricelands. *Mosq. News* 43: 59–62.

Sutherland, D. J.; Kung, S. P.; Ehrenburg, H.; Hanson, J.; and Rupp, H. 1982. Formulations of *B.t.i.* and their evaluations. *N.J. Mosquito Control Assoc. Proc.* 69: 93–101.

Yu, H. S.; Lee, D. K.; Na, J. O.; and Ban, S. J. 1983. Integrated control of mosquitoes by combined use of *Bacillus thuringiensis* var. *israelensis* and larvivorous fish *Aplocheilus latipies* in simulated rice paddies in South Korea. *Korean J. Entomol.* 13: 75–84.

Yu, H. S.; Shim, J. C.; Lee, D. K.; and Yun, Y. H. 1981. Mosquito control by *Bacillus thuringiensis israelensis* in simulated rice paddies (abstract). *Korean J. Entomol.* 11: 51–52.

10

Progress in the Biological Control of Black Flies with *Bacillus thuringiensis israelensis*, with Emphasis on Temperate Climates

DANIEL P. MOLLOY

10.1 NATURE OF THE BLACK FLY PROBLEM IN TEMPERATE CLIMATES

In contrast to the problems of onchocerciasis in the tropics, black flies in temperate climates are not known to transmit human diseases. This is not to say, however, that these flies do not cause medical problems in temperate regions. Black fly bites can induce an allergic response in sensitive individuals; skin inflammation, nausea, headache, and fever are possible consequences. Scratching the itchy scab that forms after the bite can result in localized bacterial infection. In general, however, the pest status of black flies in temperate climates does not stem from these medical problems, but rather from the annoyance caused by their flying and biting.

Their habit of incessantly flying around one's head, crawling on the skin, and flying into the eyes, ears, and nose can be quite aggravating. Thus, black flies frequently have an adverse impact on outdoor activities, including employee productivity (Olejnicek, Matha, and Weiser 1985b), recreation, and tourism. Cattle rearing can also be severely affected when black flies are abundant. Fredeen (1985) estimated that losses to beef producers during one year in east-central Saskatchewan exceeded $2.9 million. This was due to unrealized weight gains, delayed conceptions, fatalities, replacement of debilitated bulls, fence repairs, supplementary feeding, and the increased costs of labor and veterinary services. In terms of the transmission of serious disease in temperate areas, black flies are known vectors of protozoan blood parasites of fowl (*Leucocytozoon* spp.), which have been periodically responsible for severe economic losses in the poultry industry (Noblet, Kissam, and Adkins 1975; Barnett 1977). Temperate-climate black flies do not merit the significant pest status of tropical vector species, yet they are pests of considerable importance, and concerted efforts are made for their control.

10.2 ADVENT OF *B.t.i.*

Although broad-spectrum chemical insecticides have traditionally been the primary agent used for black fly control in temperate regions, their undesirable environmental effects have prompted the development of more ecologically sound control strategies, such as biological control methods. Research on one such biological agent, the pathogenic bacterium *B.t.i.* (*Bacillus thuringiensis* subsp. *israelensis*), has produced rapid and significant progress since its original isolation (Goldberg and Margalit 1977). Research and development of *B.t.i.* as a black fly control agent was previously summarized by Gaugler and Finney (1982). The present paper focuses on updating subsequent progress and highlighting recent research advancements. Discussion of the use of *B.t.i.* for large-scale control has been limited to those programs conducted in temperate regions. Analysis of the largest operational program using *B.t.i.* against black flies, the Onchocerciasis Control Programme in West Africa, appears in chapter 11.

10.3 TRENDS IN RECENT RESEARCH EFFORTS

One significant change in *B.t.i.* research in recent years is that the emphasis has moved from laboratory investigations to field trials—a logical and valuable development. In essence, the laboratory investigations of the 1978–1982 period provided the groundwork for subsequent field evaluations by identifying promising commercial formulations and elucidating a number of factors likely to play significant roles in field efficacy. Although field trials require greater resources both to conduct and analyze, they are inherently more valuable in the data they supply, for example, efficacy and impact on nontargets. Laboratory investigations of *B.t.i.* and black flies continue to be conducted, but only when field tests examining the parameter(s) under study would be premature or impractical.

One factor that facilitated the compilation of data for the present review was the general trend toward adoption of the standard field protocol suggested by Undeen and Lacey (1982). This trend is to be commended. In past years, there were almost as many field protocols as there were individuals doing *B.t.i.*/black fly research, and comparisons were difficult since data had been generated using a wide variety of testing methods.

Data interpretation was difficult for field-trial reports in which treatment concentrations were expressed solely as spores/ml and/or in which durations of treatment were not indicated (e.g., Olejnicek, Matha, and Weiser 1985a, 1985b; Olejnicek 1986). In preparing this review, it became evident that numerous field trials had been conducted without concurrent upstream controls (e.g., Car 1984; Car and de Moor 1984; Pistrang and Burger 1984; Back et al. 1985; de Moor and Car 1986). The inclusion of such upstream control data would have strengthened the conclusions of these otherwise well executed studies. This was particularly true in studies that used changes in drift intensity to monitor nontarget effects, since drift intensities can vary widely from day to day or even hour to hour. As pointed out by Dejoux, Gibon, and Yameogo (1985), concurrent upstream controls allow better data analysis by providing simultaneous data on invertebrate densities in untreated and treated areas.

10.4 FACTORS AFFECTING THE EFFICACY OF *B.t.i.* AGAINST BLACK FLIES

A variety of biotic and abiotic factors have been cited as playing a role in determining the field effectiveness of *B.t.i.* For the convenience of discussion, these factors have arbitrarily been grouped into the following four categories: Environmental, Black Fly, Formulation, and Treatment Parameters.

10.4.1 Environmental Parameters

10.4.1.1 Discharge. Discharge is well known to be a principal factor determining the *carry* of black fly larvicides (i.e., the distance downstream that an application of a larvicide produces high black fly mortality). For *B.t.i.*, this was initially confirmed in the field trials of Undeen and Colbo (1980). Since then, this relationship has been reaffirmed in numerous field trials (table 10.1). For example, the poor carry (about 50 m) of *B.t.i.* in low-discharge streams (Gaugler et al. 1983) contrasted sharply with the carry (about 5,000 m) achieved in a large river trial (de Moor and Car 1986). The overriding influence of discharge in determining carry was also demonstrated in the field trials of Lacey and Undeen (1984) in which doubling and trebling the concentration increased carry, but not in proportion to the increase in dosage.

10.4.1.2 Stream Profile. Carry has been shown to correlate strongly with stream profile, especially depth-to-width ratio (Undeen, Lacey, and Avery 1984). Decreased carry has been noted in streams with a high ratio of surface area to water volume, apparently due to increased contact and adherence of *B.t.i.* particles to stream substrates.

10.4.1.3 Turbidity. Guillet, Escaffre, et al. (1985b) found that under natural conditions the turbidity of water did not affect the efficacy of formulations with individual spores and crystals, but did affect the efficacy of formulations in which spores and crystals were clumped as large particles. The decreased efficacy noted with large-particle formulations in turbid water was considered to be a consequence of competition between large, naturally

TABLE 10.1.
Summary of Recent Field Trials Using *B.t.i.* against Black Fly Larvae

Principal black fly species	Product/ formulation	Application Rate: Conc. (ppm)	Application Rate: Duration (min)	Equivalent conc. (ppm/1 min)	Water temp. (° C)	Location	Discharge (L/min)	Results: Range in % mortality (upstream–downstream)	Results: Distance from treatment (m)	Source
Austrosimulium laticorne and *A. multicorne*	Teknar®	2.0	15.0	30.0	19	New Zealand	3,372	100–90	0–577	(1)
Austrosimulium laticorne and *A. multicorne*	R153-78 primary powder	0.2	15.0	3.0	10.5	New Zealand	1,680	100–65	0–580	(1)
Simulium ochraceum, S. metallicum, and *S. callidum*	Teknar®	45.0	0.5	22.5	19	Mexico	1,020	100–29	10–100	(2)
Simulium ochraceum, S. metallicum, and *S. callidum*	Teknar®	7.5	1.0	7.5	19	Mexico	1,080	100–24	25–100	(2)
Similium venustum/verecundum and *S. tuberosum*	Vectobac® WP	20.0	1.0	20.0	21	USA	4,100	100	20–800	(3)
Simulium venustum/verecundum and *S. tuberosum*	Teknar®	40.0	1.0	40.0	19	USA	8,000	100–5	20–1,200	(3)
Simulium venustum/verecundum and *S. tuberosum*	Teknar®	40.0	1.0	40.0	16	USA	6,800	100–45	20–1,100	(3)
Simulium hargreavesi	Powder produced by Ben Gurion University	3.0	10.0	30.0	17^b	South Africa	2,040	54–5	20–200	(4)
Simulium hargreavesi	Teknar®	1.6	10.0	16.0	17^b	South Africa	3,060	68–52	20–80	(4)
Simulium adersi	Teknar®	1.6	10.0	16.0	17^b	South Africa	3,060	78–6	80–200	(4)

Simulium adersi and *S. damnosum s.l.*	Teknar®	1.6	10.0	16.0	24	South Africa	360,000	81–33	70	(5)
Simulium adersi and *S. hargreavesi*	Teknar®	2.3	7.0	16.1	28	South Africa	30,000	100–85	35	(5)
Simulium vittatum	Bactimos® WP	10.0	1.0	10.0	20	USA	21,400	88–8	10–625	(6)
Simulium vittatum	Bactimos® WP	10.0	1.0	10.0	20	USA	17,200–21,400	61 (mean)	10–400	(6)
Simulium vittatum	Bactimos® WP	30.0	1.0	30.0	20	USA	21,400	96–72	10–625	(6)
Simulium vittatum	Vectobac® WP	10.0	1.0	10.0	11	USA	31,400	88–52	10–400	(6)
Simulium vittatum	Vectobac® WP	10.0	1.0	10.0	11–23	USA	25,000–31,400	80 (mean)	10–400	(6)
Simulium vittatum	Vectobac® WP	20.0	1.0	20.0	11	USA	31,400	96–75	10–400	(6)
Simulium vittatum	Teknar®	10.0	1.0	10.0	23	USA	25,000–27,000	85 (mean)	10–400	(6)
Simulium venustum (cytotype ACgB)	Teknar®	10.0	2.0	20.0	9–16	USA	13,500	100–0	10–800	(7)
Simulium venustum, S. tuberosum, Prosimulium mixtum, and *Stegopterna mutata*	Teknar®	5.86	15.0	87.9	11–15	Canada	114,000	95–82	50–250	(8)
Simulium vittatum	Teknar® 2X	5.0	1.0	5.0	22	USA	31,800	96	100	(9)
Simulium vittatum	Teknar® 2X oil base	5.0	1.0	5.0	22	USA	31,800	84	100	(9)
Simulium vittatum	Teknar® 2X oil base	10.0	1.0	10.0	22	USA	31,800	95	100	(9)
Simulium vittatum	Vectobac® AS	5.0	1.0	5.0	22	USA	31,800	81	100	(9)
Simulium vittatum	Vectobac® AS	10.0	1.0	10.0	22	USA	31,800	93	100	(9)
Simulium vittatum	Teknar®	5.0	1.0	5.0	22	USA	31,800	76	100	(9)
Simulium vittatum	Teknar®	10.0	1.0	10.0	22	USA	31,800	91	100	(9)
Simulium vittatum	Skeetal® F	5.0	1.0	5.0	22	USA	31,800	75	100	(9)

(*continued*)

TABLE 10.1. (Continued)

Principal black fly species	Product/ formulation	Application Rate: Conc. (ppm)	Application Rate: Duration (min)	Equivalent conc. (ppm/1 min)	Water temp. (° C)	Location	Discharge (L/min)	Results: Range in % mortality (upstream–downstream)	Results: Distance from treatment (m)	Source
Simulium vittatum	Skeetal® F	10.0	1.0	10.0	22	USA	31,800	94	100	(9)
Simulium vittatum	Bactimos® FC	5.0	1.0	5.0	22	USA	31,800	60	100	(9)
Simulium vittatum	Bactimos® FC	10.0	1.0	10.0	22	USA	31,800	92	100	(9)
Simulium aokii	Teknar®	1.5	60.0	90.0	—	Japan	6,240	100	50–100	(10)
Simulium japonicum	Teknar®	10.0	10.0	100.0	—	Japan	2,700	100	50–100	(10)
Simulium uchidai	Teknar®	100.0	1.0	100.0	—	Japan	1,260	100	50–100	(10)
Simulium uchidai	Teknar®	10.0	30.0	300.0	—	Japan	480	100	50–100	(10)
Simulium chutteri	Teknar®	1.6	10.0	16.0	17–20	South Africa	2,280,000	95–75	200–5,000	(11)
Simulium tuberosum, S. venustum, and *S. corbis*	Vectobac® AS	10.0	1.0	10.0	9.3	USA	9,300	68–23	50–600	(12)
Simulium tuberosum, S. venustum, and *S. corbis*	Vectobac® AS	10.0	5.0	50.0	12.1	USA	12,300	100–41	50–900	(12)
Simulium tuberosum, S. venustum, and *S. corbis*	Vectobac® AS	10.0	5.0	50.0	9.6	USA	1,620	100–35	150–1,250	(12)
Simulium tuberosum, S. venustum, and *S. corbis*	Vectobac® AS	10.0	5.0	50.0	—	USA	9,240	100–38	14–250	(12)

SOURCES: (1) Chilcott, Pillai, and Kalmakoff 1983; (2) Gaugler et al. 1983; (3) Horosko and Noblet 1983; (4) Car 1984; (5) Car and de Moor 1984; (6) Lacey and Undeen 1984; (7) Pistrang and Burger 1984; (8) Back et al. 1985; (9) Lacey and Heitzman 1985; (10) Nakamura et al. 1985; (11) de Moor and Car 1986; (12) Gibbs et al. 1986.

NOTE: Field trials in which treatment concentrations were expressed solely as spores/ml and/or in which durations of treatments were not indicated could not be included in this table.

[a] Possible detoxification of bacteria since aqueous suspension of powder was produced in blender (see Guillet and Duval 1985).

[b] Water contained high sewage and chloride content.

occurring particles and the *B.t.i.* clumps, thus lowering the probability of the latter's ingestion.

10.4.1.4 Pollutants. Field trials have suggested lower efficacy of *B.t.i.* in waters containing sewage and with high chloride content (Car 1984).

10.4.1.5 Water Temperature. The decreased efficacy of *B.t.i.* at lower temperature, first noted by Molloy, Gaugler, and Jamnback (1981), has been reconfirmed in a number of recent laboratory and field tests (Colbo and O'Brien 1984; Olejnicek, Matha, and Weiser 1985b; Matha et al. 1986; Olejnicek 1986; Lacoursière and Charpentier 1988). This decline in efficacy is not a simple linear one; in the extensive laboratory tests of Lacoursière and Charpentier (1988), the sharpest declines in efficacy were noted against *Simulium decorum* larvae between 18° C and 12° C and against *Prosimulium mixtum/fuscum* between 12° C and 4° C.

Streams and rivers in temperate climates require their initial treatments in the spring, when waters are often near 0° C. The achievement of acceptable levels of black fly mortality, while still possible at such temperatures, is made more difficult by *B.t.i.*'s reduced efficacy. Treatments have to be modified by increasing the exposure time and/or concentration. In either case, more *B.t.i.* is required per application. Colbo and O'Brien (1984) suggested that the reduced efficacy of *B.t.i.* at very cold temperatures may be the result of a lack of continuous feeding by the larvae. Thus, they recommended a longer application time (5 minutes vs. the usual 1 minute) to increase the probability that all larvae would ingest some *B.t.i.* during its passage downstream. Olejnicek, Matha, and Weiser (1985b), who reported lower efficacy at 0–3° C than at 17–19° C, suggested that the efficacy of the bacterium in cold water was inhibited not only by reduced ingestion, but also by reduced digestion of the toxic crystals. Likewise, Lacoursière and Charpentier (1988) have suggested that the reduced efficacy of *B.t.i.* observed at lower temperatures is a result of changes in larval feeding behavior and physiology.

10.4.1.6 pH. In the first laboratory tests examining the effect of pH on the efficacy of *B.t.i.* against black flies, Lacoursière and Charpentier (1988) observed higher mortality against *S. decorum* larvae at higher pH values (4.5 vs. 9.0). Their data support the laboratory studies of Lacey, Mulla, and Dulmage (1978), who reported a positive correlation between increasing pH and the pathogenicity of *B. thuringiensis* subsp. *kenyae* to *Simulium vittatum*. The reduced efficacy of *B.t.i.* in acidic water now remains to be confirmed in controlled field studies.

10.4.1.7 Degree of Vertical Mixing in the Water Column. Since *B.t.i.* is generally broadcast onto a stream's surface during a treatment, very little vertical mixing actually occurs at the point of application. This lack of a homogeneous mixture of *B.t.i.* in the water column is likely responsible for the anomalous results attained in some single-application field trials (Frommer et al. 1981), wherein the percentage mortality near the treatment point was less than that achieved further downstream. Likewise, the percentage mortality achieved at lentic outflows (beaver dams, lake outlets, and so on) is also often unsatisfactory because of poor mixing (Colbo and O'Brien 1984; Molloy and Struble 1989).

10.4.1.8 Reduced Water Velocity Due to Negative Relief. The settling out or dilution of *B. t. i.* caused by its passage through a slow-moving pool has often been cited as a factor decreasing carry (Molloy and Jamnback 1981; Colbo and O'Brien 1984; Lacey and Undeen 1986). In fact, Colbo and O'Brien (1984) suggested that in the stream systems of the Canadian Shield, the interruption of flow caused by low or negative relief (i.e., gullies, pools) was the most important element limiting downstream carry of Teknar®, the *B. t. i.* formulation they used.

Although certain *B. t. i.* formulations (particularly those with large particle sizes) undoubtedly do settle out to some extent in pools, it is not probable that formulations like Teknar® (i.e., with individual spores and crystals) settle out in the relatively brief period in which they are passing through a pool (0.5–5 hours). Analysis of the data of Molloy et al. (1984) indicates that the settling rate of Teknar® crystals in still water is <5 mm/hr. If Teknar® crystals do actually settle out in pools, the only explanation that appears likely is that they are binding to larger, naturally occurring particles, which subsequently accelerates settling to the bottom.

10.4.1.9 Attachment of *B.t.i.* to Benthic Substrates/Sediments. The short carry generally observed in streams with a high ratio of surface area to water volume suggests that *B.t.i.* is being removed from stream water as a result of the direct contact of *B. t. i.* particles with benthic substrates or sediments (Undeen, Lacey, and Avery 1984). Although direct benthic recovery of *B. t. i.* particles has not been reported, nontarget data do provide circumstantial evidence to support the above hypothesis. Following application of *B. t. i.* to a Quebec stream at an intentionally high dosage (5.28 ppm/15 min), mortality in two chironomid genera that normally feed on organic debris indicated that a portion of the *B. t. i.* had been retained in benthic sediments (Back et al. 1985). In the same field trial, blephariceríd larvae, which feed by scraping, were also adversely affected, thereby suggesting that *B. t. i.* was attached to periphyton they ingested. The apparent presence of *B. t. i.* on periphyton also lends strength to the theory that the presence of vegetation reduces carry (Undeen, Lacey, and Avery 1984), particularly the carry of large-particle formulations (Lacey and Undeen 1984). The field trials of Matha et al. (1986) also provide data suggesting that dense vegetation decreases *B. t. i.* efficacy.

10.4.1.10 Other Factors. Other abiotic factors, such as turbulence (Lacey and Undeen 1986) and the density of aquatic filter-feeding organisms (Undeen and Colbo 1980), have been suggested as limiting carry; but conclusive experimental evidence of their effects has not yet been presented.

10.4.2 Black Fly Parameters

10.4.2.1 Larval Age. Although *B.t.i.* is effective against all larval instars, early instars are considerably more susceptible. This negative correlation between susceptibility and larval age, first noted by Guillet and Escaffre (1979), has been repeatedly reaffirmed in recent trials (Guillet, Escaffre, and Prud'hom 1982b; Gaugler et al. 1983; Olejnicek, Matha, and Weiser 1985a; Olejnicek 1986).

10.4.2.2 Species. All simuliid species tested to date have proven susceptible to *B.t.i.* Although Molloy, Gaugler, Jamnback (1981) found last-instar *Simulium verecundum* to be three times more susceptible than last-instar *S. vittatum,* they theorized that the differences in mortality might simply have been a result of the differences in body size (i.e., larger larvae are generally less susceptible to *B.t.i.,* and last instars of *S. vittatum* are about three times the size of *S. verecundum*). Habib (1983) also reported differences in species susceptibility among two *Simulium* spp., but did not provide body-size data or otherwise speculate about the differential susceptibility. More recently, Lacoursière and Charpentier (1988) stressed the need for determining "absolute" species susceptibility by considering the optimal level of feeding activity for each species under its preferred environmental conditions (e.g., temperature).

10.4.2.3 Feeding Behavior. Since the toxicity of *B.t.i.* is due solely to its role as a stomach poison, larval feeding behavior can play a significant role in determining the efficacy of field applications. Evidence has been presented indicating that the following factors may reduce *B.t.i.* efficacy by interfering with, or otherwise causing the cessation of, normal feeding: lack of feeding during molting (Back et al. 1985); feeding inhibition due to excess particulate load (Gaugler and Molloy 1980) or formulation additives (Molloy, Gaugler, and Jamnback 1981); reduced or intermittent feeding at very cold temperatures (Colbo and O'Brien 1984; Olejnicek, Matha, and Weiser 1985b).

10.4.3 Formulation Parameters

Guillet, Hougard, et al. (1985c), in a series of laboratory trials with a primary powder and formulations produced from it, demonstrated how formulation per se could significantly increase the efficacy of a powder against black flies. At equal dosages, mortalities increased from <30% to >99%, solely through formulation of the powdered primary product. Stream trials (Lacey and Heitzman 1985) have further demonstrated that formulation can clearly play a critical role in determining field effectiveness of a product against black flies. Formulation characteristics affecting efficacy and/or carry are discussed below.

10.4.3.1 Particle Size. The laboratory trials of Molloy et al. (1984) found that a formulation's toxicity correlated positively with its particle size. Their findings further supported the work of Guillet and Escaffre (1979), who provided the initial evidence of such a relationship. Recently, Guillet, Escaffre, et al. (1985a) further demonstrated the significance of particle size; tests showed that formulation potency generally increased with particle size, up to an optimum of 35–50 μm.

10.4.3.2 Powdered vs. Liquid Formulations. Since powdered formulations generally have larger particles than liquid (flowable concentrate) formulations, it was not surprising to find that the powdered formulations generally had superior toxicities (Molloy et al. 1984). Field trials, however, have indicated that powdered formulations generally have less carry than liquid formulations (Guillet, Escaffre, and Prud'hom 1982a; Lacey and Undeen 1984). Thus, even if a liquid formulation does not have an inherently high toxicity, when applied at an appropriate dosage, it will kill black flies for greater distances downstream than would a powdered formulation. The reason for the farther carry of liquid formulations (i.e., those with smaller particles) appears to be simply a matter of physics. Stokes's law (Weast 1982) states that the settling velocity of a particle is directly proportional to the square of its radius. Therefore, among particles of equal density, the larger ones would settle with greater velocity. Molloy et al. (1984) demonstrated Stokes's law to be valid with *B. t. i.* formulations. Their data indicated that the settling rates of formulations in still water corresponded directly with their particle size; those formulations with the largest particles had the fastest settling rates. Lacey and Heitzman (1985) have suggested that, in addition to reduced carry due to settling, formulations with large particles may have reduced carry due to the increased likelihood of filtration through contact with benthic substrates.

10.4.3.3 Formulation Additives. Ingredients are commonly added to *B. t. i.* products to improve field effectiveness. For example, surfactants are included to increase miscibility, thereby accelerating the rate of spreading and mixing once a product is applied to water. Not all additives, however, improve efficacy. Guillet, Escaffre, et al. (1985a) reported that formulation efficacy was adversely affected by a cement used to prevent disintegration of clumps of *B. t. i.* particles. Similarly, Molloy, Gaugler, and Jamnback (1981) indicated that oil-like droplets in an emulsified *B. t. i.* formulation disrupted feeding behavior.

10.4.4 Treatment Parameters

10.4.4.1 Concentration. As would be expected, all tests to date have indicated a positive correlation between percentage mortality and *B. t. i.* concentration.

10.4.4.2 Duration of Application. Guillet, Escaffre, et al. (1985b) indicated that, for a given quantity of *B.t.i.*, the efficacy of a formulation with individual spores and crystals (Teknar®) was not related to exposure time, but that the efficacy of formulations with large particles was enhanced by long exposures at low concentrations. Recent tests conducted by others to examine the effect of application time on efficacy have been less conclusive (Lacey and Undeen 1984; Nakamura et al. 1985). The question of what exposure period is of optimal efficacy is of reduced importance when one realizes that, except for the farthest upstream application, *B.t.i.* slugs produced by individual treatments overlap each other during their downstream transport. Regardless of the optimal exposure period, however, long application periods at lentic outflows and at initial upstream treatment points are recommended to maximize mixing of the *B.t.i.* in the current.

10.4.4.3 Preparation of Powered Formulations. To prepare powdered *B.t.i.* formulations for experimentation, high-speed blenders have occasionally been used to produce aqueous suspensions (Molloy and Jamnback 1981; Car 1984). Extensive investigations by Guillet and Duval (1985), however, have demonstrated that the use of a blender to prepare aqueous *B.t.i.* stock suspensions can seriously reduce toxicity. Their findings are in agreement with the observation of Molloy (pers. comm. in Undeen and Lacey 1982), who reported the detoxification of a powdered *B.t.i.* formulation following blending. The results of Guillet and Duval (1985) demonstrate the necessity of avoiding any mixing technique that could have a strong mechanical impact on *B.t.i.*'s crystals. They suggest that the main factor responsible for detoxification might be the activation of proteases present in the *B.t.i.* formulations and their subsequent degradation of toxin(s). Thus, the authors recommend that the activity of proteases be taken into account in the preparation of commercial *B.t.i.* preparations.

Due to the decline in field experimentation with *B.t.i.* powdered formulations, the need for aqueous suspensions of powders has been reduced to small-scale laboratory investigations. In such cases, the suspension methods of Lacey, Undeen, and Chance (1982) are recommended.

10.5 LACK OF CORRELATION IN FORMULATION POTENCY AGAINST BLACK FLIES AND MOSQUITOES

Molloy et al. (1984) found that the potency ratings (International Toxicity Units/mg, or ITU/mg) of *B.t.i.* formulations against mosquitoes and against black flies did not correlate well ($r^2 = 0.47$). This indicated that the potency of a formulation against black flies could not be simply predicted from the *Aedes aegypti* ITU/mg rating on the label. A similar lack of correlation has been observed by others (Guillet, Escaffre, and Prud'hom 1982a; Lacey and Heitzman 1985). In recent laboratory trials,

Guillet, Hougard, et al. (1985c) have demonstrated that the potency of a formulation against black flies and mosquitoes did begin to correlate when black fly exposure periods were increased well beyond their normal 10-minute laboratory exposure; following exposures of 10 minutes, 3 hours, and 24 hours, r^2 values of 0.20, 0.72, and 0.92 were achieved, respectively. Guillet, Hougard, et al. (1985c) suggested that the changes in correlation values were related to differences in toxin activation in the gut.

10.6 EFFECT OF *B.t.i.* ON BLACK FLY POPULATIONS

Recent field trials to assess the efficacy and carry of *B.t.i.* formulations in temperate climates have been summarized in table 10.1. In general, applications at dosages equivalent to 5–30 ppm/1 min achieved good carry for at least 50–250 m in moderate-sized streams (1,000–20,000 liters/min). The relatively early commercial availability of the liquid formulation Teknar® in 1981 (Knutti and Beck 1987) contributed substantially to its frequent listing in table 10.1.

These recent field trials contributed significantly to the rapid progress made with *B.t.i.* by identifying a number of commercial formulations (e.g., Teknar®, Vectobac®), that are now proving effective in operational programs in temperate regions; and by revealing, or further substantiating, which biotic and abiotic parameters relate to field effectiveness. In addition, insight into other aspects of the interaction of *B.t.i.* and black flies has been gained from these trials. For example, it is clear that the effect of *B.t.i.* treatments on black flies is rapid and dramatic. When high black fly mortality is achieved (>95%), almost all mortality occurs within 24 hours, and benthic substrates are littered with immobile, flaccid, larval carcasses, which are detached rather easily by the current as they putrify. Thus, a large increase in black fly larval drift (cadavers) typically occurs within hours after treatment. For example, in a 2.3 ppm/7 min application in a South African river, black fly drift increased more than 60-fold within 43 minutes after treatment (Car and de Moor 1984). Likewise, a maximum increase in black fly drift of 81-fold was recorded in a 2.0 ppm/10 min Teknar® application in the Ivory Coast (Dejoux, Gibon, and Yameogo 1985). In a high-dosage trial in a Quebec lake outlet with Teknar® at 5.28 ppm/15 min, simuliid drift was 64–92 times greater than pretreatment levels (Back et al. 1985). In this Quebec trial, drift and counting-plate data also verified *B.t.i.*'s rapid action; most detachment or mortality occurred within the first 3 hours after treatment.

10.7 EFFECT OF *B.t.i.* ON NONTARGET POPULATIONS

10.7.1 Toxicity

In laboratory toxicity tests against brook trout fry using Teknar® at levels over 1,000 times greater than operational control dosages (i.e.,

3,000–6,000 ppm/45 min), the only mortality noted was attributed to formulation ingredients, not to the *B.t.i.* itself (Fortin, Lapointe, and Charpentier 1986); the authors concluded that Teknar®, when used at normal application rates, was nontoxic to brook trout fry.

Recent field trials have further substantiated the unusually high selectivity of *B.t.i.* among lotic invertebrates (table 10.2). Among the wide range of groups investigated, *B.t.i.* has been demonstrated to be toxic to only a small group of insects, almost all of which are flies in the superfamily Culicoidea (Diptera: Nematocera). As in mosquito habitats, chironomid larvae are the most likely group of nontargets to suffer some mortality (Car and de Moor 1984; Pistrang and Burger 1984; Back et al. 1985; de Moor and Car 1986). For example, in the nontarget investigations of Rutschke and Grunewald (1984), only chironomids were observed to be killed in a stream trial using a dosage 17 times higher than that used for simuliid control. Trials in New York streams have found that filter-feeding chironomids are susceptible, but at levels far below that of black flies (Molloy, unpub. data). The high-dosage Teknar® trial of Back et al. (1985) at 5.28 ppm/15 min (a dosage approximately 3–15 times greater than operational levels) demonstrated that *B.t.i.*. could also be lethal to blephariceri‍d larvae (visual estimate of 30% mortality). Back et al. (1985) judged that the nontarget impact of their treatment was against chironomids and blephariceri‍ds exposed to *B.t.i.* present on the bottom of the stream or attached to periphyton growing on rocks. They concluded that, considering the high dosage used, Teknar® was a highly selective larvicide. Similarly, the results of extensive nontarget assessments in tropical rivers (Dejoux, Gibon, and Yameogo 1985) indicated that Teknar® had a nearly total short-term innocuity and that repeated applications had no deleterious effect on nontarget populations over a nine-week period.

10.7.2 Inducement of Drift

Sampling by Dejoux, Gibon, and Yameogo (1985) in a West African river treated at normal operational levels with Teknar® (1.6 ppm/10 min) consistently revealed increases (14.5–19.7%) in the overall benthic density of nontargets. They interpreted these posttreatment benthic increases to be the result of the reattachment of nontargets that Teknar® had induced to drift from areas slightly upstream, where the nontargets had been exposed to a more concentrated slug of *B.t.i.* Thus, they hypothesized that increases in nontarget benthic densities occurred because the number of drifting insects that had colonized their sampling substrates exceeded the number of insects that had detached. In a special series of applications of Teknar® to a previously untreated river, Dejoux, Gibon, and Yameogo (1985) recorded a 4- to 5-fold increase in invertebrate drift. Such increases in drift, however, were considered very low compared to the 50- to

TABLE 10.2.
Summary of Effects on Nontarget Invertebrates during Recent Field Trials Using *B.t.i.*

Principal invertebrate groups studied	Product/ formulation	Application Rate		Equivalent conc. (ppm/1 min)	Water temp. (° C)	Location	Results	Source
		Conc. (ppm)	Duration (min)					
Ephemeroptera, Trichoptera, Plecoptera, Coleoptera, Diptera (Dixidae, Tipulidae, Chironomidae), and Gastropoda	Teknar®	2.0	15.0	30.0	19	New Zealand	No adverse effects noted	(1)
Ephemeroptera, Trichoptera, Plecoptera, Coleoptera, Diptera (Dixidae, Tipulidae, Chironomidae), and Gastropoda	R153–78 primary powder	0.2	15.0	3.0	10.5	New Zealand	No adverse effects noted	(1)
Ephemeroptera, Plecoptera, Trichoptera, Diptera (Chironomidae), Gastropoda, Platyhelminthes, mites, and Oligochaeta	Teknar®	1.6	10.0	16.0	24	South Africa	No adverse effects conclusively demonstrated	(2)
Ephemeroptera, Plecoptera, Trichoptera, Diptera (Chironomidae), Gastropoda, Platyhelminthes, mites, and Oligochaeta	Teknar®	2.3	7.0	16.1	28	South Africa	No adverse effects noted except for decrease in tanytarsine chironimds	(2)
Ephemeroptera, Plecoptera, Trichoptera, and Diptera (Chironomidae)	Teknar®	10.0	2.0	20.0	9–16	USA	Some chironomid mortality noted; temporary increases in drift of Ephemeroptera and Trichoptera observed	(3)
Collembola, Odonata, Megaloptera, Coleoptera (Elmidae, Gyrinidae, Psephenidae), Ephemeroptera, Plecoptera, Trichoptera, and Diptera (Chaoboridae, Ceratopogonidae, Empididae, Tipulidae, Psychodidae, Chironomidae, Blephariceridae)	Teknar®	5.86	15.0	87.9	11–15	Canada	No evident drift increase in taxa except for Blephariceridae, which were severely affected (drift increased 50X and mortality estimated at 30%); densities of nontargets were not reduced on artificial substrates except for 2 genera of Chironomidae (Eu-	(4)

							kiefferella and *Polypedilum* declined 26–39%; larvae of the chironomid genus *Phaenopsectra* showed signs of toxemia)	
Ephemeroptera (Baetidae, Tricorythidae, Caenidae), Odonata (Libellulidae), Coleoptera (Elmidae), Lepidoptera (Pyralidae), Trichoptera (Hydropsychidae, Leptoceridae), Diptera (Chironomidae, Ceratopogonidae, Rhagionidae), Oligochaeta, mites, and mollusks (Ancylidae)	R153-78 primary powder	0.2	10.0	2.0	—	Ivory Coast	Increase in drift of Chironomini, Trichoptera, and Ancylidae noted, but no evidence of mortality	(5)
Trichoptera (Hydropsychidae, Hydroptilidae, Philopotamidae, Leptoceridae), Diptera (Chironomidae, Rhagionidae, Ceratopogonidae), Ephemeroptera (Baetidae, Caenidae), and Lepidoptera (Pyralidae)	Teknar®	1.6	10.0	16.0	—	Ivory Coast	Increase in nontarget drift and increase in nontarget density on substrates suggested reattachment of insects that were drifting in response to the treatment	(5)
Ephemeroptera (Baetidae, Caenidae), Trichoptera (Hydropsychidae, Hydroptilidae, Philopotamidae), Diptera (Chironomidae), and Odonata (Libellulidae, Zygoptera)	Teknar®	1.6	10.0	16.0	—	Ivory Coast	After 9 consecutive weekly treatments, an increase in orthoclad chironomids and a decrease in hydropsychid caddisflies were recorded	(5)
Ephemeroptera, Diptera (Chironomidae), Trichoptera, and Platyhelminthes	Teknar®	1.6	10.0	16.0	17–20	South Africa	No adverse effects noted except for decrease in tanytarsine chironomids	(6)
Ephemeroptera (Ephemerellidae, Ephemeridae, Heptageniidae, Siphlonuridae), Plecoptera (Perlodidae), Trichoptera (Lepidostomatidae), Diptera (Blepharoceridae, Athericidae, Chironomidae), Coleoptera (Elmidae), Acarina	Vectobac® AS	10.0	1–5	10–50	9–12	USA	No evidence of impact on nontargets in either substrate or drift samples	(7)

SOURCES: (1) Chilcott, Pillai, and Kalmakoff 1983; (2) Car and de Moor 1984; (3) Pistrang and Burger 1984; (4) Back et al. 1985; (5) Dejoux, Gibon, and Yameogo 1985; (6) de Moor and Car 1986; (7) Gibbs et al. 1986.

200-fold increases in drift that typically occur after organophosphorus insecticide treatments.

Dejoux, Gibon, and Yameogo (1985) indicated that the nontarget drift they observed could well have been induced by formulation ingredients (e.g., dispersants), rather than by *B.t.i.* For example, minigutter tests that they conducted in West Africa with an unformulated *B.t.i.* product, R153-78 primary powder, produced no evidence of an increase in the overall intensity of nontarget drift. In the final analysis, however, they concluded that the inducement of drift by Teknar®, since it was relatively low and nonlethal, was likely to have no serious effect on the nontarget population balance in West African rivers.

Analysis of posttreatment drift in temperate climates has been limited. Pistrang and Burger (1984) provided evidence of temporary increases in the drift of mayflies and caddisflies following treatment of a stream with Teknar®. In contrast, however, no increases in nontarget drift, except for those nematocerous flies subjected to intoxication (discussed above), were observed following the high-dosage Teknar® application of Back et al. (1985). Similarly, Gibbs et al. (1986) and Molloy (unpub. data) in their field trials with Vectobac® AS observed no increases in the drift of nontargets.

10.8 TREATMENT AND ASSESSMENT METHODOLOGIES

10.8.1 Discharge Calculation

In order to determine the quantity of a *B.t.i.* formulation required to achieve a particular concentration in stream water, a discharge calculation is normally required. Amrine (1983) presented a calculator program for computing stream discharge from a float method. In New York, a modified float method has been devised (Molloy and Struble 1988) and is currently used as a standard for discharge calculation. A variety of other practical and accurate methods will no doubt be employed in other operational programs throughout temperate climates.

Undeen, Lacey, and Avery (1984), however, have suggested that stream width, not discharge, be the key factor used to determine the quantity of *B.t.i.* to add during an application. Their data suggest that a protocol based on stream width would be at least as good a criterion as discharge in predicting downstream carry since it has the advantage of being self-correcting for variability in stream morphology. As Undeen, Lacey, and Avery (1984) point out, a treatment protocol based on stream width is not really such a radical idea since it would be analogous to the treatment of a mosquito habitat on the basis of its surface area rather than its volume. It is true that operational use of the stream-width protocol would result in some overdosing when, for example, a wide stream with an unusually low discharge was treated. Adoption of this protocol, however, is a distinct possibility for the future, especially if nontarget field data continue to verify the relative innocuity of *B.t.i.* even in such overdose situations.

10.8.2 Treatment

Programs in New York (Molloy and Struble 1989) and elsewhere are demonstrating that streambank application of flowable concentrate formulations is a relatively simple procedure involving the manual broadcasting of *B.t.i.* across the stream. The practicality of aerial application, the standard method for treating rivers in the West African Onchocerciasis Control Programme, is currently being evaluated for use in temperate regions (G. Jones and M. Colbo, pers. comm.).

10.8.3 Assessments of Black Fly Mortality

In research programs, an accurate estimation of larval mortality requires a considerable effort, including determination of larval densities both pretreatment and posttreatment. In actual operational control programs, limited resources do not allow such a precise method to be used, and such highly quantitative methods are really unnecessary. In New York operational programs, for example, 95% is the minimum acceptable larval mortality, and simple posttreatment determination of percentage mortality among attached larvae has proven to be a practical, economical, and sufficiently accurate method of assessment. Due to detachment of black flies, this method almost invariably will underestimate the true percentage mortality (Back et al. 1985). Yet, because of the inherent underestimation of the technique, one can be confident that a 95% mortality reading among attached larvae really means that the true mortality was ≥95%.

10.9 PRODUCT/FORMULATION IMPROVEMENT

An examination of table 10.1 reveals that a number of effective commercial products now exist for use against temperate-climate black flies. In addition to the efficacy demonstrated by these formulations, the long-term storage stability (Guillet, Escaffre, and Prud'hom 1982c) and uniform high quality of mass-produced commercial products (Guillet, Doannio, et al. 1985) have also been demonstrated. Guillet, Doannio, et al. (1985), for example, indicated that the Teknar® shipped for use in the Onchocerciasis Control Programme during 1982–1983 (about 600,000 liters) had uniform toxicity, efficacy, and physical characteristics. However, interest in the evaluation of new commercial products/formulations remains, and improvements are still desired in a number of the characteristics of *B. t. i.* formulations, as discussed below.

B. t. i. provides shorter carry than chemical insecticides, thus requiring greater effort in treatment. The relatively shorter carry of *B. t. i.* appears to be a problem inherent with particulate, insoluble insecticides, but this should not discourage formulation studies to maximize *B. t. i.* carry. *B. t. i.* formulations are also considerably bulkier than chemical insecticides. Thus, further concentration of the endotoxin is needed. This problem is currently being addressed through genetic engineering studies. The problems of *B. t. i.*'s bulk and short carry have been acutely felt in the Onchocerciasis Control Programme. The quantities of Teknar® needed for applications in the wet season have been 3–20 times that of conventional insecticides (Kurtak 1986).

Although the presence of *B. t. i.* spores in commercial products has no bearing on efficacy, environmental concerns over possible recycling of *B. t. i.* (Dupont and Boisvert 1986) have encouraged industry to produce spore-free formulations.

Because of inadequate mixing in running water, *B. t. i.* formulations typically have required some dilution with water before application in order to increase effectiveness (Guillet, Escaffre, and Prud'hom 1982b), thereby further compounding the bulk problem. The need to mix the product with water prior to application could be avoided with the development of formulations with better miscibility and self-dispersion. Recent field trials have indicated that such advancements are being made and that commercial formulations may not necessarily require dilution in either manual (Lacey and Heitzman 1985) or aerial (G. Jones, pers. comm.) application.

10.10 AREA-CONTROL PROGRAMS

Although numerous single-application field trials have verified the efficacy of *B. t. i.* as a black fly larvicide (table 10.1), relatively few studies have been conducted in temperate climates to demonstrate the feasibility of using *B. t. i.* in large-scale area-control programs. These pilot studies, however, have had very encouraging results.

Reduction in the number of adult black flies has been demonstrated in both North American and European studies. In the initial temperate-climate study, at least a 70% seasonal reduction of adult flies was achieved in a Newfoundland watershed (Colbo and O'Brien 1984). Similar reductions in adult black flies have been achieved in several other North American investigations, including large-scale studies in Labrador (Colbo 1984), South Carolina

(Horosko and Noblet 1986), New York (Molloy and Struble 1989), and New Hampshire (J. Burger, pers. comm.). Similarly, *B.t.i.* treatment of a 4-km section of river in Germany was reported to have reduced the biting density from several thousand to 50–100 bites/cow (Rutschke and Grunewald 1984).

Besides demonstrating the technical feasibility of area control of black flies using *B.t.i.*, these pioneering studies also provided data on the economics of such programs. Colbo and O'Brien (1984) reported that labor, not the cost of the *B.t.i.* product, was one of the largest expenditures in an area-control program. This was confirmed in an area-control trial in the Adirondack Mountains of New York in which product cost was only about 5% of the total budget (Molloy and Struble 1989). The latter study also provided evidence that ground application programs using *B.t.i.* were actually less expensive than aerial application programs using conventional chemical insecticides. Colbo (1984), following extensive large-scale trials in Labrador, concluded that the cost of area-control *B.t.i.* programs was likely to be within the reach of many of the larger communities of northern Canada. Further information on the economic feasibility of area control using *B.t.i.* should also be forthcoming from the large-scale *B.t.i.* programs that are in progress in Pennsylvania (G. Jones, pers. comm.), Minnesota (Simmons and Sjogren 1984), New Hampshire (J. Burger, pers. comm.), and Maine (D. Hutchins, pers. comm.). The use of *B.t.i.* has been operational in New York since 1984, and the total seasonal cost of such programs is as low as $100/sq km (T. Martin, pers. comm.). Such a low cost, however, is only likely for towns that lack extensive stream/river systems and have several years of experience in operating their own *B.t.i.* programs. A cost of $400/sq km is more typical for New York communities, particularly in their first year of operational control (Molloy and Struble 1989).

In contrast to the attractive economics of *B.t.i.* use for area control in temperate climates, the operational use of *B.t.i.* in the Onchocerciasis Control Programme has been more expensive than use of conventional chemical insecticides, due primarily to the problems of *B.t.i.*'s inferior carry and greater bulk (Kurtak 1986). Taking into account the additional flying time and ground transport of insecticide, the use of *B.t.i.* is several times more expensive than temephos (Abate®) even at moderate river dosages. Only at very low discharge, where carry is similar, is *B.t.i.* more economical than temephos (World Health Organization 1989).

In summary, although area-control trials in temperate areas have been few in number, they have created a solid framework for commencement of operational programs. In New York, for example, some 10 towns with a total area of about 1,000 sq km were treated in 1989. An additional indication of the acceptance of *B.t.i.* as a practical and effective black fly control method in New York State is the availability of one-week, state-approved courses specifically designed to train technicians for ground-applied *B.t.i.* programs.

10.11 POSSIBILITY OF RESISTANCE

B.t.i. has been demonstrated to lack cross-resistance with chemical insecticides (Sun, Georghiou, and Weiss 1980; World Health Organization 1984), thus allowing it to be an effective replacement larvicide in areas where black fly populations have become resistant to chemical insecticides. The question remains, however, as to the likelihood of black flies developing resistance to *B.t.i.* per se. It has recently been demonstrated that *Bacillus thuringiensis* (probably subsp. *kurstaki*) can produce resistance in populations of *Plodia interpunctella*, a lepidopteran pest of stored grain and grain products (McGaughey 1985). Selection pressure in such grain bins, however, is ideal for the development of resistance. Use of a *B. thuringiensis* product in such a confined habitat is quite different from the typical use of *B.t.i.* in communities in temperate regions where there would be a constant influx of genetic material from nearby untreated black fly populations. Thus, the development of black fly resistance to *B.t.i.* in temperate climates has a relatively lower probability.

Because of prior experience with the development of resistance to operationally used larvicides (Kurtak 1986), a standard methodology for the monitoring of black fly resistance to *B.t.i.* has now been developed in the Onchocerciasis Control Programme (Guillet, Hougard, et al. 1985a, 1985b). This method was adapted from the standard laboratory testing methods suggested by Lacey, Undeen, and Chance (1982) and was specifically tailored for use in the field to monitor the susceptibility of *Simulium damnosum*. If resistance is to develop, it is far more likely to occur where *B.t.i.* is used year-round in an extensive geographic region. Thus, it is of value to follow the progress of these West African tests to observe the degree to which resistance can be detected.

10.12 RESEARCH PRIORITIES IN TEMPERATE CLIMATES

Because black flies are not known to be involved in the transmission of human disease in temperate regions, far greater demands are traditionally put on the environmental acceptability of control agents by governmental regulatory agencies. As a result, such agencies usually are very interested in reviewing available efficacy and nontarget data before granting permission for local use of a black fly control agent. Even though all research to date has indicated that *B.t.i.* is an effective control agent with unparalleled selectivity, it is not uncommon for regulatory agencies to request field trials in their own regions to further substantiate the effects of *B.t.i.* use. As interest in the use of *B.t.i.* as a black fly larvicide becomes more widespread in temperate regions, more short-term efficacy/nontarget studies are likely to be generated.

Considerable data are accumulating on the short-term environmental effects of *B.t.i.* application to lotic habitats, but no information is yet available on long-term effects. Since black fly habitats are so diverse (spring-fed rivulets, large rivers, and so on), the variety of ecosystems available as potential study sites is enormous. Unfortunately, the low availability of funding for such long-term studies is in direct contrast to the abundance of research opportunities. Thus, few such studies are likely to occur. Primary among the long-term concerns of regulatory agencies is the potential for an adverse impact on predators due to the removal of black flies as a prey source. Our laboratory is currently investigating such long-term effects of *B.t.i.* application by treating a third-order trout stream in New York year-round for two years (Molloy 1984). The data from this study have reconfirmed the high selectivity of *B.t.i.* (Molloy, unpub. data), and analysis of these data in regard to the ecological importance of black flies should be completed shortly.

Only a few area-control projects have been conducted to date in temperate regions. More large-scale projects are needed in a variety of terrains/watersheds to further define the technical and economic feasibility of operational use of *B.t.i.* as a black fly control agent. Questions relevant to any black fly larvicidal program should also be addressed in such large-scale programs, including the following: (1) the cost/benefit ratio of treating all black fly habitats versus just the major breeding sites of pest species; (2) the distance from the community requiring protection that the larviciding treatments must be extended (i.e., the size of the larviciding buffer zone); and (3) the most effective and economical methods of treatment (e.g., ground versus aerial).

10.13 CONCLUSIONS

Considering that the original isolation of *B.t.i.* occurred only a decade ago, the progress made in the development of this bacterium as a black fly control agent has been truly extraordinary. Its rapid transition from the laboratory to operational use has not occurred by mere chance, but rather is the result of an intensive international research effort (funded in large part by the UNDP/World Bank/WHO Special Programme for Research and Training in Tropical Diseases). As outlined below, this effort has demonstrated that *B.t.i.* has a number of characteristics considered highly desirable in a black fly control agent:

1. Efficacy—*B.t.i.*'s endotoxin is effective against all instars, against all species thus far tested, and under a wide variety of environmental conditions.

2. Selectivity—In contrast to broad-spectrum chemical insecticides, very few nontarget insects are directly affected; *B. t. i.*'s specificity as a black fly larvicide is unparalleled.
3. Lack of cross-resistance—*B. t. i.* has been demonstrated to be a useful and effective replacement insecticide in areas where black flies have become resistant to chemical insecticides.
4. Commercial availability—Industry's prior experience with other varieties of ***Bacillus thuringiensis*** has greatly facilitated the production of effective, stable, and economical *B. t. i.* formulations.

These characteristics, in combination with the recent development of practical and economical protocols for its operational use in temperate climates, have made *B. t. i.* an attractive alternative to broad-spectrum chemical insecticides. Although operational use of *B. t. i.* began in 1982 in the tropics (in the Onchocerciasis Control Programme), its use against black flies in temperate climates is only beginning to be exploited. The next decade will certainly witness a steady and widespread expansion of the role that *B. t. i.* will play in black fly population management. In their review paper several years ago, Gaugler and Finney (1982) concluded that *B. t. i.* had high potential as a black fly control agent. That potential is now beginning to be realized in temperate regions.

A final note. The successful development and use of *B. t. i.* as a black fly control agent has to some extent discouraged the search for, and experimentation on, other potentially useful biological control agents of black flies. Experience dictates that such an approach to pest control research is shortsighted and, in the final analysis, counterproductive toward guaranteeing successful, long-term, ecologically sound control of black flies. Discovery and development of an agent like *B. t. i.* is truly a rare event in the history of biological control. Unless a similar intensity of research is maintained, such an event is unlikely to occur again soon.

Acknowledgments

Special thanks to Lise Molloy, Bob Struble, and Barbara Griffin for assistance in producing this paper. Long-term financial support for *B. t. i.*/black fly studies from the UNDP/World Bank/WHO Special Programme for Research and Training in Tropical Diseases and from the National Institutes of Health is gratefully acknowledged. Contribution No. 499 of the New York State Science Service.

References

Amrine, J. W., Jr. 1983. Measuring stream discharge and calculating treatment of rates of *Bacillus thuringiensis* (H14) for black fly control. *Mosq. News* 43: 17–19.

Back, C.; Boisvert, J.; Lacoursière, J. O.; and Charpentier, G. 1985. High-dosage treatment of a Quebec stream with *Bacillus thuringiensis* serovar. *israelensis*: Efficacy against black fly larvae (Diptera: Simuliidae) and impact on non-target insects. *Can. Entomol.* 117: 1523–1534.

Barnett, B. D. 1977. Leucocytozoon disease of turkeys. *World's Poult. Sci. J.* 33: 76–87.

Car, M. 1984. Laboratory and field trials with two *Bacillus thuringiensis* var. *israelensis* products for *Simulium* (Diptera: Nematocera) control in a small polluted river in South Africa. *Onderstepoort J. Vet. Res.* 51: 141–144.

Car, M., and de Moor, F. C. 1984. The response of Vaal River drift and benthos to *Simulium* (Diptera: Nematocera) control using *Bacillus thuringiensis* var. *israelensis* (H-14). *Onderstepoort J. Vet. Res.* 51: 155–160.

Chilcott, C. N.; Pillai, J. S.; and Kalmakoff, J. 1983. Efficacy of *Bacillus thuringiensis* var. *israelensis* as a biocontrol agent against larvae of Simuliidae (Diptera) in New Zealand. *N.Z. J. Zool.* 10: 319–326.

Colbo, M. H. 1984. Control of black flies (Simuliidae) using *Bacillus thuringiensis* var. *israelensis* (Bti) as a larvicide, with emphasis on the northern programs. In *Proceedings of the Thirty-first annual meeting, Canadian Pest Management Society,* Winnipeg, Manitoba, 98–109.

Colbo, M. H., and O'Brien, H. 1984. A pilot black fly (Diptera: Simuliidae) control program using *Bacillus thuringiensis* var. *israelensis* in Newfoundland. *Can. Entomol.* 116: 1085–1096.

Dejoux, C.; Gibon, F. M.; and Yameogo, L. 1985. Toxicité pour la faune non-cible de quelques insecticides nouveaux utilisés en milieu aquatique tropical. Part 4, Le *Bacillus thuringiensis* var. *israelensis* H-14. *Rev. Hydrobiol. Trop.* (1) 18: 31–49.

de Moor, F. C., and Car, M. 1986. A field evaluation of *Bacillus thuringiensis* var. *israelensis* as a biological control agent for *Simulium chutteri* Lewis 1965 (Diptera: Nematocera) in the Middle Orange River. *Onderstepoort J. Vet. Res.* 53: 43–50.

Dupont, C., and Boisvert, J. 1986. Persistence of *Bacillus thuringiensis* serovar. *israelensis* toxic activity in the environment and interaction with natural substrates. *Water Air Soil Pollut.* 29: 425–438.

Fortin, C.; Lapointe, D.; and Charpentier, G. 1986. Susceptibility of brook trout (*Salvelinus fontinalis*) fry to a liquid formulation of *Bacillus thuringiensis* serovar. *israelensis* (Teknar®) used for blackfly control. *Can. J. Aquat. Sci.* 43: 1667–1670.

Fredeen, F.J.H. 1985. Some economic effects of outbreaks of black flies (*Simulium luggeri* Nicholson and Mickel) in Saskatchewan. *Quaest. Entomol.* 21: 175–208.

Frommer, R. L.; Hembree, S. C.; Nelson, J. H.; Remington, M. P.; and Gibbs, P. H. 1981. The evaluation of *Bacillus thuringiensis* var. *israelensis* in reducing *Simulium vittatum* (Diptera: Simuliidae) larvae in their natural habitat with no extensive aquatic vegetative growth. *Mosq. News* 41: 339–347.

Gaugler, R., and Finney, J. R. 1982. A review of *Bacillus thuringiensis* var. *israelensis* (serotype 14) as a biological control agent of black flies (Simuliidae). In *Biological control of black flies (Diptera: Simuliidae) with* Bacillus thuringiensis *var.* israelensis *(Serotype 14): A review with recommendations for laboratory and field protocol,* ed. D. Molloy, 1–17. Misc. Publ. Entomol. Soc. Am. 12, no. 4.

Gaugler, R.; Kaplan, B.; Alvarado, C.; Montoya, J.; and Ortega, M. 1983. Assessment of *Bacillus thuringiensis* serotype 14 and *Steinernema feltiae* (Nematoda: Steinernematidae) for control of the *Simulium* vectors of onchocerciasis in Mexico. *Entomophaga* 28: 309–315.

Gaugler, R., and Molloy, D. 1980. Feeding inhibition in black fly larvae (Diptera: Simuliidae) and its effects on the pathogenicity of *Bacillus thuringiensis* var. *israelensis*. *Environ. Entomol.* 9: 704–708.

Gibbs, K. E.; Brautigam, F. C.; Stubbs, C. S.; and Zibilske, L. M. 1986. *Experimental applications of B.t.i. for larval black fly control: Persistence and downstream carry, efficacy, impact on non-target invertebrates and fish feeding.* Maine Life Sci. Agric. Exp. Stn. Tech. Bull. 123.

Goldberg, L. J., and Margalit, J. 1977. A bacterial spore demonstrating rapid larvicidal activity against *Anopheles sergentii, Uranotaenia unguiculata, Culex univitattus, Aedes aegypti,* and *Culex pipiens. Mosq. News* 37: 355–358.

Guillet, P.; Doannio, J.; Hougard, J. M.; Escaffre, H.; and Duval, J. 1985. Production de masse d'un agent de lutte biologique, le *Bacillus thuringiensis* H14, pour la lutte contre les vecteurs de l'onchocercose en Afrique de l'Ouest. In *La Lutte contre les Vecteurs de l'Onchocercose en Afrique de l'Ouest: Etude de la Résistance et Recherche de Nouveaux Larvicides,* ed. P. Guillet, 339–350. Thèse Doctorat es Sciences Naturelles, Université Paris Sud-Orsay.

Guillet, P., and Duval, J. 1985. The effect of blending on the toxicity of *Bacillus thuringiensis* serotype H14 preparations against mosquito and blackfly larvae. In *La Lutte contre les Vecteurs de l'Onchocercose en Afrique de l'Ouest: Etude de la Résistance et Recherche de Nouveaux Larvicides,* ed. P. Guillet, 389–399. Thèse Doctorat es Sciences Naturelles, Uni versité Paris Sud-Orsay.

Guillet, P., and Escaffre, H. 1979. Evaluation de *Bacillus thuringiensis israelensis* de Barjac pour la lutte contre les larves de *Simulium damnosum* s.l. Part 2, *Efficacité comparée de trois formulations expérimentales.* WHO/VBC/79.735. Mimeo.

Guillet, P.; Escaffre, H.; and Prud'hom, J. M. 1982a. *Bacillus thuringiensis* H14: A biocontrol agent for onchocerciasis control in West Africa. In *Invertebrate pathology and microbial control,* 460–465. Proceedings of the Third International Colloquium on Invertebrate Pathology. Univ. Sussex, Brighton, United Kingdom.

———. 1982b. L'utilisation d'une formulation à base de *Bacillus thuringiensis* H14 dans la lutte contre l'onchocercose en Afrique de l'Ouest. Part 1, Efficacité et modalités d'application. *Cah. ORSTOM, sér. Ent. méd. et Parasitol.* 20 (3): 175–180.

———. 1982c. L'utilisation d'une formulation à base de *Bacillus thuringiensis* H14 dans la lutte contre l'onchocercose en Afrique de l'Ouest. Part 2, Stabilité dans les conditions de stockage en milieu tropical. *Cah. ORSTOM, sér. Ent. méd. et Parasitol.* 20 (3): 181–185.

Guillet, P.; Escaffre, H.; Prud'hom, J. M.; and Bakayoko, S. 1985a. Etude des facteurs conditionnant l'efficacité des préparations à base de *Bacillus thuringiensis* H14 vis-à-vis des larves du complexe *Simulium damnosum* (Diptera: Simuliidae). Part 1, Influence de la nature et de la taille des particules. *Cah. ORSTOM, Sér. Ent. méd. et Parasitol.* 23 (4): 257–264.

———. 1985b. Etude des facteurs conditionnant l'efficacité des préparations à base de *Bacillus thuringiensis* H14 vis-à-vis des larves du complexe *Simulium damnosum* (Diptera: Simuliidae). Part 2, Influence du temps de contact et de la quantité de particules naturelles en suspension dans l'eau. *Cah. ORSTOM, sér. Ent. méd. et Parasitol.* 23 (4): 265–271.

Guillet, P.; Hougard, J. M.; Doannio, J.; Escaffre, H.; and Duval, J. 1985a. Evaluation de la sensibilité des larves du complexe *Simulium damnosum* à la toxine du *Bacillus thuringiensis* H14. Part 1, Méthodologie. *Cah. ORSTOM, sér. Ent. méd. et Parasitol.* 23 (4): 241–250.

———. 1985b. Evaluation de la sensibilité des larves du complexe *Simulium damnosum* à la toxine du *Bacillus thuringiensis* H14. Part 2, Sensibilité relative de quelques groupes d'espèces et possibilités d'utilisation de doses diagnostiques. *Cah. ORSTOM, sér. Ent. méd. et Parasitol.* 23 (4): 251–255.

———. 1985c. Toxicité comparée du *Bacillus thuringiensis* H14 pour les larves de moustiques et de simulies. In *La Lutte contre les Vecteurs de l'Onchocercose en Afrique de l'Ouest: Etude de la Résistance et Recherche de Nouveaux Larvicides,* ed. P. Guillet, 373–385. Thèse Doctorat es Sciences Naturelles, Université Paris Sud-Orsay.

Habib, M.E.M. 1983. Potency of *Bacillus thuringiensis* var. *israelensis* (H-14) against some aquatic dipterous insects. *Z. Ang. Entomol.* 95: 368–376.

Horosko, S., III, and Noblet, R. 1983. Efficacy of *Bacillus thuringiensis* var. *israelensis* for control of black fly larvae in South Carolina. *J. Ga. Entomol. Soc.* 18: 531–537.

———. 1986. Black fly control and suppression of leucocytozoonosis in turkeys. *J. Agric. Entomol.* 3: 10–24.

Knutti, H. J., and Beck, W. R. 1987. The control of black fly larvae with Teknar®. In *Black flies: ecology, population management, and annotated world list,* ed. K. C. Kim and R. W. Merritt, 409–418. University Park, Penn.: Pennsylvania State University.

Kurtak, D. C. 1986. Insecticide resistance in the Onchocerciasis Control Programme. *Parasitol. Today* 2 (1): 20–21.

Lacey, L. A., and Heitzman, C. M. 1985. Efficacy of flowable concentrate formulations of *Bacillus thuringiensis* var. *israelensis* against black flies (Diptera: Simuliidae). *J. Amer. Mosq. Control Assoc.* 1: 493–497.

Lacey, L. A.; Mulla, M. S.; and Dulmage, H. T. 1978. Some factors affecting the pathogenicity of *Bacillus thuringiensis* Berliner against blackflies. *Environ. Entomol.* 7: 583–588.

Lacey, L. A., and Undeen, A. H. 1984. Effect of formulation, concentration, and application time on the efficacy of *Bacillus thuringiensis* (H-14) against black fly (Diptera: Simuliidae) larvae under natural conditions. *J. Econ. Entomol.* 77: 412–418.

———. 1986. Microbial control of black flies and mosquitoes. *Ann. Rev. Entomol.* 31: 265–296.

Lacey, L. A.; Undeen, A. H.; and Chance, M. M. 1982. Laboratory procedures for the bioassay and comparative efficacy evaluation of *Bacillus thuringiensis* var. *israelensis* (serotype 14) against black flies (Simuliidae). In *Biological control of black flies (Diptera: Simuliidae) with* Bacillus thuringiensis *var.* israelensis *(serotype 14): A review with recommendations for laboratory and field protocol,* ed. D. Molloy, 19–23. Misc. Publ. Entomol. Soc. Amer. 12, no. 4.

Lacoursière, J. O., and Charpentier, G. 1988. Laboratory study of the influence of water temperature and pH on *Bacillus thuringiensis* var. *israelensis* efficacy against black fly larvae (Diptera: Simuliidae). *J. Amer. Mosq. Control Assoc.* 4: 64–72.

McGaughey, W. H. 1985. Insect resistance to the biological insecticide *Bacillus thuringiensis. Science* 229: 193–195.

Matha, V.; Weiser, J.; Olejnicek, J.; and Tonner, M. 1986. Possibilities of application of preparations based on *Bacillus thuringiensis* in the biological control of black fly larvae. In *Dipterologica Bohemoslavaca IV,* 135–138. Proceedings of the Eighth Dipterological Seminar at Ceske Budejovice. Jihoceske Museum, Czechoslovakia.

Molloy, D. 1984. The black fly debate. NAHO (N.Y. State Museum, Albany) 17: 7–10.

Molloy, D.; Gaugler, R.; and Jamnback, H. 1981. Factors influencing efficacy of *Bacillus thuringiensis* var. *israelensis* as a biological control agent of black fly larvae. *J. Econ. Entomol.* 74: 61–64.

Molloy, D., and Jamnback, H. 1981. Field evaluation of *Bacillus thuringiensis* var. *israelensis* as a black fly biocontrol agent and its effect on nontarget stream insects. *J. Econ. Entomol.* 74: 314–318.

Molloy, D. P., and Struble, R. H. 1988. A simple and inexpensive method for determining stream discharge from a streambank. *J. Freshw. Ecol.* 4: 477–481.

———. 1989. Investigation of the feasibility of the microbial control of black flies (Diptera: Simuliidae) with *Bacillus thuringiensis* var. *israelensis* in the Adirondack Mountains of New York. *Bull. Soc. Vector Ecol.* 14: 266–276.

Molloy, D.; Wraight, S. P.; Kaplan, B.; Gerardi, J.; and Peterson, P. 1984. Laboratory evaluation of commercial formulations of *Bacillus thuringiensis* var. *israelensis* against mosquito and black fly larvae. *J. Agric. Entomol.* 1: 161–168.

Nakamura, Y.; Kanayama, A.; Sato, H.; Fujita, K.; Tabaru, Y.; and Shimada, A. 1985. Field trials of *Bacillus thuringiensis* var. *israelensis* against blackfly larvae in Japan. *Jpn. J. Sanit. Zool.* 36: 371–373.

Noblet, R.; Kissam, J. B.; and Adkins, T. R., Jr. 1975. *Leucocytozoon smithi:* Incidence of transmission by black flies in South Carolina (Diptera: Simuliidae). *J. Med. Entomol.* 12: 111–114.

Olejnicek, J. 1986. The use of *Bacillus thuringiensis* var. *israelensis* in the biological control of blackflies in Czechoslovakia. *Wiad. Parazytol.* 32: 539–542.

Olejnicek, J.; Matha, V.; and Weiser, J. 1985a. The effectiveness of *Bacillus thuringiensis* var. *israelensis* against black fly larvae under the field conditions of South Bohemia. *Vet. Med.* (Prague) 30: 433–441.

———. 1985b. The efficacy of *Bacillus thuringiensis* var. *israelensis* against larvae of the blackfly *Odagmia ornata* (Meig.) (Simuliidae) at low temperatures. *Folia Parasitol.* (Prague) 32: 271–277.

Pistrang, L. A.; and Burger, J. F. 1984. Effect of *Bacillus thuringiensis* var. *israelensis* on a genetically-defined population of black flies (Diptera: Simuliidae) and associated insects in a montane New Hampshire stream. *Can. Entomol.* 116: 975–981.

Rutschke, J., and Grunewald, J. 1984. The control of blackflies (Diptera: Simuliidae) as cattle pest with *Bacillus thuringiensis* H-14. *Zentralbl. Bakteriol. Mikrobiol. Hyg.,* ser. A, 258: 413.

Simmons, K. R., and Sjogren, R. D. 1984. Black fly (Diptera: Simuliidae) problems and their control strategies in Minnesota. *Bull. Soc. Vector Ecol.* 9: 21–22.

Sun, C. N.; Georghiou, G. P.; and Weiss, K. 1980. Toxicity of *Bacillus thuringiensis* var. *israelensis* to mosquito larvae variously resistant to conventional insecticides. *Mosq. News* 40: 614–618.

Undeen, A. H., and Colbo, M. H. 1980. The efficacy of *Bacillus thuringiensis* var. *israelensis* against blackfly larvae (Diptera: Simuliidae) in their natural habitat. *Mosq. News* 40: 181–184.

Undeen, A. H., and Lacey, L. A. 1982. Field procedures for the evaluation of *Bacillus thuringiensis* var. *israelensis* (serotype 14) against black flies (Simuliidae) and nontarget organisms in streams. In *Biological control of black flies (Diptera: Simuliidae) with* Bacillus thuringiensis *var.* israelensis *(Serotype 14): A review with recommendations for laboratory and field protocol,* ed. D. Molloy, 25–30. Misc. Publ. Entomol. Soc. Am. 12, no. 4.

Undeen, A. H.; Lacey, L. A.; and Avery, S. W. 1984. A system for recommending dosage of *Bacillus thuringiensis* (H-14) for control of simuliid larvae in small streams based upon stream width. *Mosq. News* 44: 553–559.

Weast, R. C., ed. 1982. *CRC handbook of chemistry and physics,* 113–114. Boca Raton, Fla.: CRC Press.

World Health Organization. 1984. Report of the seventh meeting of the scientific working group on biological control of vectors. TDR/BCV/SWG-7/84.3. Mimeo.

_______. 1989. Informal consultation on bacterial formulations for cost-effective vector control in endemic areas. WHO/VBC/89.979. Mimeo.

11

Use of *Bacillus thuringiensis israelensis* for Onchocerciasis Control in West Africa

PIERRE GUILLET
DANIEL C. KURTAK
BERNARD PHILIPPON
ROLF MEYER

11.1 INTRODUCTION

Onchocerciasis, or river blindness, constitutes a major public health problem in savanna areas of West Africa and an obstacle to the socioeconomic development of the countries concerned.

The ecology of the vectors, black flies belonging to the *Simulium damnosum* complex (8 species and several forms of the *S. damnosum* complex recorded in West Africa), as well as the biology of the parasite they transmit, the filarial worm *Onchocerca volvulus*, determine the strategy for controlling the disease. The vectors breed year-round with 25 to 30 generations per year (Le Berre 1966). Larval breeding sites are located in fast-flowing water of medium and large rivers with an average larval life of 8 to 10 days. Females have a long flight range and can migrate over considerable distances (over 300 km) when assisted by winds (Le Berre et al. 1979; Walsh, Davies, and Garms 1981). The life of the parasite in humans is longer than 10 years, and, up to now, no drug has been available for mass chemotherapy.

The only effective strategy for controlling onchocerciasis is to interrupt the transmission through vector control. This control is presently achieved using larvicides exclusively because black fly larvae are much easier to locate and reach than adults. Eggs and pupae are unaffected by most insecticides.

For a campaign to be effective, treatments must be repeated weekly to avoid outbreaks of indigenous flies and must be conducted over a very wide area in order to limit or prevent the effects of reinvasion by flies from untreated areas. Transmission has to be virtually interrupted for at least 15 years until the parasite reservoir in humans is exhausted.

The vast Onchocerciasis Control Programme (OCP) was launched in West Africa in 1974 under the auspices of the World Health Organization, largely through the efforts of the international community (fig. 11.1). The

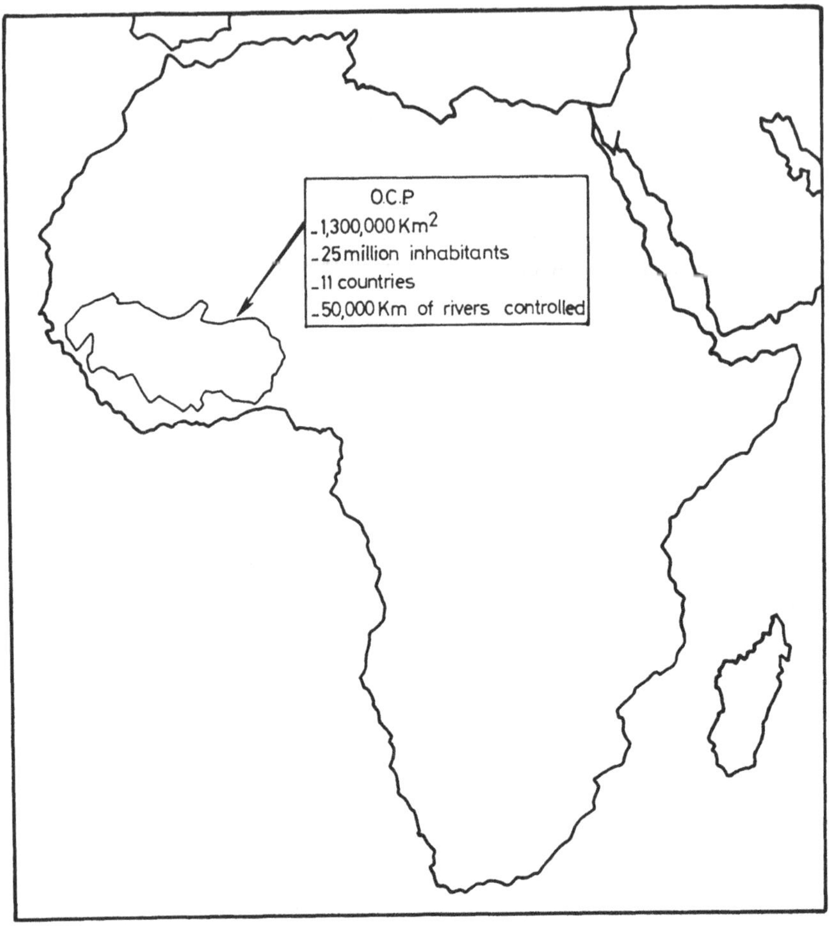

FIGURE 11.1 OCP in Africa

objectives of OCP are not to eradicate the vectors or even the parasite, which is unrealistic, but to put an end to onchocerciasis as a disease of public health and socioeconomic importance and to ensure that there will be no recrudescence of the disease thereafter.

From 1974 to 1987, the treated area covered 764,000 km^2 in seven countries (Benin, Burkina Faso, Côte d'Ivoire, Ghana, Mali, Niger, and Togo; see fig. 11.2). In this initial area, during the rainy season, up to 6,000 selective insecticide treatments might be applied to a maximum river length of 23,000 km every week using nine helicopters and two fixed-wing aircraft.

From 1987 to 1989, OCP was extended to the southeast and west to include new areas in southern Benin, southern Ghana, southern Togo, western Mali, and four additional countries (Guinea, Guinea Bissau, Senegal, and

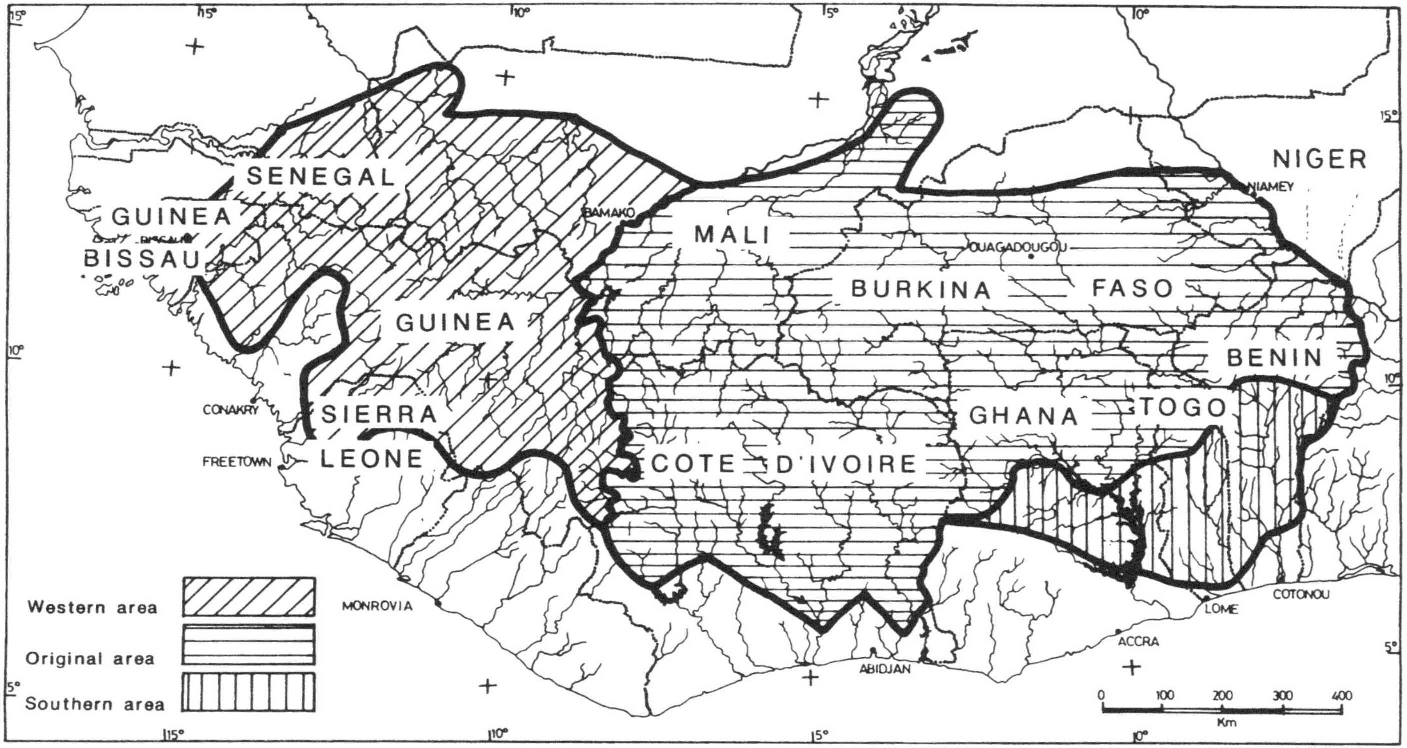

FIGURE 11.2 OCP in West Africa

Sierra Leone). The coverage of rivers to be treated thus increased from 23,000 to 50,000 km in an area of 1.3 million km^2.

After 12 years of control, the results are considered to be satisfactory and even impressive. Transmission in more than 600,000 km^2 (over 80% of the initial treated area) has been interrupted or brought below the level at which the ocular forms of the disease do not reappear. So far, all outbreaks of autochthonous flies have been controlled despite the rapid emergence of resistance and local larviciding failures. Intolerable transmission is still prevalent in border areas because of seasonal waves of immigrant infective flies. It has been demonstrated that this invasion can be controlled by appropriate treatments of external sources of breeding (Philippon et al. 1986).

The decreasing trends in human infection confirm the entomological results. In well-protected areas, all the indicator parameters (prevalence and community microfilarial loads) show a sharp decline, enabling prediction of the virtual extinction of infection 12 to 15 years after the treatments started. There is no incidence in the age group of children born after the implementation of vector control, which means that virtually none of the 3.5 million children born in the protected area will be at risk of ocular onchocerciasis during their life. The decline in ocular disease is still more impressive. Ocular infection has almost disappeared, and lesion development has been virtually interrupted everywhere.

Although a marked improvement is noticeable in reinvaded areas, the parasitological and ophthalmological results are not satisfactory and do not allow us to foresee the elimination of infection in the same time frame (Remme 1984).

In protected valleys that were previously uninhabited or deserted, vast spontaneous or organized repopulation movements have already been initiated, allowing the cultivation of new, fertile riparian lands (Hervouet et al. 1984).

Up to now, owing to a close standardized monitoring of the treated rivers under the supervision of an independent ecological group, no irreversible and intolerable damage has been demonstrated amongst the aquatic invertebrate fauna. No detectable effect attributable to OCP larviciding has been observed on fish populations (Leveque 1985).

For an insecticide to be used in such a large program, it has to be properly formulated and fulfill many requirements, including the following:

1. Efficacy of 99.9% against black fly larvae at low dosage and over a long distance downstream from the treatment point (the carry), considering the volume of product to be transported and applied by the aircraft
2. Spontaneous and rapid dispersion in river water when applied in a liquid form without prior dilution
3. Good storage stability in the field for at least one year in the open in a tropical environment

TABLE 11.1.
Comparative Performances of the Formulations Presently Used in the OCP with Reference to Abate®

Formulation	Dosage (mg ai/1/10 min.)	Volume per m^3/s discharge	Carry (km) and Relative Cost: Low discharge Carry	Low discharge Cost	High discharge Carry	High discharge Cost
Abate® EC	0.05–0.1	0.15–0.3	1–3	1	40–50	1
Teknar®	1.6	960 + 20% water	1–2	1.4	5–7	18
HPD or Vectobac® 12AS	1.2	720	1–2	1.4	10–12	4.3
Chlorphoxim EC	0.05	0.15	1–2	0.7	15	1.3
Permethrin EC	0.015	0.045	na	na	15	0.7
Carbosulfan	0.05	0.12	na	na	10–15	2.3

4. Good selectivity and low toxicity against humans, mammals, fish, and other nontarget organisms
5. Reasonable cost-effectiveness

During the dry season, when river discharges are low in the rivers, the carry of any insecticide is very poor because of its dilution in ponds between breeding sites; therefore, practically every larval site has to be treated individually. When the discharge increases during the rainy season, insecticide is carried downstream. Its carry is then generally proportional to the discharge rate, but it also depends a great deal on the insecticide, its percentage of active ingredient, and the formulation used. Abate®, a 20% emulsifiable concentrate of the organophosphate (OP) temephos, has been used since the beginning of the OCP and represents the best compromise of these requirements. Its average carry during the rainy season ranges from 40 to 50 km (table 11.1).

The application of larvicides that affect the entire target population, the high frequency of treatments, and the large surface area treated at the same time exert a very strong selection pressure. This pressure, together with the high reproductive rate of the female black fly, has offered significant prospects for the development of resistance. The first resistance to temephos in the OCP area was demonstrated in 1980 (Guillet et al. 1980) in the *Simulium soubrense/sanctipauli* Vajime & Dunbar subcomplex (which now appears to be only *S. sanctipauli* s.n., Post 1986; Post, pers. comm.), a forest species originally distributed only in the southern part of the OCP area. Because of the complete failure of operational treatments and the impossibility of increasing the dosage rate due to the environmental impact and cost of treatments, temephos had to be replaced immediately by another OP, chlorphoxim. Unfortunately, resistance to this compound also quickly developed (Kurtak et al. 1982), followed by cross-resistance to most other OP compounds. More recently, resistance to temephos has extended to the savannah species *S. damnosum* s.str. and *Simulium sirbanum* Vajime & Dunbar, the major vectors of severe onchocerciasis, and, at present, most of the OCP area is affected.

Despite considerable efforts to test new alternative chemical larvicides, only two compounds have been selected, permethrin (a pyrethroid) and carbosulfan (a carbamate), neither of which exhibited any cross-resistance to temephos (Kurtak, Jamnback, et al. 1987). Unfortunately, these two larvicides are more toxic than temephos to nontarget organisms and can be used only for rainy season treatments when high flow occurs in the rivers (discharge >70m^3/s). Moreover, resistance to pyrethroids is generally rapidly acquired, and permethrin, which is also extensively used for crop protection in the OCP area, should be regarded as a stopgap insecticide rather than a replacement larvicide. At the moment, in areas of organophosphate resistance, no chemical larvicides are available for continuous treatment during the six to eight consecutive months of the dry season, when river discharges are at their lowest.

Bacillus thuringiensis subsp. *israelensis* (*B.t.i.*) was first tested against *S. damnosum* complex larvae in West Africa in 1979 in the laboratory (Undeen and Berl 1979) and in the field (Guillet and de Barjac 1979). A commercial formulation was available for limited operational use by 1981 (Guillet, Escaffre, and Prud'hom 1982c; Lacey et al. 1982). Although *B.t.i.* formulations had severe limitations, they immediately appeared as one solution, if not the only one, to the problems created by OP resistance. Therefore, research was intensified to select more suitable formulations, to develop proper application equipment, and to study the factors affecting field efficacy. In this way, *B.t.i.* has become progressively the key larvicide for the OCP.

11.2 THE SEARCH FOR SUITABLE *B.t.i.* FORMULATIONS

The first trials carried out with experimental formulations showed, as is commonly observed with chemical larvicides, that the efficacy depends both on the formulation itself and on its active ingredient (toxin) content (Guillet and Escaffre 1979; Molloy, Gaugler, and Jamnback, 1981; Molloy et al. 1984).

Basically, two types of formulations were produced by industry, powders and liquid suspensions with large spherical aggregates (mean particle 30 to 50 μ) or suspensions of very fine particles (0.5 to 5 μ). The efficacy of formulations with large particles soon appeared to be significantly better than those with small ones. Several experimental samples were then produced in order to investigate the influence of the nature and the size of particles. Depending on the formulation, the optimum size was found to vary from 30 to 50 μ. The efficacy against young-instar larvae decreased sharply with increasing particle size (Guillet, Escaffre, et al. 1985a).

Subsequently, it was shown that formulations with large particles lost their efficacy in turbid water during the rainy season or after floods, due to competition between *B.t.i.* and natural particles in their capture and ingestion (Guillet, Escaffre, et al. 1985b). Moreover, their efficacy was also dependent on exposure time. Brief exposures, commonly observed in operational treatments, result in significantly reduced efficacy (Guillet, Escaffre, et al. 1985b). On the contrary, the efficacy of formulations with fine particles is not as dependent on water turbidity in a range covering most of the natural conditions found in West Africa and also is not dependent on exposure time within a range of 1 to 60 minutes. Thus, the latter formulations appeared to be much more suitable to the conditions of operational use prevailing in the OCP. Other types of formulations were also tested, such as oil suspensions, emulsions, or surface films, but none gave as satisfactory results as the water-dispersible concentrates of very fine particles.

A formulation of this type was rapidly selected (Guillet, Escaffre, and Prud'hom 1982a; Lacey et al. 1982), and the limited studies of its performance allowed its operational use in emergency situations caused by OP resistance to be considered. This formulation was very stable under storage in the open in tropical environments (Guillet, Escaffre, and Prud'hom 1982a). Quality-control tests were performed on a 600,000-liter industrial production of Teknar®. From batch to batch, its efficacy against black fly larvae and its *Aedes aegypti* potency were satisfactory and reasonably constant (Guillet 1985).

However, the physical characteristics of different batches varied significantly, causing serious complications in treatment operations. Prior to application, Teknar® had to be diluted 20% with water, and sometimes the dilution rate had to be adjusted depending on the viscosity. Moreover, its efficacy and its operational carry (about 5 km under favorable conditions of discharge) were relatively limited compared to Abate® (table 11.1). Thus, the relatively low cost-effectiveness of Teknar® stimulated the search for improved formulations.

Through a close collaboration between industry and WHO, several hundred experimental formulations were produced by a small number of firms and then tested in West Africa. The objective was to develop a self-dispersible liquid formulation with both suitable and constant physical characteristics and a higher toxin content. This was necessary in order to limit the bulk of product to be applied to a maximum of 1.5 times that for Abate® (maximum $450 cc/m^3/s$).

After four years of intensive screening, the level of efficacy of the formulation used operationally increased by 33%, and, due to improved viscosity characteristics, dilution became unnecessary. Consequently, the volume to be transported and applied decreased by 60%. In addition, two formulations from different companies with similar performances (Teknar HPD® and Vectobac 12AS®) have become available for operational use.

One of the serious difficulties encountered in the improvement of the *B.t.i.* preparations was the lack of a systematic relationship between *Ae. aegypti* potency and toxicity to larvae of the *S. damnosum* complex. This was observed with formulations as well as with unformulated primary products (Guillet, Escaffre, and Prud'hom 1982b; Molloy et al. 1984; Guillet 1985). When preparations were tested against mosquito and black fly larvae under similar conditions (24-hour exposure in distilled water), a close correlation between results was observed ($r^2 = 0.92$). When the exposure time for black fly larvae was reduced to 3 hours (with proportionally increased concentrations) in distilled water, the correlation was much lower ($r^2 = 0.72$). Finally, with a 10-minute exposure of black fly larvae in river water no correlation at all was observed ($r^2 = 0.2$; Guillet 1985). This indicates that *Ae. aegypti* potency is not, therefore, a reliable criterion for predicting the efficacy of *B.t.i.* preparations, including primary products. Consequently, they must be tested against *Simulium damnosum* complex larvae, which increases the delay and cost in the evaluation process.

Other difficulties lie in the assessment of the physical characteristics and potency of the formulations. Their dispersibility is generally related to their viscosity; viscous products usually have poor dispersibility and a high rate of settling, and they create pumping problems when refilling aircraft and spraying.

Viscosity is measured either with a No. 3 Zahn cup or a Brookfield LVT viscometer. With the first method, 44 ml of formulation should be discharged in less than 10 seconds at 25° C. With the second method, the viscosity at 30° C with a No. 3 spindle at 30 rpm should not exceed 500 cps. Subsequently, a third method was developed in the OCP to measure the dispersibility and settling rate in terms of the average suspension index (ASI) based on the dispersion of 1 ml of formulation in 1 liter of water in a glass cylinder and assessment with a Hach® Ratioturbidimeter in terms of nephelometric turbidity units (NTU). The ASI units vary from less than 0.7 (poor dispersion) to more than 0.95 (excellent dispersion). This routine procedure does not constitute an official method. Visually, a formulation is considered to be suitable if the drops break up when they hit the water and disperse well into clouds of fine particles. No large clumps should sediment out and settle to the bottom. These very simple criteria also provide valuable firsthand information.

The active ingredient content of formulations is guaranteed in terms of *Ae. aegypti* units and cannot be controlled using standard procedures as for chemical pesticides. A standard bioassay method has been developed and proposed by WHO using a standard preparation produced by the Pasteur Institute Paris. However, this method, which gives reliable results expressed in International Toxicity Units (ITU/*Ae. aegypti*/mg), is not systematically used by firms and does not represent an official method at present. In addition, because of the delay required to carry out such potency tests in a reference laboratory, batches are generally purchased before any quality control can be finalized. At the moment, no serious quality-control difficulties have appeared after six years of intensive use of *B.t.i.* formulations. However, the acceptance of formulation specifications and a standard bioassay procedure by firms, laboratories, and users is still highly desirable.

11.3 OPERATIONAL USE OF *B.t.i.* IN THE OCP

Since 1985, Teknar HPD®, a new formulation produced from an asporogenic strain, has been used by the OCP. It exhibits much-improved characteristics when compared with the previous Teknar® formulation (table 11.1). In addition to a reduced volume of application and a better carry, HPD disperses spontaneously and thus can be applied with fixed-wing aircraft; with these attributes, HPD can be used to treat rivers with discharge rates up to $200m^3/s$ if necessary. From current research on the dispersion and carry of HPD it appears that, by increasing application dosages, carry could be improved in a cost-effective way due to reductions in application cost, and that a high dosage of *B.t.i.* is acceptable due to its remarkable selectivity. Mathematical models based on field experiments and bioassays both in Canada and in West Africa should be available in the near future, and their use will optimize the application of *B.t.i.* formulations in control operations (Back, Chalifour, and Boisvert, pers. comm.).

However, HPD is still more difficult to apply than Abate®. HPD must be sprayed carefully from bank to bank, and the number of helicopter hours needed to treat a given area with HPD is usually 1.5 to 2 times that with Abate®. The change from Abate® to *B.t.i.* created many problems, which have led to the development of improved spraying equipment compatible with the five larvicides currently used in the OCP. *B.t.i.* formulations were corrosive to aluminum parts like pumps, valves, and booms. As much as possible, they were replaced by fiberglass, rubber, copper, or stainless steel parts. New fiberglass holding tanks were designed without baffles and with a sloped bottom to improve the flow of the product. A jet agitation system was installed to enable mixing in the tank as well as a second flow meter for viscous products. The spray boom was equipped with 5 BX50 nozzles, which resulted in a higher application pressure, a finer spray, and a better product dispersion.

Some losses of efficacy of both Teknar® and HPD were observed in rivers with high concentrations of suspended algae such as the Comoe, Bandama, and Sassandra during the height of the dry season (low river levels and warm water) and also in stretches of rivers polluted with organic material. When the algal density reaches 1,500 to 3,000 cells/ml, the operational dose must be doubled.

Due to the extension of OP resistance to the savannah species and its increasing geographical spread, the use of *B. t. i.* in the OCP is increasing. The strategy for the rotational use of different larvicides can be summarized as follows:

In areas free of OP resistance. *B. t. i.* is used during the dry season. A change to Abate® is made when increased stream flow gives good transport and carry of the larvicide. The flow-rate threshold depends on the size of the river.

In areas with temephos resistance. Chlorphoxim replaces Abate® in rotation with permethrin (usually 8 cycles with chlorphoxim followed by permethrin for 4 cycles when the discharge reaches 70 m^3/s). In this case, permethrin is used to slow down the development of resistance to chlorphoxim (Kurtak, Meyer, et al. 1987).

In areas with OP resistance (both temephos and chlorphoxim). *B. t. i.* is used up to 70 m^3/s and then replaced by carbosulfan or permethrin. In this case, during the rainy season, the main river stretches are treated with compounds more toxic than Abate®, but the tributaries still remain under treatment with *B. t. i.* In the Sankarani basin in Guinea, for example, where no OP compounds are used to prevent the geographical extension of the resistance to the west (over a total of 2,300 km of rivers), only 30% (660 km) are treated with chemicals during the rainy season. Along these 660 km, *B. t. i.* is used below 70 m^3/s, which constitutes 60 to 65% of the treatments. In this way, stretches treated with chemicals are quickly recolonized by nontarget organisms drifting in from tributaries and also during the dry season when *B. t. i.* treatment is taking place.

The rotation between *B. t. i* and chemical larvicides is adjusted week by week, basin by basin, according to entomological, hydrological, epidemiological, and environmental factors. At present, it is hoped that judicious rotation will delay the appearance of resistance to temephos and chlorphoxim and, to some extent, together with the extensive movements and seasonal replacements of fly populations in dry savannah area, will help recover quasi-normal susceptibility levels once resistance has developed. This strategy is not based on theoretical models or strong scientific arguments but on a pragmatic and empirical attitude based on experience progressively gained in the field. Results obtained so far are very encouraging.

B. t. i. is now by far the main larvicide for the OCP, the use of which allows the management of resistance to insecticides in the *S. damnosum* complex. From 1981 to 1988, more than 2 million liters of *B. t. i.* were used (table 11.2).

TABLE 11.2.
Volume of Different Larvicides Used in the OCP since the Appearance of Resistance to Temephos in 1980

Year	Abate®	*B.t.i.*	Chlorophoxim	Permethrin	Carbosulfan
1980	184,517	0	5,700	0	0
1981	132,497	7,987	81,462	0	0
1982	162,750	233,000	6,700	0	0
1983	74,760	310,000	35,782	0	0
1984	77,849	257,000	56,685	0	0
1985	130,117	210,563	6,306	2,968	8,916
1986	120,000	385,000	15,400	10,200	20,000
1987	70,000	299,150	29,600	10,400	8,125
TOTAL	952,490	1,702,700	237,635	23,568	37,041

With the extension of the OCP, it was estimated that in 1988 about 400,000 liters of formulation were used, accounting for 68% of the total volume of larvicides applied. Abate® then accounted for 19%, and other chemicals accounted for only 13% of the total.

B.t.i. is well known for its remarkable selectivity. River treatments in West Africa have only very limited short-term effects on nontarget invertebrate communities. Though invertebrate drift is 20 to 40 times lower than with Abate®, the effect continues for 24 hours, which indicates some light but prolonged traumatism of treated invertebrates (Dejoux, Gibon, and Yameogo 1985). A significant increase in Ecnomidae and Hydropsychidae is obtained in the medium term, while Tricorythidae show a drop in number compared with pretreatment data (Elouard, pers. comm.). In the short term, therefore, *B.t.i.* is virtually innocuous to nontarget organisms, and in the medium term there is no evidence of deleterious action after regular applications.

11.4 DISCUSSION AND PROSPECTS

A *B.t.i.* formulation was first tested in the field under operational conditions during the same year that resistance to temephos was detected in *S. damnosum* in the OCP area. Though the formulation had a limited performance, its use as a microbial larvicide was immediately envisioned as a practical solution to problems generated by resistance to temephos. Without this entomopathogen, the OCP would have faced very serious difficulties, especially following the extension of resistance to savannah species and its geographical spread. Ten years ago, a similar situation would have been considered a dead end, recalling to some extent the failure of the Malaria Eradication Programme.

Because of its remarkable characteristics and the immediate availability of a commercial formulation with proven feasibility of use, a great deal of effort was immediately devoted by industry, WHO, and various scientific institutions to improve the performance of existing formulations. These efforts led to significant improvement, which has allowed *B.t.i.* to become the key larvicide for the OCP since 1985. A strategy was developed for the rotational use of the four chemical larvicides around *B.t.i.* The development of this strategy has already provided very encouraging results in the effective management of resistance.

Since the introduction of *B.t.i.* on a large operational scale, no significant difference has been recorded in vector biting rates and transmission potentials between river basins controlled with *B.t.i.* and rivers formerly or currently treated with chemicals. Although epidemiological trends take time to be detected, there are no indications of any change in the decline of infection in humans in areas that have been seasonally exposed to *B.t.i.* applications during the last five years.

The extensive use of *B.t.i.* creates a strong selective pressure. Though prospects for the development of resistance are, a priori, limited, they are not negligible. Resistance of agricultural pests to the ***kurstaki*** subspecies has already occurred under field conditions (McGaughey 1985), although there are few similarities with black fly control in the OCP. A method has been developed to measure the susceptibility of the *S.* ***damnosum*** complex larvae to the H14 toxin adapted to working conditions in the field (Guillet, Hougard, et al. 1985a). This method is derived from the standard bioassay procedure proposed by Lacey, Undeen, and Chance (1982). Based on a brief exposure time, it gives reliable results and is now routinely used to monitor the susceptibility of different populations of the *S.* ***damnosum*** complex–species (Guillet, Hougard, et al. 1985b).

However, continuous improvements to achieve a better cost-effectiveness for *B.t.i.* are of paramount importance to the OCP and one of its first priorities. Such improvements embrace the following:

1. The selection of more effective formulations with better physical characteristics. From recent bioassays and field trials it appears that HPD efficacy can be doubled. Although preliminary results obtained with industrial batches need to be confirmed, they are extremely encouraging for the near future. In addition, competition among firms producing formulations has resulted in substantial cost reductions.
2. The study of factors affecting field efficacy, such as the concentration of algae in river water as well as its suspended matters content
3. The mastery of several hydrological parameters, such as the transport and loss of *B.t.i.* in running water, in order to devel-

op at a practical level a predictive model for insecticide carry in relation to river discharge. In addition, the teletransmission of hydrological data through the Argos satellite system provides very accurate discharge data, thus helping to avoid costly overdosing or underdosing. This is a great improvement over the previous manual reading of river gauges, which would often be read several days before treatment.

The improvements in the use of *B.t.i.* are one component of the search for new tools actively carried out or stimulated by the OCP. These tools also include new chemical larvicides and drugs that can be used for mass treatment, such as Ivermectin®, which has reached the stage for large-scale operational use. These tools will be integrated into the OCP strategy so that the program will accomplish its objectives within the proposed time period.

Acknowledgments

In OCP, all published results are the fruit of collaboration among a large number of personnel at all levels. We thank them all heartily without attempting to mention everyone by name. We also gratefully acknowledge the support of the director of the OCP and thank him for approval to publish this document. The authors are also greatly indebted to Dr. R.H.A. Baker for correcting the draft.

References

Dejoux, C.; Gibon, F. M., and Yameogo, L. 1985. Toxicité pour la faune non-cible de quelques insecticides nouveaux utilisés en milieu aquatique tropical. Part 4, Le *Bacillus thuringiensis* var. *israelensis* H14. *Rev. Hydrobiol. Trop.* 18 (1): 31–49.

Guillet, P. 1985. La lutte contre les vecteurs de l'onchocercose en Afrique de l'Ouest: Etude de la resistance et recherches de nouveaux larvicides. These Doctorat, Universite de Paris-Sud.

Guillet, P., and de Barjac, H. 1979. Toxicite de *Bacillus thuringiensis* var. *israelensis* pour les larves de Simulies vectrices de l'onchocercose. *C. R. Acad. Sciences* 289: 549–552.

Guillet, P., and Escaffre, H. 1979. Evaluation de *Bacillus thuringiensis israelensis* de Barjac pour la lutte contre les larves de *Simulium damnosum* s.l. Part 2, *Efficacité comparée de trois formulations expérimentales.* WHO/VBC/79.735. Mimeo.

Guillet, P.; Escaffre, H.; Ouedraogo, M.; and Quillevere, D. 1980. Mise en evidence d'une resistance au temephos dans le complexe *S. damnosum* (*S. sanctipauli* et *S. soubrense*) en Côte d'Ivoire (zone du Programme de lutte contre l'onchocercose dans la region du bassin de la Volta). *Cah. ORSTOM, sér. Ent. Méd. et Parasitol.* 23 (3): 291–299.

Guillet, P.; Escaffre, H.; and Prud'hom, J. M. 1982a. *Bacillus thuringiensis* H14: a biocontrol agent for onchocerciasis control in West Africa. In *Invertebrate pathology and microbial control,* 460–465. Proceedings of the Third International Colloquium on Invertebrate Pathology. Univ. Sussex, Brighton, United Kingdom.

———. 1982b. L'utilisation d'une formulation à base de *Bacillus thuringiensis* H14 dans la lutte contre l'onchocercose en Afrique de l'Ouest. Part 1, Efficacité et modalités d'application. *Cah. ORSTOM, sér. Ent. méd. et Parasitol.* 20 (3): 175–180.

_______. 1982c. L'utilisation d'une formulation à base de *Bacillus thuringiensis* H14 dans la lutte contre l'onchocercose en Afrique de l'Ouest. Part 2, Stabilité dans les conditions de stockage en milieu tropical. *Cah. ORSTOM, sér. Ent. méd. et Parasitol.* 20 (3): 181–185.

Guillet, P.; Escaffre, H.; Prud'hom, J. M.; and Bakayoko, S. 1985a. Etude des facteurs conditionnant l'efficacité des préparations à base de *Bacillus thuringiensis* H14 vis-à-vis des larves du complexe *Simulium damnosum* (Diptera: Simuliidae). Part 1, Influence de la nature et de la taille des particules. *Cah. ORSTOM, sér. Ent. med. et Parasitol.* 23 (4): 257–264.

_______. 1985b. Etude des facteurs conditionnant l'efficacité des préparations à base de *Bacillus thuringiensis* H14 vis-à-vis des larves du complexe *Simulium damnosum* (Diptera: Simuliidae). Part 2, Influence du temps de contact et de la quantité de particules naturelles en suspension dans l'eau. *Cah. ORSTOM, sér. Ent. méd. et Parasitol.* 23 (4): 265–271.

Guillet, P.; Hougard, J. M.; Doannio, J.; Escaffre, H.; and Duval, J. 1985a. Evaluation de la sensibilité des larves du complexe *Simulium damnosum* à la toxine du *Bacillus thuringiensis* H14. Part 1, Methodologie. *Cah. ORSTOM, sér. Ent. méd. et Parasitol.* 23 (4): 241–250.

_______. 1985b. Evaluation de la sensibilité des larves du complexe *Simulium damnosum* à la toxine du *Bacillus thuringiensis* H14. Part 2, Sensibilité relative de quelques groupes d'espèces et possibilités d'utilisation de doses diagnostiques. *Cah. ORSTOM, sér. Ent. méd. et Parasitol.* 23 (4): 251–255.

Hervouet, J. P.; Clanet, J. C.; Paris, F.; and Some, H. 1984. Peuplement des vallees protegees de l'onchocercose apres 10 ans de lutte antivectorielle au Burkina Faso. WHO/OCP/GVA/84.5. Mimeo.

Kurtak, D. C. 1986. Insecticide resistance in the Onchocerciasis Control Programme. *Parasitol. Today* 2 (1): 20–21.

Kurtak, D.; Jamnback, H.; Meyer, R.; Ocran, M.; and Renaud, P. 1987. Evaluation of larvicides for the control of *Simulium damnosum* s.l. (Diptera: Simuliidae) in West Africa. *J. Amer. Mosq. Control Assoc.* 3: 201–210.

Kurtak, D.; Meyer, R.; Ocran, M.; Ouédraogo, M.; Renaud, P.; Sawadogo, R. O.; and Télé, B. 1987. Management of insecticide resistance in control of the *Simulium damnosum* complex by the Onchocerciasis Control Programme, West Africa: Potential use of negative correlation between organophosphate resistance and pyrethroid susceptibility. *Med. Vet. Entomol.* 1: 137–146.

Kurtak, D. C.; Ouedraogo, M.; Ocran, M.; Barro, T.; and Guillet, P. 1982. Preliminary note on the appearance in Ivory Coast of resistance to chlorphoxim in *Simulium soubrense-sanctipauli* larvae already resistant to temephos (Abate ®). WHO/VBC/82.850. Mimeo.

Lacey, L. A.; Escaffre, H.; Philippon, B.; Sékétéli, A.; and Guillet, P. 1982. A large river treatment with *Bacillus thuringiensis* (H-14) for the control of *Simulium damnosum* s.l. in the Onchoceriasis Control Programme. *Tropenmed. Parasitol.* 33: 97–101.

Lacey, L. A.; Undeen, A. H.; and Chance, M. M. 1982. Laboratory procedures for the bioassay and comparative efficacy evaluation of *Bacillus thuringiensis* var. *israelensis* (serotype H14) against black flies (Simuliidae). In *Biological control of black flies (Diptera: Simuliidae) with* Bacillus thuringiensis *var.* israelensis *(serotype 14): A review with recommendations for laboratory and field protocol*, ed. D. Molloy, 19–23. Misc. Publ. Entomol. Soc. Amer. 12, no. 4.

Le Berre, R. 1966. Contribution a l'etude biologique et ecologique de *Simulium damnosum:* Theobald, 1903. *Mem. ORSTOM* 17: 1–204.

Le Berre, R.; Garms, R.; Davies, J. B.; Walsh, J. F.; and Philippon, B. 1979. Displacements of *Simulium* and strategy of control against onchocerciasis. *Phil. Trans. R. Soc. London* 287 B: 277–288.

Leveque, C. 1985. The use of insecticides in the Onchocerciasis Control Programme and the aquatic monitoring in West Africa.

McGaughey, W. H. 1985. Insect resistance to the biological insecticide *Bacillus thuringiensis. Science* 229: 193–195.

Molloy, D.; Gaugler, R.; and Jamnback, H. 1981. Factors influencing efficacy of *Bacillus thuringiensis* var. *israelensis* as a biological control agent of black fly larvae. *J. Econ. Entomol.* 74: 61–64.

Molloy, D.; Wraight, S. P.; Kaplan, B.; Gerardi, J.; and Peterson, P. 1984. Laboratory evaluation of commercial formulations of *Bacillus thuringiensis* var. *israelensis* against mosquito and black fly larvae. *J. Agric. Entomol.* 1: 161–168.

Philippon, B.; Zerbo, D. G.; Kurtak, D. C.; and Remme, J. H. 1986. Lutte contre l'onchocercose en Afrique de l'Ouest: Bilan et perspectives entomologiques apres 11 ans du lutte antivectorielle. C.R. 2eme conf. int. Entom. Expr. fse, Universite du Quebec a Trois-Rivieres, Canada.

Post, R. J. 1986. The cytotaxonomy of *Simulium sanctipauli* and *Simulium soubrense* (Diptera: Simuliidae). *Genetica* 69: 191–207.

Remme, J. H. 1984. Trends in epidemiology of onchocerciasis. *Proc. 11th int. Congr. trop. Med.* Calgary, Canada.

Undeen, A., and Berl, D. 1979. Laboratory studies on the effectiveness of *Bacillus thuringiensis* de Barjac against *Simulium damnosum* larvae. *Mosquito News* 39: 742–745.

Walsh, J. F.; Davies, J. B.; and Garms, R. 1981. Further studies on the reinvasion of the Onchocerciasis Control Programme by *Simulium damnosum* s.l.: The effects of an extension of control activities into Southern Ivory Coast during 1979. *Tropenmed. Parasit.* 32: 269–273.

12

Mammalian Safety of *Bacillus thuringiensis israelensis*

JOEL P. SIEGEL
JOHN A. SHADDUCK

12.1 INTRODUCTION

There is considerable interest in the commercial development of microbial pesticides for use as supplements or alternatives to chemicals for the control of medically important insects. Like their chemical counterparts, entomopathogens must be evaluated for their safety to both animals and humans, although the tests used to evaluate their safety differ from the chemical safety protocols. Microbial safety tests concentrate on acute toxicity and vertebrate infectivity, while chemical safety tests focus on acute toxicity, neurotoxicity, and carcinogenicity. The ocular and dermal irritancy of both chemical and microbial agents are additional concerns.

The underlying assumption of chemical safety testing is that every chemical has a biological effect if the dose administered is high enough. Toxicity is typically described by the median lethal dose, or LD_{50}. The chemical administered to an animal may be metabolized, excreted, or both; the chemical or its breakdown products may persist in the animal's body and exert an effect over time. Therefore, long-term evaluation is an important component of chemical testing. In contrast, microbial agents affect the target through infection or the action of toxins, and they may be evaluated over a much shorter time period (two to four weeks). However, they are not expected to affect nonsusceptible species, and an LD_{50} cannot be determined by conventional modes of exposure such as dermal or oral administration (Shadduck 1983).

In an attempt to create a vulnerable mammalian system, microbial safety tests follow a parallel approach to the LD_{50} tests but vary the route of exposure rather than the dose in order to produce a detectable effect. Methods of administration include conventional routes such as oral, dermal, subcutaneous, intraperitoneal, and inhalation, as well as unconventional methods of exposure such as intraocular and intracerebral injection. This testing philosophy is known as maximum challenge, and it seeks the route that most severely compromises an animal's natural defenses. Various maximum-challenge protocols have been developed by the United States Environmental Protection Agency and by European nations. Our laboratory has used this approach to evaluate the mammalian pathogenicity of *Bacillus sphaericus*, *Bacillus thuringiensis* subsp. *israelensis*, *Metarhizium anisopliae*, and *Nosema algerae* (Shadduck, Singer, and Lause 1980; Siegel, Shadduck, and Szabo 1987; Shadduck, Roberts, and Lause 1982; Shadduck 1983).

The primary difficulty associated with maximum-challenge testing involves the interpretation of mortality. One must be wary of prematurely rejecting a candidate agent, because even nonpathogenic bacteria can produce mortality when injected in massive quantities into a vulnerable site such as the brain. Mortality after administration of a high dose does not automatically lead to a candidate's rejection; several routes of exposure must be evaluated. According to Burges (1981), there is no such thing as a no-risk situation; one cannot absolutely prove a negative. He further states that registration of a chemical or microbial is essentially a statement of usage in which risks are acceptable.

There are concerns about the mammalian pathogenicity of the genus *Bacillus* because one member, *B. anthracis*, is a virulent mammalian pathogen, and various species of *Bacillus* have been associated with infections following traumatic wounds (Pearson 1970). For these reasons, *Bacillus thuringiensis* underwent careful testing during its commercial development, including full chemical safety tests and infectivity studies. Rats fed as many as 2×10^{12} viable spores per kg showed no illness, and human volunteers fed 3×10^9 spores per day for five days reported no adverse effects (Fisher and Rosner 1959). However, Lamanna and Jones (1963) reported that six strains of *B. thuringiensis* maintained on nutrient agar killed mice when injected intraperitoneally; the LD_{50} per mouse for bacilli ranged from 6.3×10^5 to 1.6×10^7, and the LD_{50} per mouse for spores ranged from 4.1×10^7 to 1.1×10^9. The existence of β-exotoxin was not known at that time and may have confounded their results; the use of strains containing β-exotoxin has been prohibited since 1977 in the United States.

There are two cases in the literature associating *B. thuringiensis* with mammalian infection outside the laboratory. *B. thuringiensis* was isolated from a fatal case of bovine mastitis in 1971 (Gordon 1977) and from a human ocular lesion following exposure to a liquid formulation (Samples and Buettner 1983). This latter case is controversial because vegetative bacilli were not observed in the corneal ulcer. *B. thuringiensis* was identified as the causal agent following its recovery from a swab of the corneal lesion; spores may have persisted from the initial exposure and then germinated in the culture medium. *B. thuringiensis* has an excellent safety record, especially when it is compared to chemical pesticides used in the same environment. Because of the close relationship between *B. thuringiensis* H3a,3b, and *B. thuringiensis* subsp. *israelensis* (serotype H14), we would expect that *B.t.i.*

is nonpathogenic to vertebrates as well. In this chapter, we report the results of oral, intraperitoneal, subcutaneous, aerosol, intracerebral, and ocular administration of six isolates of *B. thuringiensis* (H14) to mice, rats, and rabbits.

12.2 SOURCE AND PREPARATION OF CULTURES

Preparations of *B.t.i.* were received from four sources. Dry powders (ABG-6108, 6406-125) and suspension (*B.t.i.* 82161, ABG-6193) were received from Abbott Laboratories. A liquid suspension was received from Sandoz Pharmaceuticals (SAN 402 WDC 60), and lyophilized powder (R153-78) and spore-impregnated paper strips were received from Dr. H. de Barjac, Institute Pasteur. R. Bellon Laboratories, France, furnished lyophilized powders of both *B. thuringiensis* H3a,3b and *B.t.i.*

The Abbott powders were suspended in Hank's Buffered Salt Solution (HBSS) and washed five times before injection; 6406-125 was also seeded onto Difco bacto agar plates enriched with Difco Brain Heart Infusion (BHI) and then incubated at 30° C for 24 hours. The bacterial colonies were gently scraped from the plates, resuspended in HBSS, and washed three times before injection. This agar culture procedure was also performed on the lyophilized powder from R. Bellon Laboratories and the spore-impregnated paper strips from Dr. H. de Barjac.

R153-78 powder suspended in sterile distilled water was centrifuged at 15,000 rpm for ten minutes. The supernatant was discarded, and the pellet was resuspended in sterile distilled water; this procedure was repeated four times. The pellet was resuspended after the fifth wash in 30 ml of HBSS with sodium bicarbonate buffer before injection. A similar centrifugation and washing procedure, using 25-minute intervals and phosphate buffered saline (PBS), was conducted on the suspension SAN 402 WDC 60; this suspension was also applied mixed with the supplied inert carrier and PBS.

Fifteen ml of *B.t.i.* 82161 and *B.t.i.* ABG-6193 were each mixed with an equal volume of a stock solution (pH 7.2, 0.15 M) of sterile PBS for a final volume of 30 ml. The suspension was then centrifuged at 1,000 × g for 30 minutes in order to pellet the bacteria, and the supernatant was discarded. The pellet was resuspended in PBS to the original volume of 30 ml., then centrifuged at 1,000 × g; this washing procedure was repeated three more times for a total of four washes. The number of colony-forming units (cfu) per ml for each suspension was determined by serial dilution and then plating the dilutions on BHI agar plates, after which the plates were incubated from 24 hours at 30° C. Individual colonies were then counted, and the number of cfu per ml determined.

TABLE 12.1.
Mortality Resulting from Oral and Intraperitoneal Administration of *B.t.i.* to Rats

		Route	
Isolate[a]	Dose Per Rat	Oral	Intraperitoneal
A	6.9×10^7	0/6	0/6
A[b]	6.9×10^7	0/6	0/6
B	9.0×10^6	1/6	1/6
B[b]	9.0×10^6	0/6	0/6
Controls			
C	1.4×10^7	0/6	0/6
C[b]	1.4×10^7	0/6	0/6
HBSS	—	0/30	0/30

NOTE: Mortality data are presented as number dead/number exposed.
[a]A = lyophilized powder (R. Bellon); B = R153-78 (H. de Barjac); C = *Bacillus thuringiensis* H3a,3b (R. Bellon); HBSS = Hanks Buffered Salt Solution.
[b]Autoclaved.

12.3 RESULTS AND DISCUSSION

12.3.1 Oral and Intraperitoneal Administration

Forty-eight young rats (Hsd: Sprague Dawley [SD] BR) received viable and autoclaved inocula (0.1 ml) of two isolates of *B.t.i.* R153-78 either orally or intraperitoneally (table 12.1). The animals were observed daily for signs of illness or mortality for one week; *B. thuringiensis* H3a,3b and HBSS served as controls. Mortality occurred in the rats receiving R153-78 (one-sixth died after oral exposure, and one-sixth died after intraperitoneal injection), but this mortality was not statistically significant. Trauma may have occurred during administration of the inoculum or while handling the rats; mortality cannot be attributed to *B.t.i.*

12.3.2 Subcutaneous Injection

Twenty mice (BALB/cAnNHsd BR) received subcutaneous injections (10^9 organisms) of isolate R153-78; 15 mice received viable inocula, and 5 mice received autoclaved inocula. The mice were observed daily for signs of illness or death for two weeks. Subcutaneous injection of both viable and autoclaved inocula resulted in abscesses at the injection site. Because abscesses formed after injection with autoclaved organisms, they most likely resulted from the presence of heat-stable foreign material. There was no

TABLE 12.2.
Recovery of *B.t.i.* Isolate R153-78 from Mice and Rats

Species	Route	Dose Per Animal	Recovery Site	
Mouse	Subcutaneous	1.0×10^9	injection site	= 15/15[a]
			spleen	= 15/15
	Intracerebral	6.5×10^8	brain	= 10/10
			spleen	= 10/10
Rat	Intraperitoneal	7.4×10^7	peritoneal swab	= 0/15
			spleen	= 4/15
	Intracerebral	1.9×10^7	brain	= 2/10
			spleen	= 2/10

NOTE: All animals were sacrificed 14 days after injection, and the data are reported as the number positive/number injected.
[a]Abscess at injection site; also present in those receiving autoclaved inoculum.

mortality or illness as a result of injection, but the inoculum apparently entered the bloodstream since viable organisms were recovered from the spleens of all mice receiving viable *B.t.i.* (table 12.2).

12.3.3 Aerosol Exposure

Young rats (Hsd: Sprague Dawley [SD] BR) were exposed to an aerosol (2.05×10^6 viable spores per ml) of SAN 402 WDC 60 in a dynamic inhalation chamber for 30 minutes. Greater than 45% of the solution nebulized had particles ranging from 0.4 to 7 μm at 27.2°C and 44% humidity. The rats were killed at intervals from 3 hours to 27 days after exposure, and the lungs were removed, weighed, minced, and cultured on BHI agar plates for recovery of *B.t.i.*

The maximum number of bacteria was recovered from the lung 3 hours after exposure (1.18×10^4 bacteria per lung), and by 2.5 days after exposure the bacteria recovered dropped to 3.01×10^3 bacteria per lung. From days 3 to 27 after exposure, no further *B.t.i.* was recovered. There was no evidence of bacterial replication in the lungs, and no gross lesions were observed. These data indicate that clearance of *B.t.i.* is rapid—3 logs within 3 days.

12.3.4 Intracerebral Injection

Three experiments were conducted to evaluate the effects of culture medium and incubation temperature as well as isolate on the intracerebral toxicity of *B.t.i.* to rats; the persistence of the organism inside the brain was also determined. Young rats (Hsd: Sprague Dawley [SD] BR) were

TABLE 12.3.
Intracerebral Toxicity of 3 Isolates of *B.t.i.* and 1 Isolate of *Bacillus thuringiensis* H3a,3b in Weanling Rats

		Exponent			
Source	Multiple	10^7	10^6	10^5	10^4
R. Bellon	3.5	5/6	1/6	0/6	0/6
R. Bellon[a]	3.5	0/6			
Abbot 6406-125	20.0	5/6	0/6	0/6	0/6
Abbott 6406-125[a]	20.0	1/6			
Abbott 6406-125[b]	1.2	4/6	0/6	0/6	0/6
Abbott 6406-125[ab]	1.2	0/6			
H. de Barjac R153-78[b]	1.5	5/6	0/6	0/6	0/6
H. de Barjac R153-78[ab]	1.5	0/6			
Controls					
B. thuringiensis H3a,3b	0.7	5/6	2/6		
B. thuringiensis H3a,3b	0.7	0/6			

NOTE: Mortality data are presented as the number dead/number injected.
[a]Autoclaved inoculum.
[b]Grown on agar 24 hours at 30° C.

injected intracerebrally with 0.05 ml of inoculum; *B. thuringiensis* H3a,3b served as the control in two experiments. Mortality occurring during the first hour after injection was attributed to trauma, and only mortality occurring after the first hour was recorded; all mortality occurred during the first day following injection. Affected rats had severe convulsions and assumed a hunched posture; they also moved abnormally and lay down frequently, making paddling motions with their limbs prior to death.

The effects of incubation temperature and culture medium were evaluated using BHI agar plates and BHI broth incubated at 30° and 37° C. *B. thuringiensis* H3a,3b was evaluated under the same conditions; both isolates came from R. Bellon Laboratories. The intracerebral toxicity of *B.t.i.* was influenced by both medium and incubation temperature; mortality was greatest when the culture was incubated at 30° C (88% vs. 54%, $p < 0.01$, two-way ANOVA), and mortality also increased when BHI broth was used (83% vs. 58%, $0.05 > p > 0.025$, two-way ANOVA). Incubation temperature did not affect the intracerebral toxicity of *B. thuringiensis* H3a,3b, but culture medium did. Toxicity was greater when grown on BHI broth (86% vs. 40%, $p = 0.01$, one-way ANOVA).

In order to quantify the toxicity of three isolates of *B.t.i.*, a series of 10-fold dilutions of viable and autoclaved bacteria were injected intracerebrally, and mortality was evaluated over one week (table 12.3). Viable and autoclaved *B. thuringiensis* H3a,3b served as controls. Viable inocula caused significant mortality when 10^7 organisms were injected, but toxicity fell sharply when the concentration of bacteria decreased; 10^6 organisms caused no significant mortality. Abbott 6406-125 grown at 30° C did not appear more toxic than the original powder preparation, but the commercial growth conditions are unknown. The intracerebral toxicity of *B. thuringiensis* H3a,3b and *B. t. i.* was comparable; autoclaved inocula of either variety were not toxic, so intracerebral toxicity may result from the presence of bacterial metabolites and/or heat-labile toxins at high concentrations.

The ability of *B. t. i.* to replicate or persist in brain tissue was evaluated. Twenty-six young rats (Hsd: Sprague Dawley [SD] BR) were intracerebrally injected with 1.15×10^5 organisms, and the animals were sacrificed 10, 13, 17, 20, 24, and 27 days after exposure. The right cerebral hemisphere was saved sterilely, weighed, ground, and suspended in 10 ml of HBSS. The suspension was then titrated in order to determine the number of viable organisms per 100 mg of wet tissue.

Bacterial recovery fell rapidly, which indicated that no bacterial replication occurred. The maximum number of bacteria was recovered on the first day of collection, day 10 (750 bacteria per 100 mg wet tissue); by day 22, only 100 bacteria per 100 mg wet tissue were recovered. The rate of clearance was logarithmic and can be described by the following equation:

$$Y = 2207 - 712 \text{ LnX} \ (r^2 = 0.58, p < 0.001).$$

Although there was no replication, viable bacteria persisted for as long as 3 weeks, most likely as spores.

The data from these three experiments indicate that the intracerebral toxicity of *B. thuringiensis* H3a,3b and *B. t. i.* may be influenced by the growth medium; incubation temperature influenced the toxicity of *B. t. i.* as well. There was no evidence of replication, and no gross lesions were observed, although there was persistence of *B. t. i.* for 3 weeks in brain tissue. *B. thuringiensis* H3a,3b and *B. t. i.* were comparable in terms of their toxicity, and a concentration of 10^7 organisms injected caused significant mortality. Since *B. thuringiensis* has a history of safe use, mortality produced at this concentration is not worrisome. Additionally, United States federal regulations specify that injection of 10^6 spores of *B. thuringiensis* per mouse gives an ample safety margin (Burges 1981), and the route of administration specified in the regulations was not as challenging as intracerebral injection.

12.3.5 Clearance

The spread of *B. t. i.* from injection sites into the general circulation was monitored by removing and mincing the spleen and then culturing the tissue suspension. Bacteria injected into tissue may be redistributed by the flow of tissue fluid through the lymphatic system and may eventually reach the bloodstream. Subsequently, they may be filtered by the spleen or liver. *B. t. i.* was recovered from the spleens of rats and mice following subcutaneous, intraperitoneal, and intracerebral injection (table 12.2). There was no evidence of multiplication, though viable organisms were recovered as late as 3 weeks after injection.

Two experiments were conducted in order to determine the importance of an intact immune system in preventing infection and to quantify the rate of clearance or disappearance of bacteria from the spleen. Immune suppression was achieved by two means. General suppression of the immune system was accomplished by administration of 1.25 mg cortisone acetate (Cortone® Acetate; Merck, Sharp, and Dohme) intramuscularly twice a week, starting 1 week prior to injection with *B.t.i.,* to 40 inbred mice (BALB/cAnNHsd BR); 42 inbred euthymic mice without corticosteroid injections served as controls. The mice were sacrificed at regular intervals beginning 4 days after administration of *B.t.i.,* and the experiment ended 49 days after administration. The spleens were collected aseptically and minced, and the number of bacteria per gram spleen was calculated by serial dilution. The need for an intact thymic-dependent lymphoid system for protection against *B.t.i.* was evaluated by injecting the bacterium into 30 athymic mice (Hsd: Athymic Nude nu AF), which lack T-lymphocytes and also cannot mount T-cell–dependent antibody response. Thirty euthymic inbred mice served as controls. Three mice per group were sacrificed at selected intervals beginning day 1 and ending 36 days after injection with *B.t.i.;* their spleens were collected and the number of cfu per gram spleen determined as in the previous experiment.

Recovery of *B.t.i.* declined over time from all experimental groups, including immunosuppressed and immunodeficient (athymic) mice, although there was considerable variation. *B.t.i.* SAN 402 WDC 60 cleared logarithmically from euthymic mice according to the formula $Y = 2{,}068{,}860 - 515{,}079 \text{ Ln } X$, $r^2 = 0.39$, $p < 0.001$. The same suspension cleared logarithmically from euthymic mice injected with corticosteroids according to the formula $Y = 3{,}266{,}344 - 777{,}778 \text{ Ln } X$, $r^2 = 0.31$, $p < 0.001$. *B.t.i.*-82161 clearance from athymic mice was best described by a power curve, $\text{Ln } Y = \text{Ln } 8{,}952{,}128 - 0.769 X$, $r^2 = 0.51$, $p < 0.001$; while the same suspension cleared logarithmically from euthymic mice according to the formula $Y = 2{,}407{,}101 - 580{,}856 \text{ Ln } X$, $r^2 = 0.18$, $p < 0.001$. The data from this last experiment are reported in table 12.4 and illustrate the extent of variance as well as the difficulty in predicting clearance in an individual mouse. The rate of clearance of *B.t.i.* was similar for euthymic mice receiving SAN 402 WDC 60 and *B.t.i.* 82161 because the slopes of the two curves could be pooled, but the curves themselves cannot be pooled because the intercepts are different, which may reflect the different concentrations of bacteria injected (SAN 402 $= 3.4 \times 10^7$ cfu per gram spleen, 82161 $= 2.6 \times 10^7$ cfu per gram spleen). Clearance rates differed for the coricosteroid injected and T-cell deficient mice, since the slopes of their curves could not be pooled. Athymic mice had the highest initial levels of bacteria present in their spleens, and the euthymic mice had the lowest.

TABLE 12.4.
Clearance of *B.t.i.* from the Spleens of Inbred and Athymic Mice

Days After Injection	Inbred Mice	Athymic Mice
1	1,106,857 ± 535,089	7,686,667 ± 558,000
3	4,309,091 ± 1,701,319	4,080,235 ± 2,118,113
7	393,345 ± 70,810	1,451,632 ± 61,310
10	828,952 ± 335,402	1,205,558 ± 482,465
14	322,991 ± 168,692	565,827 ± 92,788
17	1,011,686 ± 645,473	460,896 ± 30,340
21	429,466 ± 51,912	717,319 ± 82,894
24	635,572 ± 260,305	1,274,231 ± 447,298
28	613,916 ± 156,846	1,400,171 ± 994,637
36	208,866 ± 32,457	232,388 ± 15,096

NOTE: Data are given as mean colony-forming units (cfu) per gram spleen ± the standard error of the mean.

The literature states that athymic mice have elevated macrophage activity (Sharp and Colston 1984; Tamura et al. 1980), so the elevated bacterial level in athymic mice is puzzling. Phagocytosis should be enhanced in athymic mice, and this increased phagocytic activity should initially result in lower levels of bacteria in athymic mice than in euthymic mice until the immune system of euthymic mice becomes activated. We repeated this intraperitoneal injection experiment using *B. thuringiensis* H3a,3b; 30 athymic and 30 euthymic mice each received 1.9×10^6 spores, and their spleens were titrated in the manner already described. Athymic mice receiving *B. thuringiensis* H3a,3b had lower initial levels of bacteria than their conventional counterparts. Our experimental design could not determine whether this was due to enhanced phagocytosis or to lysis. More importantly, both varieties of bacteria persisted in the spleen for as long as 7 weeks in conventional and immunosuppressed mice, although replication did not occur.

An additional experiment was conducted in order to determine the maximum length of persistence of *B.t.i.*. Outbred mice were injected intraperitoneally with a washed suspension of *B.t.i.* 82161 containing 2.72×10^7 cfu in 0.5 ml. Mice were selected at intervals over 80 days and sacrificed; their spleens were collected and titrated as previously described. The clearance rate for *B.t.i.* was calculated by regression techniques, and heart blood was sampled on days 67 and 80.

There was no decline in cfu recovery over the 80-day period (table 12.5), and a regression could not be calculated. Heart blood from the mice was positive for *B.t.i.* on days 67 and 80, indicating that the mice were bacteremic. Many spleens were enlarged, and spleen weights as high as 860 mg were recorded. Analysis of the original innoculum using heat treatment at 65°

TABLE 12.5.
Recovery of *B.t.i.* Colony-forming Units (cfu) from the Spleens of Outbred Female Euthymic Mice

Days After Injection	N[a]	CFU Per Gram Spleen (± Standard Error)
11	4	1,323,295 ± 518,995
15	4	4,937,782 ± 1,288,348
25	4	3,091,309 ± 1,456,307
31	4	2,559,625 ± 1,080,284
45	4	2,241,752 ± 727,558
53	4	1,098,541 ± 310,108
58	3	3,321,789 ± 184,893
67	3	2,058,438 ± 745,282
80	3	1,612,387 ± 10,279

NOTE: Initial inoculum was 2.72×10^7 cfu per mouse.
[a]Number of spleens sampled.

C for 30 minutes indicated that the mice were injected with an innoculum containing 80% heat-susceptible cfu (germinated spores and/or bacilli), and this may have resulted in the failure of *B.t.i.* to clear. We note that the *B.t.i.* 82161 had been kept at 4° C for several years, and germination may have occurred under these conditions.

New material (ABG-6193) was received and tested for the percentage of spores present after the washing procedure. This material contained 75% spores based on response to heat treatment, and did not differ significantly from the spore composition of unwashed material. The washed material was then split into two aliquots, with one aliquot heat-treated to kill all vegetative forms and the other kept at room temperature. Equal cfu from both aliquots (1.5×10^7) were injected into new outbred mice and the short-term (15-day) clearance rate determined.

Both the mice receiving the heat-treated inoculum containing 100% spores (spores are unaffected by the heat-treatment regimen) and the mice injected with the mixed inoculum successfully cleared *B.t.i.* from their spleens over the 15-day period. Clearance for both groups could be described by an exponential function: the mice that received the 100% spore innoculum cleared *B.t.i.* at twice the rate as the mice receiving the mixed innoculum. However, these slopes did not differ significantly due to considerable variation in the mixed innoculum group (table 12.6).

An additional clearance experiment was conducted using euthymic and athymic mice, which have an impaired immune response since they lack T-lymphocytes. Both groups of mice were injected with 2×10^7 cfu of a mixed innoculum of *B.t.i.* in order to determine if athymic mice were more susceptible to infection than euthymic mice when vegetative stages of *B.t.i.* were

TABLE 12.6.
Recovery of *B.t.i.* Colony-forming Units (cfu) from Outbred Female Euthymic Mice

Days After Injection	*100% Spore Inoculum* cfu per Gram Spleen ± Standard Error	*Mixed Inoculum* cfu per Gram Spleen ± Standard Error
1	10,959,596 ± 2,071,323	1,597,578 ± 153,425
3	9,672,749 ± 6,475,596	1,282,011 ± 717,115
6	1,858,975 ± 498,601	879,062 ± 330,688
12	202,778 ± 57,200	1,026,437 ± 512,491
15	329,791 ± 132,595	464,896 ± 375,018

NOTE: Initial inoculum was 1.50×10^7 cfu per mouse.

present in the inoculum. Mice from both groups were killed at intervals over 27 days; the rate of clearance was determined, and the strain differences were evaluated using both regression and ANOVA techniques.

B.t.i. did not clear during the first 10 days from either euthymic or athymic mice, but cfu recovery declined after that point in time for both groups. Although the regressions of cfu recovery were significant, considerable variation produced very low r^2 values. Analysis of clearance using a two-factor ANOVA indicated that clearance differed significantly between athymic and euthymic mice—athymic mice had higher levels of *B.t.i.* in their spleens—but the levels in both groups declined over time.

Lamanna and Jones (1963) stated that "when a spore is injected, three opportunities in time are presented for phagocytosis preceding multiplication: injected spore phase, emergent germination phase, and fully developed vegetative bacillus phase. With injection of vegetative bacilli, only one stage of the life cycle is available for effective phagocytosis before vegetative proliferation proceeds to successful infection." Therefore, one would predict that a 100% spore innoculum would clear more rapidly from a test animal than a 100% bacilli innoculum because there would be less likelihood of replacement of those bacteria removed by host defenses.

Our laboratory clearance data confirm this statement. Mice injected with 100% *B.t.i.* spores cleared the innoculum more rapidly than did mice injected with vegetative (heat-susceptible) stages. The one experiment in which *B.t.i.* failed to clear inadvertently used an innoculum containing 80% vegetative stages, and, based on the successful clearance of *B.t.i.* 82161 in earlier experiments, it is reasonable to assume that considerable germination had taken place during storage. The new commercial suspension tested (ABG-6193) did not contain 100% spores, but heat-susceptible stages only comprised 25% of the suspension. Exposure of euthymic and athymic mice to this material, which more accurately reflects the end-use circumstances, resulted in successful clearance, although there was considerable variation in cfu recovery. We stress that, although there was evidence of multiplication in these studies, such as failure of *B.t.i.* to clear, there was no evidence of disease. Representative tissue from lung, spleen, liver, heart, and kidneys was examined, and no lesions were present. Therefore, we found no evidence that *B.t.i.* was infectious and feel that our studies underscore the importance of innoculum composition, as well as route of administration, in assessing the results of clearance and safety studies.

12.3.6 Ocular Irritancy

Two isolates of *B.t.i.* (R153-78, ABG-6108) were administered daily as powders (50 mg) to six rabbit eyes (2 to 3 kg, New Zealand White) for a total of 10 times over a two-week period. Each treated eye received the preparation in the conjunctival cul-de-sac, and an untreated eye in each rabbit served as a control. Glove powder (Bio-Sorb lot B208) was administered daily to three eyes as a control treatment for ABG-6108 as well. The conjunctiva was scored according to Draize, Woodward, and Calvery (1944), and the eyes were also observed with a slit lamp and the corneas were scored according to a standard system of McDonald and Shadduck (1977).

ABG-6108 was not an irritant, and the congestion it caused was minimal; its effects were indistinguishable from glove powder (table 12.7). In contrast, R153-78 was quite irritating, and it caused significant conjunctival congestion, swelling, and discharge, as well as severe corneal irritation following

TABLE 12.7.
Ocular Irritancy of *B.t.i.* Isolate ABG-6108

	Days After Application																			
	1		2		3		4		5		6		7		11		12		13	
	A	B	A	B	A	B	A	B	A	B	A	B	A	B	A	B	A	B	A	B
Conjunctiva																				
Congestion	6	0	6	1	5	1	4	1	4	1	6	3	6	3	6	3	6	3	6	3
Swelling	0	0	0	0	0	0	0	0	0	0	0	0	0	0	0	0	0	0	0	0
Discharge	0	0	1	0	0	0	0	0	0	0	0	0	4	0	4	0	4	0	2	0
Iritis	0	0	0	0	0	0	0	0	0	0	0	0	0	0	0	0	0	0	0	0
Corneal																				
Opacity	0	0	0	0	0	0	0	0	0	0	1	0	2	0	2	0	2	0	2	0

NOTE: Treated eyes were dosed daily with 50 mg material, 5 times per week over a 2-week period. The number of treated eyes differing from normal eyes is given; a total of 6 eyes were treated. Viable *B.t.i.* persisted up to 41 days in the eye. A = Isolate ABG-6108; B = Glove powder: Bio-Sorb lot B208. Arbrook Inc., Arlington, Texas. Three eyes were exposed.

TABLE 12.8.
Ocular Irritancy of *B.t.i.* Isolate R153-78

	Days After Application								
	1	2	3	4	7	9	10	11	14
Conjunctiva									
Congestion	8	8	8	8	8	8	8	6	8
Swelling	5	4	8	4	0	5	5	4	0
Discharge	8	8	0	8	0	6	6	6	0
Iritis	0	0	2	1	0	0	0	0	0
Corneal									
Opacity	2	0	3	3	5	7	7	5	5

NOTE: Treated eyes were dosed daily with 50 mg material, 5 times per week over a 2-week period. The number of treated eyes differing from normal eyes is given; a total of 8 eyes were treated.

repeated administration (table 12.8). The conjunctival symptoms rapidly decreased following cessation of administration.

The rapidity of the clearance suggests that the ocular irritancy of R153-78 was primarily due to its physical characteristics; when exposure ceased, the effects disappeared. R153-78 had been stored in a desiccator at 4° C and had the consistency of dry paste when it was applied. There were large, sharp-edged clumps that were retained in the conjunctival cul-de-sac. In contrast, ABG-6108 was a dry and dusty powder with a much smaller particle size, and less material was retained in the conjunctival cul-de-sac. Neither isolate caused infection, although viable spores persisted for as long as 41 days in the eyes of the rabbits.

12.4 SUMMARY

Oral, intraperitoneal, and aerosol exposure of *B. t. i.* produced no significant illness or mortality. The entomopathogen disappeared rapidly from the lungs of rats, and there was no evidence of multiplication. Subcutaneous injection of *B. t. i.* produced abscesses, but the abscesses appeared following injection with autoclaved material as well. These abscesses most likely arose from the presence of heat-stable foreign material in the injection site. The ocular irritancy of *B. t. i.* depended on the physical state of the preparation; fine powders produced no irritation, and large clumps caused considerable corneal irritation, conjunctival swelling, congestion, and discharge.

Intracerebral injection of 10^7 cfu resulted in mortality, and both culture medium and incubation temperature affected the toxicity of *B. t. i.* Mortality may have resulted from the direct deposition of large quantities of bacteria and/or bacterial metabolites into a vulnerable area of the brain; there were no gross lesions or signs of infection, however. The intracerebral toxicity of *B. t. i.* was comparable to that of *B. thuringiensis*, so we feel that mortality resulting from injection of this high concentration of bacteria is not worrisome.

B. t. i. was not confined to the injection site and entered the general circulation following subcutaneous, intraperitoneal, and intracerebral injection. Viable spores persisted in the brain for three weeks, and in one clearance study, injection of mice with a predominantly vegetative innoculum (determined by susceptibility to heat) resulted in a failure of *B. t. i.* to clear. Multiplication may have occurred in mice, because *B. t. i.* remained at the same level despite removal by host defenses, and follow-up studies suggested that inoculum composition played an important role in determining the rate of clearance. An intact immune system was not necessary for successful clearance of *B. t. i.*, although athymic mice had higher levels of *B. t. i.* in their spleens than did their euthymic counterparts.

As use of *B. t. i.* increases, so will human exposure to this entomopathogen; predictably, isolation of *B. t. i.* from humans will increase. The ability of the spores of both *B. t. i.* and *B. thuringiensis* H3a,3b to remain viable in mammalian tissue for extended periods of time may lead to their incorrect identification as causal agents of mammalian infection, which underscores the importance of identifying vegetative stages of a bacterium in a lesion before labeling a particular organism as the cause of disease. This issue is further complicated by our finding that commercial suspensions of *B. t. i.* may contain vegetative forms and that the clearance rate of *B. t. i.* in laboratory animals is influenced by innoculum composition. Consequently, even if vegetative stages of *B. t. i.* are recovered from humans or animals, it does not mean that this organism caused disease and may be a simple function of the elapsed time since exposure. In our studies, intraperitoneal injection of *B. t. i.* was the only route that caused significant mortality; since it only occurred at high concentrations (10^7 cfu per rat), we conclude that *B. t. i.* is not a significant mammalian pathogen. *B. t. i.* appears comparable to *B. thuringiensis* H3a,3b with regard to safety to laboratory animals, and it can be used safely in environments where human and mammalian exposure is likely to occur.

We have not dealt in depth with the toxicity of alkali-solubilized (pH 12) crystal δ-endotoxin of *B. t. i.* to mammals because it is not activated in mammals. However, *B. t. i.* endotoxin activated in the laboratory has been demonstrated to be toxic to mice and cytolytic to human erythrocytes. Thomas and Ellar (1983) reported that intravenous injection of 15–30 μg of solubilized δ-endotoxin per gram body weight of BALB/c suckling mice resulted in death within two to three hours. This solubilized crystal endotoxin was not toxic when administered *per os*, and this was independently confirmed by H. de Barjac (pers. comm.). The purified endotoxin was also hemolytic to rat, sheep, mouse, and horse erythrocytes in vitro. Armstrong, Rohrmann, and Beaudreau (1985) reported that a 25-kDa polypeptide was responsible for the mouse toxicity and that 100 μg of this polypeptide injected intraperitoneally into six-week-old BALB/c mice caused mortality within six hours. Gill, Singh, and Hornung (1987) noted that two toxic polypeptides of 24- and 25-kDa could be purified from parasporal proteinaceous crystals of *B.t.i.* and that both of these polypeptides lysed human erythrocytes in vitro. The authors concluded that the cytolytic polypeptides did not act as a nonspecific detergent on the cell membrane, but rather interacted with phospholipid receptors on the cell membrane to increase permeability.

12.5 SUMMARY OF OTHER STUDIES

As cited by permission of the authors (de Barjac et al. 1980), the following are the conclusions of other experiments on the safety of *B.t.i.*, which further confirm the innocuousness of this bacterium to mammals.

The toxicological studies were performed by experimental exposure of various animals (mice, rats, guinea pigs, and rabbits), and indicated the absence of acute and prolonged toxicity. All acute toxicity tests gave negative results through a variety of pathways—subcutaneous and intraperitoneal injection, gavage, inhalation, percutaneous application, scarification of the skin, and ocular inoculations—using mean doses of about 10^7 to 10^8 bacteria per animal. There was also no anaphylactic shock obtained in guinea pigs, and successive passages in mice of *B.t.i.* did not induce any virulence. In investigating subacute oral toxicity in mice and rats, through repeated administration of *B.t.i.* for a period of three weeks, during which the equivalent of about 10^{11} to 10^{12} bacteria per animal was ingested, results were negative. In summary, neither pathological symptoms, nor diseases, nor mortality have been observed. The behavior and the weight gain of the experimentally exposed animals were within the normal range and did not significantly differ from the controls. At the end of each study, necropsies were conducted. No gross lesions or microscopic lesions were observed, and reisolation tests from various organs were negative. *B. thuringiensis* serotype H14 appears to be well tolerated by each species used in the tests. The bacteria did not multiply in mammals and were eliminated quite rapidly, very likely by natural defenses. These results confirm the innocuity of *B. thuringiensis*, already widely demonstrated for the other serotypes, H1 and 3a,3b, which for many years have been subject to large-scale use in agricultural and forestry areas.

Acknowledgments

Supported by the World Health Organization Special Program for Research and Training in Tropical Diseases. Equipment and guidance for the aerosol exposures were provided by Dr. Stanley Curtis, Department of Animal Sciences, University of Illinois, and Rose Ann Meccoli provided significant technical assistance.

References

Armstrong, J. L.; Rohrmann, G. F.; and Beaudreau, G. S. 1985. Delta endotoxin of *Bacillus thuringiensis* subsp. *israelensis*. *J. Bacteriol.* 161: (1): 39–46.

Barjac, H. de; Larget, I.; Benichou, L.; Cosmao Dumanoir, V.; Viviani, G.; Ripouteau, H.; and Papion, S. 1980. Innocuity tests on mammals with serotype H14 of *Bacillus thuringiensis*. WHO Document WHO/VBC/80.761. Mimeo.

Burges, H. D. 1981. Safety, safety testing, and quality control of microbial pesticides. In *Microbial control of pests and plant diseases*, ed. H. D. Burges, 738–767. London: Academic Press.

Draize, J. H.; Woodard, G.; and Calvery, H. O. 1944. Methods for the study of irritation and toxicity of substances applied topically to the skin and mucous membranes. *J. Pharmacol. Exp. Therap.* 82: 377–390.

Fisher, R., and Rosner, L. 1959. Toxicology of the microbial insecticide, Thuricide. *Agricul. Food Safety* 10: 686–688.

Gill, S. S.; Singh, G. J. P.; and Hornung, J. M. 1987. Cell membrane interaction of *Bacillus thuringiensis* subsp. *israelensis* cytolytic toxin. *Infect. Immun.* 55: 1300–1308.

Gordon, R. E. 1977. Some taxonomic observations on the genus *Bacillus*. In *Biological regulation of vectors: The saprophytic and aerobic bacteria and fungi*, ed. J. B. Briggs, 67–82. U.S. Department of Health, Education, and Welfare, Publ. NIH 77-1180.

Lamanna, C., and Jones, L. 1963. Lethality for mice of vegetative and spore forms of *Bacillus cereus* and *Bacillus cereus*-like insect pathogens injected intraperitoneally and subcutaneously. *J. Bacteriol.* 85: 532–535.

McDonald, T., and Shadduck, J. A. 1977. Ocular irritation testing. In *Advances in modern toxicology*, ed. H. I. Maiback and F. N. Marzulli, 4: 130–191.

Pearson, H. E. 1970. Human infections caused by organisms of the *Bacillus* species. *Amer. J. Clin. Path.* 53: 506–515.

Samples, J. R., and Buettner, H. 1983. Corneal ulcer caused by a biological insecticide (*Bacillus thuringiensis*). *Am. J. Ophthalmol.* 95: 258–260.

Shadduck, J. A. 1983. Some considerations on the safety evaluation of non-viral microbial pesticides. *WHO Bull.* 61 (1): 117–128.

Shadduck, J. A.; Roberts, D. W.; and Lause, S. 1982. Mammalian safety tests of *Metarhizium anisopliae*: Preliminary results. *Environ. Entomol.* 11: 189–192.

Shadduck, J. A.; Singer, S.; and Lause, S. 1980. Lack of mammalian pathogenicity of entomicidal isolates of *Bacillus sphaericus*. *Environ. Entomol.* 9: 403–407.

Sharp, A. K., and Colston, M. J. 1984. Elevated macrophage activity in nude mice. *Exp. Cell Biol.* 51 (1–2): 44–47.

Siegel, J. P.; Shadduck, J. A.; and Szabo, J. 1987. Safety of the entomopathogen *Bacillus thuringiensis* var. *israelensis* for mammals. *J. Econ. Entomol.* 80: 717–723.

Tamura, T.; Sakaguchi, A.; Kai, C.; and Fujiwara, K. 1980. The role of macrophages in mouse hepatitus virus infection nude mice. *Microbiol. Immunol.* 23 (10): 965–974.

Thomas, W. E., and Ellar, D. J. 1983. *Bacillus thuringiensis* var. *israelensis* crystal δ-endotoxin: Effects on insect and mammalian cells in vitro and in vivo. *J. Cell Sci.* 60: 181–197.

PART 2

Bacillus sphaericus

13

Introduction to the Study of *Bacillus sphaericus* as a Mosquito Control Agent

SAMUEL SINGER

13.1 INTRODUCTION

Some of us are *Bacillus sphaericus* watchers, including John Briggs, Betty Davidson, Al Yousten, Larry Lacey, and Huguette de Barjac, to name but a few. We were present either during the early isolation and development of this bacterium or, having worked with the microorganism, were convinced early on of its great potential as a field agent for the control of mosquito larvae. Needless to say, all of us are pleased by the recent progress in the field—in the isolation of improved strains and in the energy presently evident in this most interesting research area. We are all even more pleased that industry (Abbott Laboratories in the United States and the Biochem-Solvay group in Belgium) has decided to consider, seriously, the development of material for field use.

The early history of *B. sphaericus* is marked by the isolation of key strains, which may have not been as active as the present candidates, but did play an important role during the development of the *B. sphaericus* array of cultures (table 13.1). Unlike *Bacillus thuringiensis* subsp. *israelensis*, which is the result of only one documented isolation (Margalit and Dean 1985), the history of *B. sphaericus* is the history of the isolation of a series. An examination of the history of the development of this ubiquitous microorganism is the theme of this chapter. Lessons from this history should point out the directions that we should go in the future.

13.2 HISTORY OF EXTANT STRAINS

Kellen, during a routine surveillance of rock holes near Big Creek, Fresno, California, collected several moribund fourth-instar larvae of *Culiseta incidens* (Kellen and Meyers 1964). From these larvae he isolated several bacteria, among which was a *B. sphaericus*. Because of work at Bioferm with agricultural *B. thuringiensis*, it was decided to investigate the

TABLE 13.1.
History of the Development of Key *Bacillus sphaericus* Strains

Key Strains	Historical Significance of Key Strains	Other Strains and Relationship to Key Strains
K (USA)	First reported active isolate work discontinued after brief program	Q(USA); Phage group 1; low larvicidal activity
SSII-1 (India)	First active isolate universally available; fermentation and population stability problems	1404 (Phillipines); 1883, 1885–1893, 1895, 1896 (Israel); Phage group 2; moderate larvicidal activity
1593 (Indonesia)	First fermentation and population stable strain; one of key field candidates	1691, 1881 (El Salvador); 2013-4, 2013-6 (Roumania); 2117-1 (Phillipines); 2500, 2501 (Thailand); Phage group 3; high larvicidal activity
2362 (Nigeria)	First highly active African strain; the prime field candidate	
Lysenko isolates: 2537-2, 2533-1 (K1), 2533-1 (K2) (Guyana); 2601 (Hungary); 2602 (Czeckoslovakia)	First active strains isolated from nonmosquito terrestrial sources	
2297 (Sri Lanka)	Crystal first noted in this strain of key field candidates	2173, 2377 (India); 2317-3 (Thailand); Phage group 4; mixed larvicidal activity 1894 (Israel); Phage group 5; 2115 (Phillipines); Phage group 6; 2315 (Thailand); Phage group 7; moderate larvicidal activity

potential of this strain. Kellen's strain was later designated as strain K. It was the first reported active *B. sphaericus* isolate (Kellen and Meyers 1964; Kellen et al. 1965). After a modest effort and for various commercial reasons, work on this strain was discontinued in 1967. The LC_{50} values of strain K were of a low order of larvicidal activity. Being originally of proprietary interest this strain was not generally available (except from Kellen), and work with it was no longer pursued by Kellen or by us.

In 1970, through affiliation with John Briggs and his Collaborating Center for Biological Control (WHO/CCBC), a part of a new World Health Organization (WHO) effort (Cantwell and Laird 1966; Singer 1973) to examine potential agents obtained from WHO field stations, I was sent a series of dead larvae from which many of the subsequent candidates were isolated. The isolation of strain SSII-1 (Singer 1973) renewed interest in *B. sphaericus*. This and subsequent strains were made generally available. Strain SSII-1 demonstrated that strain K was not a singular event. Up to this point, and in a few isolated instances as late as 1976, many workers felt that the *Bacilli* they had isolated from moribund larvae were opportunistic saprophytic microorganisms not worth a further look.

The isolation of SSII-1 also afforded an opportunity to study the pathogenesis of the organism (Davidson, Singer, and Briggs 1975), its safety and ecological impact (Davidson, Singer, and Briggs 1975; Davidson et al. 1977; Singer 1975; Shadduck, Singer, and Lause 1980), and tests in the field (Ramoska, Singer, and Levy 1977; Ramoska, Burgess, and Singer 1978). In light of the primitive nature of our fermentation efforts when compared to present-day biotechnology, culture preparations of strain SSII-1 were unstable; as a result, preliminary field tests gave a mixed efficacy signal. Only broth cultures were available, and shipping even refrigerated materials any distance, more often than not, gave spurious results. It required heroic efforts—Briggs hand-carried refrigerated samples to Nigeria (via Geneva) for testing—to get any information from endogenous malarial areas.

In early 1975, material received from WHO/CCBC changed all of this. Strain 1593, derived from this material, performed better from a fermentation point of view. It gave LC_{50} values that one could consider superior (Singer 1977). None of the strains isolated to date show mammalian toxicity, and none show any effects against nontarget species, so they do not perturb the environment. Early commercial fermentation material (dry powders) had some problems, but for a first time, commercial preparation of the material was quite useful. The production and availability of international standard powders by de Barjac greatly ameliorated the immediate problems.

Contrary to popular belief, the isolation by Goldberg and Margalit (1977) of what later came to be called *B.t.i.* (*Bacillus thuringiensis* subsp. *israelensis*, de Barjac 1978a) had a salutary effect on *B. sphaericus* work. Most of the effort with *B. sphaericus* to that point (and to a large extent even now) had been carried out in academic laboratories and institutions in the United States and in Europe. With the isolation of *B.t.i.*, industry's interest was aroused. Here was a situation where the technology was in place. Industrial companies had been producing *B. thuringiensis* material for agricultural use for over thirty years. It took only a slight modification of their technology to be able to produce *B.t.i.* Interest in microbial insecticides soon spread. This can be seen in the vast increase in the number of publications relating to these microorganisms since 1980.

Wickremesinghe and Mendis (1980) isolated what they called strain MR-4, which has since been given the designation 2297. This strain, along with strains 1593 and 2362, remains one of the principal candidates of field interest. What is particularly interesting about strain 2297 is the presence of a very large parasporal body (crystal) similar to but not identical to the *B.t.*

crystal (Davidson and Myers 1981; de Barjac and Charles 1983). The most highly active strains of *B. sphaericus* were all found to possess the "crystal" upon close inspection under phase contrast microscopy. This is not to say that this parasporal body is the site of all of the activity; many insecticidal cells that had not yet sporulated and therefore did not contain the crystal were still insecticidal (Davidson and Myers 1981).

Weiser (1984) isolated strain 2362, another highly active field candidate, which is reported to be as good or better than strains 1593 and 2297 (Yousten 1984). What is particularly interesting about this strain is that it was the first active *B. sphaericus* isolated in Africa (Nigeria) and was isolated from a nonmosquito source, namely *Simulium* adults. Along these same lines, Lysenko et al. (1985) isolated a series of five *B. sphaericus* cultures from nonmosquito sources (caterpillars and grasshoppers) having larvicidal activity similar to strains 1593 and 2362. As with 2362, which is not active against *Simulium* from which it was isolated, Lysenko's strains are only active against mosquito larvae. We therefore have the isolation of mosquito larvicidal strains from aquatic (2362) and terrestrial (Lysenko's strains) nonmosquito sources.

Of the 30 to 50 strains of *B. sphaericus* usually cited as being available for use (and the list is growing), a majority of the research effort has involved the few strains cited above. In most instances the remaining strains, such as strains Q, 1404, 1691, 1881, 2013-4, 2013-6, 2117-2, and 2500, were isolated during the above-cited course of events and usually did not offer properties that were different from the extant cultures of that time (for a list of these cultures, see Singer 1980; Yousten 1984; de Barjac et al. 1985; and chapter 14). One can judge the relative dates of the isolations by their WHO accession numbers, which subsequently became their strain number. These other cultures are worth consideration for more than historical reasons. A whole series of *B. sphaericus* cultures (strains 1883, 1885–1894, 1896) was isolated in Israel from the same source material from which *B.t.i.* (WHO/CCBC no. 1896) was isolated. Many of these cultures, isolated from specific malaria-endogenous areas have been reported to have larvicidal activity equivalent to the principal candidate cultures 1593, 2297, and 2362 (Singer 1980). These other strains would be most useful to satisfy the local national desire for endogenously derived candidates. In addition, all 30–50 cultures constitute a framework for taxonomic considerations. These cultures are also a genetic pool from which material for future genetic manipulation may be derived.

As the many larvicidal strains of *B. sphaericus* were isolated and examined in the late 1970s and 1980s, it became apparent that there was a need to differentiate among the strains. From just a fermentation point of view, one can compare only a limited number of strains at any one time or else the cost becomes prohibitive. There are few phenotypic differences between insecticidal and noninsecticidal strains other than insecticidal activity (Singer 1980; de Barjac, Véron, and Cosmao Dumanoir 1980; Yousten 1984). Other approaches had to be developed to differentiate between these strains and to verify the identity of the strains. A big step was taken in this direction when Yousten (Yousten et al. 1980; Yousten 1984) developed a combination of bacteriophage for this purpose and de Barjac used H-antigens to serotype the strains (de Barjac, Véron, and Cosmao Dumanoir 1980). There is an excellent correlation between the two methods. An interesting consequence of these typing methods has been the information that all of the most active strains, with the exception of strain 2297, belong to phage-type group 3 (Yousten et al. 1980) and serotype H5a,5b (de Barjac et al. 1985). A word of caution, however: neither method can distinguish within the phage-type or serotype group. This indicates that either the members of the phage-type group, or serotype group, are identical or that there is a limit to the methods' sensitivity. The latter is most probable. There is much more to be done in this area, particularly if the gene pool represented by these strains is to be used in genetic manipulation.

13.3 FUTURE STRAINS

With 30–50 strains of *B. sphaericus* presently available, why search for more strains? The answer lies in the spectrum of activity against the target insects. Unlike *B. t. i., B. sphaericus* is limited to effects against mosquito larvae. Even among the mosquito larvae certain genera (e.g., *Aedes*) are less susceptible to *B. sphaericus* toxin (de Barjac 1978b; Singer 1980). The prime advantage of the *B. sphaericus* group of strains lies in their ability to perform for a longer period of time in the environment (Singer 1980; see chapters 18 and 19).

There are at least two approaches that can be used to obtain more desirable strains: continued isolation from the wild and genetic manipulation of the available strains (through recombinant DNA technology or protoplast fusion). In the first approach, the number of sources from which isolates were obtained in the past is small compared to what can yet be sampled. The question could be asked, however, are we dipping too often into the same well? We should know when we reach this point of diminishing returns because we will increasingly start to isolate the same strains as measured by their phage-type or serotype. This point may already have been reached. It would seem to me that one of the lessons from Weiser's and Lysenko's work is that it may be more profitable to explore nonmosquito sources for this next generation of isolations while perhaps still isolating in the more classical manner from moribund mosquito larvae to satisfy the local need for endogenously derived candidates.

In the second approach, at present there appears to be some effort to understand the genetics of *B. sphaericus* and to obtain vectors suitable for transferring DNA material into *B. sphaericus* recipient strains (McDonald and Burke 1984). Genetic efforts with *B.t.* have progressed remarkably the last few years (Aronson, Beckman, and Dunn 1986). But if we expect direct dividends from this *B.t.* work to be immediately applicable to *B. sphaericus*, I am afraid we are bound to be disappointed. We must keep in mind that differences in biology, cell structure, and enzyme complement between these two groups may preclude direct and immediate application of information from one to the other. A separate, distinct effort will be needed to gain sufficient genetic information for future genetic manipulation of the *B. sphaericus* group. We must also ask what the goal(s) of genetic manipulation should be. I believe that most *B. sphaericus* watchers would agree that the ultimate aim is to place *B.t.* toxin(s) into *B. sphaericus* recipient cells. This would give us the benefits of both worlds: the wider spectrum of *B.t.i.* in a bacterium capable of surviving, persisting, and recycling in the environment. Perhaps it is naive to visualize that synergisms will accrue by this combination of *B.t.* and *B. sphaericus* and that new activities will develop.

Achievement of this goal by direct genetic manipulation via appropriate vector systems awaits the results of further genetic work with *B. sphaericus.* A more direct approach favored by my laboratory is that of protoplast fusion of appropriate *B. thuringiensis* and *B. sphaericus* parents. Protoplast fusion is a versatile general technique to induce genetic recombination in a variety of prokaryotes and eukaryotes. Protoplast fusion is particularly useful for procaryotes that have not been subjected to extensive genetic analysis because it does not require transducing phage, plasmid sex factors, or competency development. The technique has its own problems, however, in terms of generating and fusing stable protoplasts and regenerating viable cells from the fusants (Matsushima and Baltz 1986). Whether one proceeds via recombinant DNA techniques or protoplast fusion, one must understand that the bottom line of this particular area of bacterial insecticide research is an agent that would have a direct bearing on the reduction of the incidence of malaria and related tropical diseases.

References

Aronson, W. I.; Beckman, W.; and Dunn, P. 1986. *Bacillus thuringiensis* and related insect pathogens. *Microbiol. Rev.* 50: 1–24.

Barjac, H. de. 1978a. Une nouvelle variété de *Bacillus thuringiensis* très toxique pour les moustiques: *B. thuringiensis* var. *israelensis* sérotype 14. *C. R. Acad. Sci.* (Paris) 286D: 797–800.

———. 1978b. Un nouveau candidat à la lutte biologique contre les moustiques: *Bacillus thuringiensis* var. *israelensis. Entomophaga* 23: 309–319.

Barjac, H. de, and Charles, J.-F. 1983. Une nouvelle toxine active sur les moustiques presente dans des inclusions crystallines produites par *Bacillus sphaericus. C. R. Acad. Sci.* (Paris) 296D: 905–910.

Barjac, H. de; Larget, I.; Cosmao Dumanoir, V.; Bénichou, L.; and Viviani, G.; Ripouteau, H.; and Papion, S. 1979. Innocuité de *Bacillus sphaericus* souche 1593, pour les mammifères. WHO/VBC/79.731. Mimeo.

Barjac, H. de; Larget-Thiery, I.; Cosmao Dumanoir, V.; and Ripouteau, H. 1985. Serological classification of *Bacillus sphaericus* strains on the basis of toxicity to mosquito larvae. *Appl. Microbiol. Biotechnol.* 21: 85–90.

Barjac, H. de; Véron, M.; and Cosmao Dumanoir, V. 1980. Caractérisation biochimique et sérologique de souches de *Bacillus sphaericus* pathogènes ou non pour les moustiques. *Ann. Microbiol.* (Inst. Pasteur). 131B: 191–201.

Cantwell, G. E., and Laird, M. 1966. The World Health Organization kit for the collection and shipment of pathogens and parasites of diseased vectors. *J. Invertebr. Pathol.* 8: 442–451.

Davidson, E. W.; Martin, H. L.; Moffett, J. D.; and Singer, S. 1977. Effect of *Bacillus sphaericus* strain SSII-1 on honey bees, *Apis melifera. J. Invertebr. Pathol.* 29: 344–346.

Davidson, E. W., and Myers, P. 1981. Parasporal inclusions in *Bacillus sphaericus. FEMS Microbiol. Lett.* 10: 261–265.

Davidson, E. W.; Singer, S.; and Briggs, J. D. 1975. Pathogenesis of *Bacillus sphaericus* strain SSII-1 infections in *Culex pipiens quinquefasciatus* larvae. *J. Invertebr. Pathol.* 25: 179–184.

Goldberg, L. J., and Margalit, J. 1977. A bacterial spore demonstrating rapid larvicidal activity against *Anopheles sergentii, Uranotaenia unguiculata, Culex inivitattus, Aedes aegypti,* and *Culex pipiens. Mosq. News* 37: 355–358.

Kellen, W. R.; Clark, T. B.; Lindegren, J. E.; Ho, B. C.; Rogoff, M. H.; and Singer, S. 1965. *Bacillus sphaericus* Neide as a pathogen of mosquitoes. *J. Invertebr. Pathol.* 6: 442–448.

Kellen, W. R., and Meyers, C. M. 1964. *Bacillus sphaericus* Neide as a pathogen of mosquitoes. *J. Invertebr. Pathol.* 7: 442–448.

Lysenko, O.; Davidson, E. W.; Lacey, L. A.; and Yousten, A. A. 1985. Five new mosquito larvicidal strains of *Bacillus sphaericus* from non-mosquito origins. *J. Amer. Mosq. Control Assoc.* 1: 369–371.

McDonald, K. O., and Burke, W. F., Jr. 1984. Plasmid transformation of *Bacillus sphaericus* 1593. *J. Gen. Microbiol.* 130: 203–208.

Margalit, J., and Dean, D. 1985. The story of *Bacillus thuringiensis* var. *israelensis* (*B.t.i.*). *J. Amer. Mosq. Control Assoc.* 1: 1–7.

Matsushima, O., and Baltz, R. H. 1986. Protoplast fusion. In *Manual of industrial microbiology and biotechnology,* ed. A. L. Demain and N. A. Solomon, 170–183. Washington, D.C.: American Society for Microbiology.

Ramoska, W. A.; Burgess, J.; and Singer, S. 1978. Field applications of a bacterial insecticide. *Mosq. News* 38: 57–60.

Ramoska, W. A.; Singer, S.; and Levy, R. 1977. Bioassay of three strains of *Bacillus sphaericus* on field collected mosquito larvae. *J. Invertebr. Pathol.* 30: 151–154.

Shadduck, J. A.; Singer, S.; and Lause, S. 1980. Lack of mammalian pathogenicity of entomocidal isolates of *Bacillus sphaericus. Environ. Entomol.* 9: 403–407.

Singer, S. 1973. Insecticidal activity of recent bacterial isolates and their toxins against mosquito larvae. *Nature* 244: 110–111.

———. 1974. Entomogenous bacilli against mosquito larvae. *Dev. Industr. Microbiol.* 15: 187–194.

———. 1975. Use of bacteria for control of aquatic insect pects. In *Impact of the use of microorganisms on the aquatic environment,* 5–22. National E.P.A. Ecological Research Series, no. 600-3-75-001.

———. 1977. Isolation and development of bacterial pathogens of vectors. In *Biological regulation of vectors,* 3–18. DHEW Publication no. (NIH) 77-1180.

———. 1980. *Bacillus sphaericus* for the control of mosquitoes. *Biotechnol. Bioeng.* 22: 1335–1355.

Weiser, J. 1984. A mosquito-virulent *Bacillus sphaericus* in adult *Simulium damnosum* from Northern Nigeria. *Zbl. Mikrobiol.* 139: 57–60.

Wickremesinghe, R.S.B., and Mendis, C. L. 1980. *Bacillus sphaericus* spore from Sri Lanka demonstrating rapid larvicidal activity on *Culex quinquefasciatus. Mosq. News* 40: 387–389.

Yousten, A. A. 1984. *Bacillus sphaericus:* Microbiological factors related to its potential as a mosquito larvicide. *Adv. Biotechnol. Processes* 3: 315–343.

Yousten, A. A.; Barjac, H. de; Hedrick, J.; Cosmao Dumanoir, V.; and Myers, P. 1980. Comparison between bacteriophage typing and serotyping for the differentiation of *Bacillus sphaericus* strains. *Ann. Microbiol.* (Inst. Pasteur) 131B: 297–308.

14

Classification of *Bacillus sphaericus* Strains and Comparative Toxicity to Mosquito Larvae

HUGUETTE DE BARJAC

14.1 INTRODUCTION

Unlike *Bacillus thuringiensis*, *Bacillus sphaericus* was first recognized as a saprophytic microorganism, the spores of which currently contaminate soil and aquatic environments. Since the first reports of strains moderately toxic to mosquito larvae (strain K of Kellen and Meyers 1964; strain SSII-1 [= 1321] of Singer 1973), researchers have isolated a series of very potent *B. sphaericus* strains, beginning in 1977 with strain 1593 (Singer 1977). Strain 1593 was followed by strain 2297 (= MR-4) (Wickremesinghe and Mendis 1980), strain 2362 (= 290-8) (Weiser 1984), and strain Iab 59 (de Barjac et al. 1988). In addition to the numerous atoxic strains, in 1989 more than 150 mosquito pathogenic isolates with various toxicity levels have become available. This has created problems in their identification and differentiation—first from the atoxic *B. sphaericus*, and second from each other within the toxic group.

14.2 DIFFERENT APPROACHES TO THE CLASSIFICATION AND CHARACTERIZATION OF TOXIC STRAINS

Identification of *B. sphaericus* species in current bacteriology is typically made on the basis of its morphology: round spores located terminally in swollen sporangia. In addition to the 40 validly recognized *Bacillus* species in the 1986 Bergey's *Manual of Systematic Bacteriology*, 8 other species produce similar spores (only one of them with nonswollen sporangia). But *B. sphaericus* species can be differentiated by several phenotypic characters (Claus and Berkeley 1986). Very briefly, *B. sphaericus* is a strict aerobic round spore-former, unable to use sugars as carbon sources for growth (Russell, Jelley, and Yousten 1989).

B. sphaericus is considered to be a heterogenous species according to DNA-DNA hybridization studies by Krych, Johnson, and Yousten (1980) on 63 strains with a similar GC content (34–37%). Only 3 strains showed a significant percentage of homology (>70%) to the type-strain, thus composing Group 1. Four other homology groups were recognized, the subgroup IIA containing all of the 7 studied strains toxic to *Culex quinquefasciatus* larvae: K, Q, SSII-1, 1404, 1593, 1691, and 1881.

In a related work (de Barjac, Véron, and Cosmao Dumanoir 1980), these same strains plus two other toxic ones, 1593-4 and 1930, were grouped into class A by numerical analysis (computer assisted) of auxanograms given by 35 strains on 160 substrates. Two other classes containing nonpathogenic strains, B and C, were separated on the dendrogram, which also confirmed the heterogeneous composition of this population. On the other hand, the 78 phenotypic characters that we routinely use in *Bacillus* taxonomy and which have proven useful for differentiation of *B. thuringiensis* strains, failed to distinguish between *B. sphaericus* isolates, mainly because of a majority of negative reactions. Similar results on phenotypic traits were reported (Krych, Johnson, and Yousten 1980).

Both techniques, DNA-DNA hybridization and auxanograms, tend to group the mosquito pathogenic strains together, without discrimination between them. Now, due to the wide range in toxicity of these pathogens to mosquito larvae and to the resulting practical consequences, differentiation between these strains is essential. This can be achieved by two systems, also developed simultaneously and in close correlation, H-serotyping (de Barjac, Véron, and Cosmao Dumanoir 1980) and bacteriophage-typing (Yousten et al. 1980).

In the same work on auxanograms as cited above, we submitted the 35 *B. sphaericus* strains, including 9 pathogens, to the flagellar (H) agglutination method, first successfully adapted for differentiation of *B. thuringiensis* strains. Twenty H-serotypes were determined, 17 of which contain the atoxic strains. The pathogenic strains were grouped into 3 H-serotypes, each of them corresponding to a different level of toxicity: serotype H1a, containing the oldest known strains, K and Q, with a very low toxicity; serotype H2, containing the intermediate strains SSII-1, 14040, and 1930, with a moderate toxicity; and serotype H5, containing the more recent strains 1593, 1593-4, 1691, and 1881, exhibiting a very high toxicity. This concordance between H antigen and pathogenicity has been partially confirmed by further studies.

Comparative phage-typing of these toxic *B. sphaericus* strains together with other ones, using a set of 11 bacteriophages, gave similar subgroups (Yousten et al. 1980). The 3 toxic serotypes H1a, H2, and H5 paralleled three main patterns of response to the set of phages, which were later designated as bacteriophage groups 1, 2, and 3, among the 7 individualized groups (Yousten 1984a, 1984b). Neither single phage nor single serum specificity for one pathogenic strain was described.

The convergency of serotyping and phage-typing in the characterization of *B. sphaericus* strains is remarkable, at least for the first three phage groups. Divergence appears within phage-group 4, which contains, in addition to strain 2297 belonging to serotype H25, other strains either belonging to a different serotype or being atoxic. Serotype H25 and serotype H26a,26b were further characterized as containing pathogenic *B. sphaericus* strains (de Barjac et al. 1985). More recently, isolates from Ghana identified as belonging to serotype H6 bring to six the number of toxic serotypes (de Barjac et al. 1988).

Besides serology and phage lysis, fatty-acid patterns revealed by gas chromatography could be another way to recognize *B. sphaericus* strains larvicidal to mosquitoes. (Franchon and de Barjac, unpub. data).

14.3 COMPARATIVE TOXICITY OF VARIOUS SEROTYPES

An interesting feature is the frequent correlation between serotypes or phage-types, so far as we know, and the level of toxicity to mosquito larvae. The most toxic *B. sphaericus* strains are found in three subgroups: serotype H5a,5b (ex H5), phage-group 3, with strains 1593, 1691, 1881, and 2362 as main components; serotype H25, phage-group 4, with the main strain 2297; and serotype H6 with strain Iab 59. Strains showing an average toxicity are found in serotype H26a,26b with strains ISPC5 and 2173 and serotype H2a,2b (ex H2), phage-group 2, with strains SSII-1, 1887, and 1890 as main constituents. The less-toxic strains are found in serotype H1a, phage-group 1, with the first known toxic isolates, K and Q. The only exception to the correlation is the phage-group 4 with the less-toxic strains 2173 and 2377 and the highly toxic 2297.

It should be noted that the level of toxicity of a strain depends not only on the strain's intrinsic properties, but also on the test insect, on the bacterial growth conditions, and on the bioassay method. Every comparative evaluation must take into account all these factors. For instance, bioassays should be conducted on *Culex pipiens* or *Culex quinquefasciatus,* the most sensitive species, using homogenous early L4 larvae and with mortality data read until 48 or 72 hours; this is necessary because of the usually more progressive and slower action of *B. sphaericus* in comparison with *B.t.i.* Also, bacterial culture media giving a high sporulation rate and high yield are preferred, such as the NYSM broth of Krych, Johnson, and Yousten (1980) or MBS liquid medium of Kalfon et al. (1983).

The situation has been complicated by the fact that very often workers use somewhat different techniques and express the toxicity values in different ways (e.g., LC_{50} in whole culture dilution, or in powder weight, or in cell or spore numbers). This makes the comparison of data from different authors very difficult and often questionable. However, despite these difficulties, the following general consensus has been drawn about the toxicity of *B. sphaericus*.

The *B. sphaericus* toxic strains known so far are not pathogenic to black fly larvae, as is *B.t.i.* Among the mosquito species susceptible to *B. sphaericus*, *Culex* species are the most sensitive, followed by *Anopheles* species, which are usually less sensitive, and *Aedes* species, the least sensitive. However, variation can exist within each genus. For instance, *Aedes aegypti* (L.) is nearly insensitive to most *B. sphaericus* strains, while *Aedes melanimon* (Dyar) is sensitive. *Psorophora* sp. and *Mansonia* sp. have also proven to be sensitive.

The comparative study of the toxicity of various *B. sphaericus* strains, which permitted the above-mentioned classification of the toxic serotypes according to their mean range of larvicidal power on three mosquito species, was made with 54 strains isolated from mosquitoes, black flies, or their breeding sites (de Barjac et al. 1985). Second, the study showed that not every strain isolated from insect vectors is toxic, since nine such isolates classified into five serotypes were inactive. Inversely, toxic strains have been isolated from other insects, such as Orthoptera and Lepidoptera (Lysenko et al. 1985), or from soil.

Primarily, this study led us to subdivide the 45 studied mosquito pathogenic strains into five groups according to their toxicity level, expressed either by their LC_{50} or by their mortality percentage for *Cx. pipiens* and *Anopheles stephensi* L3-L4 larvae. The five groups were established according to five progressive levels of toxicity to *Cx. pipiens* larvae, expressed respectively as an LC_{50} in 48 hours of about 10^{-3}, 10^{-4}, 10^{-5}, 10^{-6}, and 10^{-7} dilution of final whole culture. In order to establish a clear-cut boundary between the so-called toxic strains and the atoxic ones, a limit for toxicity was arbitrarily fixed at 100% mortality in 48 hours with a 10^{-2} dilution of final whole culture.

14.4 DISTRIBUTION OF THE LARVICIDAL STRAINS AND ACTIVITY RATIOS

As a result of *B. sphaericus* research and development, more than 300 strains or isolates are now available in our laboratory collection, a part of the WHO Collaborating Center for entomopathogenic *Bacillus*. A catalogue can be sent on request. These strains, derived from different parts of the world, are grouped into 48 H-serotypes. Only the 6 serotypes mentioned above (H1a; H2a,2b; H5a,5b; H6; H25, and H26a,26b) contain the mosquito pathogenic isolates, which now number more than 150; such a figure looks inflated because of various reisolations of the same strain.

Among the larvicidal strains, the most studied and obviously most toxic ones come from tropical climates: 1593 from Indonesia, 1691 and 1881 from

El Salvador, 2297 from Sri Lanka, and 2362 from Nigeria. The last identified isolate Iab 59 from Ghana, is no exception.

A survey of the larvicidal activity of nearly all the available toxic *B. sphaericus* strains, in MBS final whole culture (= an average of 10^8 cells/ml), on *Cx. pipiens, An. stephensi,* and *Ae. aegypti,* was conducted recently in our laboratory (Thiery and de Barjac 1989). The conclusion is that the three major groups defined earlier (de Barjac, Véron, and Cosmao Dumanoir 1980) still summarize the gross pathogenicity levels: a group of low toxicity corresponding to a 100% mortality of *Cx. pipiens* early L4 larvae in 48 hours with a FWC (final whole culture) 10^{-2} dilution or an LC_{50} around 10^{-2} to 10^{-3} (serotypes H1a and H2a,2b); a group of moderate toxicity corresponding to an LC_{50} under the same conditions around 10^{-4} (serotype H26a,26b); a group of high toxicity corresponding to an LC_{50} around 10^{-6} (serotypes H5a,5b; H6; and H25).

A few examples (table 14.1) indicate that such toxicity grouping parallels *grosso modo* the serotype grouping, even if some variation in toxicity exists between the strains of the same serotype. It also indicates that within the same toxigroup, as within the same serogroup, strains with a similar toxicity level to *Cx. pipiens* larvae can differ in their larvicidal activity for *An. stephensi* or *Ae. aegypti.* So, for more accuracy in the comparative evaluation of the toxicity of *B. sphaericus* strains, one should take into account the data not only for one mosquito species, but for several. Table 14.1 shows that the most larvicidal strains, which belong to serotype H5a,5b, are very toxic to *Cx. pipiens,* a little less toxic to *An. stephensi,* and much less toxic to *Ae. aegypti* larvae. Comparatively, strains of serotypes H6 and H25 are very toxic to *Cx. pipiens,* less toxic to *An. stephensi,* and not toxic to *Ae. aegypti* larvae. Most strains of serotypes H26a,26b and H2a,2b are moderately toxic only to *Cx. pipiens* larvae.

Therefore, a better indication of the potency of a *B. sphaericus* strain could be achieved by the means of activity ratios similar to those introduced by Dulmage (1981) for *B. thuringiensis* strains (see chapter 8). An example is given in table 14.2, where, along with the activity level on *Cx. pipiens* larvae, activity ratios are calculated on the basis of the LC_{50}s in FWC dilutions. One can see that a given strain is well characterized by this kind of ratio. A ratio near 1 means a similar relative activity for the numerator/ denominator species. The higher the ratio, the less activity there is on the denominator species and vice versa. The comparison between such activity ratios for various strains of *B. sphaericus* and *B. thuringiensis* is meaningful. These ratios could also be calculated in Toxic Units, by reference to a standard, instead of culture dilutions.

The classification of *B. sphaericus* strains by H-serotypes is again interesting because usually the toxic strains of the same serotype are given a comparable pathogenic power (table 14.1). Other authors obtained similar data (Ren Gaixin et al. 1986). That does not exclude the normal occurrence

TABLE 14.1.
Toxicity of *B. sphaericus* Strains of Six Different H-serotypes to Larvae of Three Mosquito Species

Serotype	Strains	Means and ranges of LC_{50}s[a] (48 h)		
		Cx. pipiens	*An. stephensi*	*Ae. aegypti*
H5a,5b	In 1989 = 112 strains. Among the best known and most toxic ones: 1593 (or 1593-4), 1691 (or 1691-56), 1881 (or 1881-48), 2117-2, 2362 (or 290-8), k20-11, k6-3, 2533, 2601, 2602, B42, LB24, Hertl	2.5×10^{-6} (from 5.5 to 0.6×10^{-6})	1.6×10^{-5} (from 5 to 0.2×10^{-5})	9.8×10^{-4} (from 51 to 0.1×10^{-4})
H6	In 1989 = 12 toxic strains, among them Iab 59, Iab 482, Iab 509, Iab 611, Iab 620, Iab 622	3.8×10^{-6} (from 2.5 to 5.2×10^{-6})	2.2×10^{-5} (from 7 to 0.3×10^{-5})	—
H25	In 1989 = 11 toxic strains, among them 2297, 2626, 2627, 2628, 2629, 2632	6.6×10^{-6} (from 16 to 1.1×10^{-6})	3×10^{-4} (from 6.4 to 0.3×10^{-4})	—
H26a,26b	In 1989 = 5 toxic strains, among them 2173, 2315, ISPC5, LB29	3.5×10^{-4} (from 2.3 to 6.5×10^{-4})	—	—
H2a,2b	In 1989 = 9 toxic strains, among them SSII, 1886, 1887, 1889, 1890, 1891	4.9×10^{-3} (from 1.9 to 8.7×10^{-3})	—	—
H1a	In 1989 = 2 strains: K, Q	1×10^{-2}	—	—

[a] LC_{50}s are expressed in dilutions of final whole culture in MBS medium. A dash indicates the impossibility to calculate an LC_{50}, the larval mortality being less than 100% with the 10^{-2} FWC dilution.

TABLE 14.2.
Relative Activity of Various *B. sphaericus* and *B. thuringiensis* Strains on Three Mosquito Species

H-serotype	Strains	$LC_{50}s$[a] *Cx. pipiens*	$LC_{50}s$[a] *An. stephensi*	$LC_{50}s$[a] *Ae. aegypti*	Activity Level on *Cx. pipiens*	Activity Ratios[b] *Cx./An.*	Activity Ratios[b] *Cx./Ae.*	Activity Ratios[b] *An./Ae.*
B. sphaericus								
H5a,5b	1593	5.7×10^{-6}	5×10^{-6}	7.2×10^{-4}	10^{-6}	1.14	0.007	0.0069
H5a,5b	2362	2×10^{-6}	6.8×10^{-6}	8.5×10^{-4}	10^{-6}	0.29	0.002	0.008
H6	Iab 59	2.5×10^{-6}	5.2×10^{-6}	—	10^{-6}	0.48		
H25	2297	1.4×10^{-6}	1.4×10^{-6}	—	10^{-6}	1		
H26a,26b	ISPC5	2.3×10^{-4}	—	—	10^{-4}			
H2a,2b	SSII-1	4.3×10^{-3}	—	—	10^{-3}			
B. thuringiensis								
H14	*B.t.i.*	2.61×10^{-6}	1.2×10^{-6}	1.99×10^{-6}	10^{-6}	2.17	1.31	0.60
H10	73E 10–16	6×10^{-6}	1.1×10^{-3}	6.4×10^{-4}	10^{-6}	0.005	0.009	1.71
H1	BA068	1.3×10^{-3}	8.8×10^{-3}	9.1×10^{-4}	10^{-3}	0.14	1.42	9.67

[a] $LC_{50}s$ are expressed in dilutions of final whole cultures and calculated at 48 h, except for *B.t.i.*, where 24 h is used. A dash indicates that an LC_{50} cannot be determined because of a lack of pathogenicity toward the target larvae, giving less than 100% mortality with the 10^{-2} culture dilution.
[b] The mosquitoes used in the ratios are indicated by the abbreviations of the genus.

of quantitative differences in toxicity between strains of a same serotype or even, sometimes, qualitative differences, as have been recently found with the *B. thuringiensis* group. Exceptions in biology are the rule. That does not hamper the usefulness of the H antigens for classifying *B. sphaericus* and *B. thuringiensis* isolates. It has become a need, or a must, for their identification and remains as a first approach to anticipate their pathogenic potentialities.

References

Barjac, H. de, and Bonnefoi, A. 1972. Essai de classification biochimique de 64 *Bacillus* des groups II et III représentant 11 espèces différentes. *Ann. Microbiol.* (Inst. Pasteur) 122: 463–473.

Barjac, H. de; Larget-Thiery, I.; Cosmao Dumanoir, V. C.; and Ripouteau, H. 1985. Serological classification of *Bacillus sphaericus* strains in relation with toxicity to mosquito larvae. *Appl. Microbiol. Biotechnol.* 21: 85–90.

Barjac, H. de; Thiery, I.; Cosmao Dumanoir, V.; Frachon, E.; Laurent, P.; Charles, J.-F.; Hamon, S.; and Ofori, J. 1988. Another *B. sphaericus* serotype harboring strains very toxic to mosquito larvae: Serotype H6. (Inst. Pasteur) *Ann. Microbiol.* 139: 363–377.

Barjac, H. de; Véron, M.; and Cosmao Dumanoir, V. C. Caractérisation biochimique et sérologique de souches de *Bacillus sphaericus* pathogènes ou non pour les moustiques. *Ann. Microbiol.* (Inst. Pasteur) 131B: 191–201.

Claus, D., and Berkeley, R.C.W. 1986. *Genus Bacillus.* In *Bergey's manual of systematic bacteriology,* vol. 2, ed. P.H.A. Sneath, N. S. Mair, M. E. Sharpe, and J. G. Holt. Baltimore: Williams and Wilkins.

Dulmage, H. T., and Cooperators. 1981. Insecticidal activity of isolates of *Bacillus thuringiensis* and their potential for pest control. In *Microbial control of pest and plant diseases, 1970–1980,* ed. H. D. Burges, 191–220. London: Academic Press.

Kalfon, A.; Larget-Thiery, I.; Charles, J.-F.; and Barjac, H. de. 1983. Growth, sporulation, and larvicidal activity of *Bacillus sphaericus. Eur. J. Appl. Microbiol. Biotechnol.* 18: 168–173.

Kellen, W. R., and Meyers, C. M. 1964. *Bacillus sphaericus* Neide as a pathogen of mosquitoes. *Proc. Calif. Mosq. Control Assoc.* 32: 37.

Krych, V. K.; Johnson, J. L.; and Yousten, A. A. 1980. Deoxyribonucleic acid homologies among strains of *Bacillus sphaericus. International Systematic Bacteriol.* 30: 476–484.

Lysenko, O.; Davidson, E. W.; Lacey, L. A.; and Yousten, A. A. 1985. Five new mosquito larvicidal strains of *Bacillus sphaericus* from non-mosquito origins. *J. Amer. Mosq. Control Assoc.* 1: 369–371.

Ren Gaixin; Chen Wan; Chen Jinru; Liu Binzhang; and Xing Jiaqi. 1986. Biochemical and serological characterisation of *Bacillus sphaericus* strains and their toxicity to mosquito larvae. Acta Scientiarum Naturalium Universitatis Nan Kaiensis 2: 89–96.

Russell, B. L.; Jelley, S. A.; and Yousten, A. A. 1989. Carbohydrate metabolism in the mosquito pathogen *Bacillus sphaericus* 2362. *Appl. Environ. Microbiol.* 55: 294–297.

Singer, S. 1973. Insecticidal activity of recent bacterial isolates and their toxins against mosquito larvae. *Nature* 244: 110–111.

———. 1977. Isolation and development of bacterial pathogens of vectors. In *Biological regulation of vectors,* 3–18. DHEW Publication no. (NIH) 77-1180.

Thiery, I., and Barjac, H. de. 1989. Selection of the most potent *B. sphaericus* strains, based on susceptibility ratios determined on three mosquito species. *Appl. Microbiol. Biotechnol* 31: 577–581.

Weiser, J. 1984. A mosquito-virulent *Bacillus sphaericus* in adult *Simulium damnosum* from Northern Nigeria. *Zbl. Mikrobiol.* 139: 57–60.

Wickremesinghe, R.S.B., and Mendis, C. L. 1980. *Bacillus sphaericus* spore from Sri Lanka demonstrating rapid larvicidal activity on *Culex quinquefasciatus. Mosq. News* 40: 387–389.

Yousten, A. A. 1984a. *Bacillus sphaericus:* Microbiological factors related to its potential as a mosquito larvicide. *Adv. Biotechnol. Processes* 3: 315–343.

———. 1984b. Bacteriophage typing of mosquito pathogenic strains of *Bacillus sphaericus*. *J. Invertebr. Pathol.* 43: 124–125.

Yousten, A. A.; Barjac, H. de; Hedrick, J.; Cosmao-Dumanoir, V.; and Myers, P. 1980. Comparison between bacteriophage typing and serotyping for the differentiation of *Bacillus sphaericus* strains. *Ann. Microbiol.* (Inst. Pasteur) 131B: 297–308.

15

The Mosquito Larval Toxin of *Bacillus sphaericus*

ELIZABETH W. DAVIDSON
ALLAN A. YOUSTEN

15.1 LOCATION OF THE TOXIN IN THE BACTERIAL CELL

Although the pathogenicity of a strain of *Bacillus sphaericus* was described by Kellen et al. (1965), the role of a toxin in the pathology was first suggested by Singer (1973). His suggestion was verified by Myers and Yousten (1978). Early studies focused on strain SSII-1 (phage-group 2, serotype H2), which was the most toxic strain available at that time. Refer to Yousten (1984) for a list of strains and their phage types and to de Barjac et al. (1985), de Barjac et al. (1988), and Thiery and de Barjac (1989) for a list of strains and their serotypes. In strain SSII-1 the toxicity did not increase as the cells completed vegetative growth and entered sporulation, the toxin seemed to be quite easily lost during efforts at purification, and no parasporal inclusion body was observed by phase contrast or by electron microscopy (Myers and Yousten 1978, 1981). The chemical nature and location of the toxin in strain SSII-1 and in other low-toxicity strains is presently unknown. The isolation of strains having a higher level of toxicity has directed research toward these strains, which have a greater potential for practical application in mosquito control.

The isolation by Singer of the highly toxic strain 1593 provided new opportunities for the study of the *B. sphaericus* toxin. In contrast to SSII-1, the toxicity of 1593 increased dramatically at the onset of sporulation. The cell mass that resulted from the growth of oligosporogenous mutants was much less toxic than that produced by the sporulating parental strain (Myers, Yousten, and Davidson, 1979). Furthermore, the toxin was unaffected by treatments that might be used in toxin isolation such as sonication, French pressure-cell disruption, refrigerator storage, and freezing. Such treatments had severely degraded the toxicity of strain SSII-1.

The first studies of strain 1593 failed to demonstrate the presence of a parasporal body that might be related to toxicity. However, subsequent studies by Davidson (1981b) and by Davidson and Myers (1981) revealed the

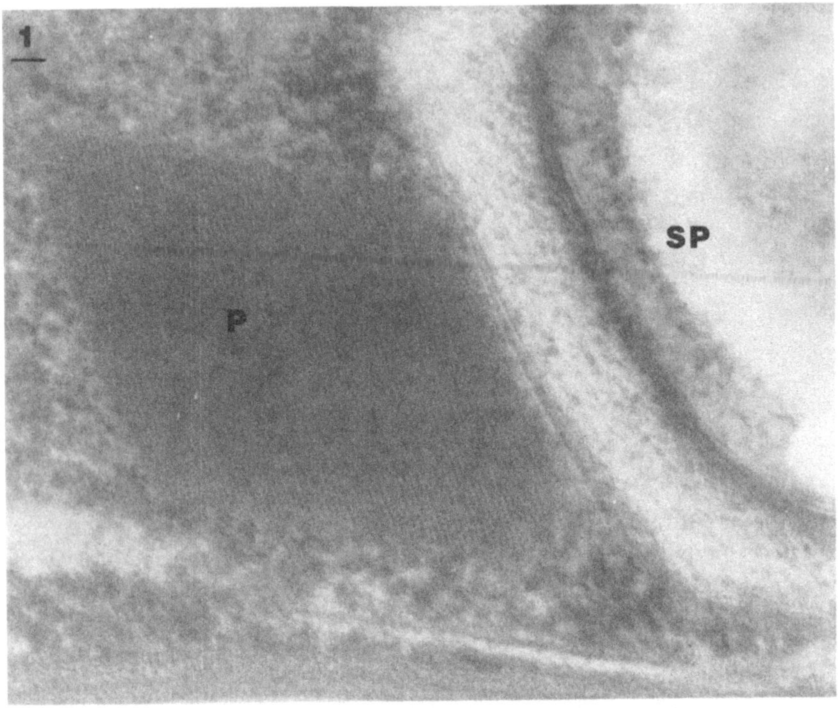

FIGURE 15.1 Electron micrograph of a thin section of *B. sphaericus* 2297 showing the spore (*SP*) and the paraspore (*P*). Bar = 0.1 μm. (From Yousten and Davidson (1982). Reprinted by permission of the American Society for Microbiology.)

presence of a variety of inclusions. There were elliptical and oval bodies that had no obvious crystal lattice and did not change in appearance during passage through the larval gut. These were probably not related to toxicity because they were found in some nonpathogenic strains as well as in pathogenic strains (Davidson and Myers 1981; Davidson 1981b). However, parasporal inclusion bodies that did display a crystal lattice (fig. 15.1) and did dissolve in the larval gut (fig. 15.2) were found in highly toxic strains 1593, 2013-4, and 1691 (all of phage-group 3, serotype H5a,5b) and in 2297 (= MR-4) (serotype H25). The paraspore present in strain 2297 is particularly large and easily seen even by light microscopy (Wickremesinghe and Mendis 1980). The paraspore of strain 2362 (Weiser 1984) is also readily visible by light microscopy, particularly when the cell is undergoing autolysis. Strains of low toxicity such as SSII-1 and Kellen Q and the nontoxic strains do not contain crystalline paraspores (Davidson and Myers 1981).

The formation of the paraspore in strain 2297 was studied by Yousten and Davidson (1982) and by Kalfon et al. (1984). The results of the two studies were in good agreement and showed that the paraspore appeared in the

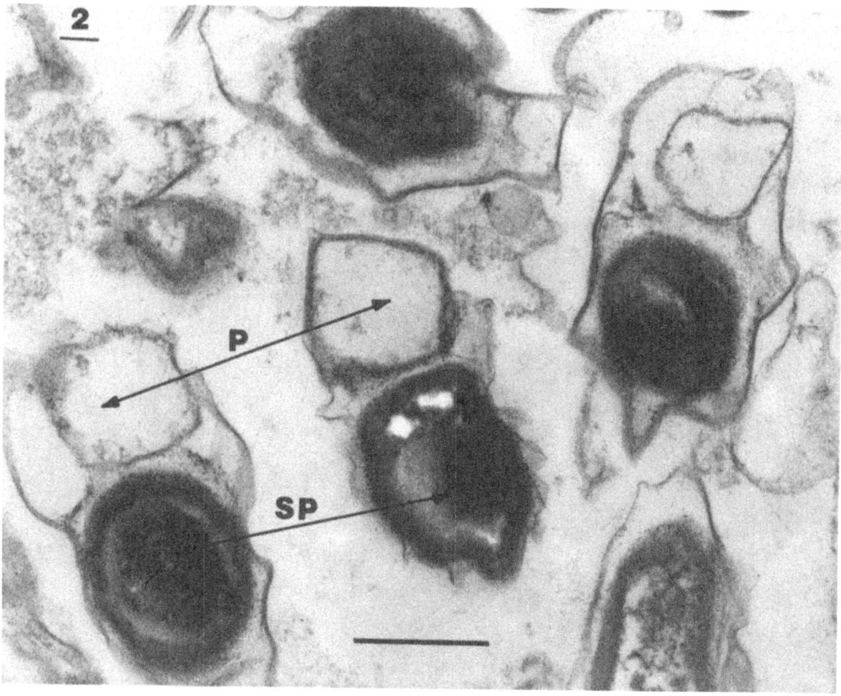

FIGURE 15.2 Spores (*SP*) and paraspores (*P*) of *B. sphaericus* 2297 after 15 minutes in the midgut of a *Cx. quinquefasciatus* larva. The matrix of the parasporal inclusions has been dissolved leaving behind an envelope. Bar = 0.5 μm. (From Yousten and Davidson (1982). Reprinted by permission of the American Society for Microbiology.)

cell after the completion of the forespore septum. It enlarged rapidly during the period when the forespore septum was engulfing one end of the cell to produce the forespore. The paraspore of *B. sphaericus* appears to have the shape of a parallelogram in thin sections, and the interior of the structure shows the striations of a lattice arrangement (fig. 15.1) as are seen in the paraspores of *Bacillus thuringiensis*. Measurement of the distance between the rows of the lattice gave a value of 6.3 nm (Yousten and Davidson 1982; de Barjac and Charles 1983). The paraspore of *B. sphaericus* appears to be composed of a single object, in contrast to the paraspore of *B. t. i.*, which has a composite structure composed of bodies of various sizes and shapes.

An envelope or membrane surrounding the paraspore was first observed at stage 3 of sporulation as the forespore engulfment was completed (Kalfon et al. 1984). This parasporal envelope is most readily observed in thin sections of cells that have been extracted with NaOH (de Barjac and Charles 1983) or have been fed to mosquito larvae (Yousten and Davidson 1982).

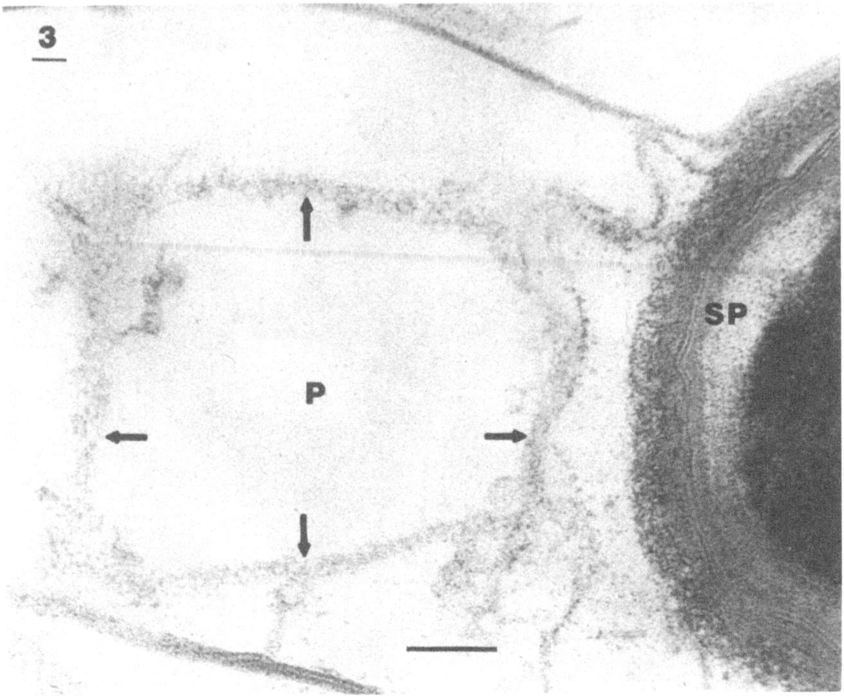

FIGURE 15.3 Spore (*S*) and paraspore (*P*) of *B. sphaericus* 2297. The matrix of the paraspore has been dissolved. The envelope (arrows) is composed of a meshlike structure. Bar = 0.1 μm. (From Yousten and Davidson (1982). Reprinted by permission of the American Society for Microbiology.)

Under these conditions the paraspore matrix is dissolved and the envelope remains intact as a shell revealing the outline of the paraspore (fig. 15.3). The envelope is not a unit membrane of the phospholipid type, but seems to be composed of a mesh or network having a thickness of about 20 nm at its narrowest (Yousten and Davidson 1982). Its appearance is very similar to the net or mesh that covers the surface of the parasporal inclusions of *B. thuringiensis* (Bulla et al. 1980). Bulla et al. (1980) suggested that the net covering *B. thuringiensis* paraspores might be related to the exosporium. The paraspore in *B. sphaericus* lies within an elongated exosporium and appears to be in contact with it at various points. The meshlike appearance of the parasporal envelope seen in strain 2297 (fig. 15.3) (Yousten and Davidson 1982) bears a striking similarity to the hexagonally arranged subunits seen on the inner surface of the exosporium of nontoxic *B. sphaericus* 9602 (Holt, Gauthier, and Tipper 1975). The parasporal envelopes of *B. sphaericus* and *B. thuringiensis* have not been isolated or chemically characterized.

The presence of an envelope on the paraspores of both of these species of *Bacillus* suggests a role in the assembly of the paraspore, but this would be difficult to demonstrate. The parasporal matrix protein may simply offer a suitable surface on which exosporium subunits assemble. Proteinaceous membranes surround certain other procaryotic inclusions, such as the poly-beta-hydroxybutyrate (PHB) granules of many bacteria and the gas vacuoles of cyanobacteria (Shively 1974). In the case of PHB granules, the surrounding membrane is the site of the enzyme that brings about polymerization of beta-hydroxybutyric acid. However, these protein membranes are only about 2–4 nm wide and have a different appearance than the meshlike form of the *B. sphaericus* parasporal inclusion envelope.

The parasporal bodies of *B. sphaericus* are the sites of much of the toxin produced by this bacterium. This conclusion is supported by both indirect and direct evidence. First, the toxicity of the cells increases as the cells begin sporulation and as the paraspores appear (Yousten and Davidson 1982; Kalfon et al. 1984; Charles et al. 1988). Second, the matrix of ingested paraspores is rapidly dissolved in the larval gut; this step could release the toxin in the appropriate site. Toxin has been extracted from a mixture of spores and paraspores by Davidson (1983), who worked with strain 1593, and by de Barjac and Charles (1983), who used strain 2297 (discussed in detail below). Both research groups used 0.05M NaOH to dissolve the matrix of the paraspores and to obtain a solution that was toxic to *Culex* larvae.

Since mixtures of spores and paraspores were used in these experiments, it is possible that some toxin was extracted from spores as well as from paraspores. To avoid this problem, Payne and Davidson (1984) dissociated spores from paraspores by passing them through a French pressure cell and by fractionation on NaBr gradients. The fraction enriched in paraspores was more toxic than any of the other fractions. It was unclear whether other cell components actually contained toxin or whether they were contaminated by the presence of a small number of paraspores. It was clear, however, that the paraspores were the most toxic component. A similar purification scheme was used by Baumann et al. (1985) to prepare toxic paraspores from strain 2362.

Although the study of *B. sphaericus* toxin presently emphasizes toxin derived from paraspores, many strains have readily detectable toxicity but do not appear to form paraspores. Since the toxicity of strains such as Kellen K and SSII-1 (phage-groups 1 and 2, serotypes 1 and 2) is low and apparently unstable, little effort has been made to determine its location in the cell or even whether it is the same protein as is found in the highly toxic, paraspore-forming strains. An early effort at toxin characterization detected toxin present on the walls of cells of strain 1593 that were in the course of sporulating (Myers and Yousten 1980). If the toxin is shown to have an affinity for other cell structures, it might appear in several cell fractions.

15.2 BIOCHEMICAL NATURE OF THE TOXIN

A small amount of soluble toxic activity was detected in the cytoplasmic fraction of broken, sporulating strain 1593 cells (Myers and Yousten 1980). This toxin was partially purified by ammonium sulfate precipitation, ion exchange, and gel filtration, and was found to have a molecular weight of around 100 kDa. Toxicity was resistant to degradation by trypsin, α-chymotrypsin, protease K, thermolysin, β-glucuronidase, lysozyme, or subtilisin, but was reduced by Pronase and subtilisin (Davidson 1982).

Soluble toxin has been extracted from the spore/crystal complex of *B. sphaericus* strains 1593, 2362, and 2297 by two methods. Tinelli, de Barjac, and Bourgouin (1980) produced a soluble extract by repeated freeze-thaw of 1593 spores. Davidson (1983) found that a toxic extract could be produced from strain 1593 spores by 0.05M NaOH extraction followed by acetate precipitation, a technique based upon an early method of extracting *B. thuringiensis* toxin (Angus 1956). The *B. sphaericus* toxin was found to have a molecular weight of 35 to 54 kDa depending upon the method used to calculate this value. The alkaline extract was quite toxic to first-instar *Culex quinquefasciatus* larvae, killing 50% of the larvae at 4.5×10^{-3} μg protein/ml. Bourgouin et al. (1984) compared toxins produced by freeze-thaw and alkaline extraction from 1593 spores. The toxin isolated from freeze-thawed spores appeared to be 110 kDa after gel filtration, but 50 kDa after ion exchange chromatography. Alkaline extracts contained two major proteins of 39 kDa and 65 kDa. These authors proposed that either two different toxins were present, or that a higher molecular weight toxin was degraded to a smaller protein during purification. A toxic extract was produced from strain 2297 by de Barjac and Charles (1983). As with strain 1593 (Davidson 1983), crystals of 2297 were found to be dissolved by alkali, and determination of the amino acid content of this extract revealed an apparent similarity to that of *B.t.i.* toxic extracts (de Barjac and Charles 1983).

Baumann et al. (1985) and Broadwell and Baumann (1986, 1987) have examined the synthesis and chemistry of the toxin from strain 2362 and a few related strains. In confirmation of the earlier studies, they found that crystal protein synthesis was initiated at the end of exponential growth. The first protein detected in the isolated crystals from strain 2362 had a mass of 125 kDa, and its appearance was followed within an hour by a protein of 110 kDa. About 7 hours later, smaller proteins of 63 and 43 kDa were detected in the crystals. As the 63- and 43-kDa proteins appeared in the crystal, the amounts of 125- and 110-kDa protein decreased. This observation led to the hypothesis that the 125-kDa protein was degraded to the 110 kDa, which in turn was degraded to the 63- and 43-kDa proteins. However, this interpretation was shown to be untenable when separate genes coding for the 43- and 63-kDa proteins were cloned in *Escherichia coli* (Berry and Hindley 1987; Hindley and Berry 1987; Baumann, Broadwell, and Baumann 1988). Also, the 125-

kDa protein, which was originally thought to be an integral part of the crystal, was shown to be identical to the cell surface layer (S-layer) protein described by Lewis, Yousten, and Murray (1987). The S-layer protein had apparently been a contaminant of earlier crystal preparations (Bowditch, Baumann, and Yousten 1989).

The gene coding for the 125-kDa protein was cloned into *E. coli* and shown to have an open-reading-frame coding for 1176 amino acids, yielding a calculated mass of 125,085 Daltons (Bowditch, Baumann, and Yousten 1989). The 110-kDa protein appears to be derived from the 125-kDa protein during sporulation. The 43- and 63-kDa proteins were shown to share considerable amino acid sequence similarity when the two open reading frames contained on a 3479-base pair *Hind*III DNA fragment from strain 2362 cloned in *E. coli* were compared. The hydrophobic regions were most conserved between these two proteins, but the two proteins did not cross-react immunologically (Baumann et al. 1985; Baumann, Broadwell, and Baumann 1988). The molecular masses of the two small crystal proteins as derived from nucleotide sequences are 41.9 kDa (Berry and Hindley 1987; Baumann, Broadwell, and Baumann 1988) and 51.4 kDa (Baumann, Broadwell, and Baumann 1988; Arapinis, de la Torre, and Szulmajster 1988). Hereafter, the two proteins in the *B. sphaericus* crystal will be referred to by the masses determined by sequencing: 41.9 kDa and 51.4 kDa. When the N-terminal portion of the deduced sequence of the 41.9-kDa protein was compared to that of the protein purified from crystal, it was found that the crystal protein lacked 4 of the N-terminal amino acids encoded by the gene (Hindley and Berry 1987; Baumann, Broadwell, and Baumann 1988). Apparently a tetrapeptide is removed from the N-terminus of the protein as a posttranslational step as it is deposited in the crystal.

There is only very limited information available about the distribution of the crystal proteins among the various strains of *B. sphaericus.* Most information has come from studies of strain 2362 where the 51.4 and 41.9 proteins have been demonstrated. The toxic 41.9-kDa protein has been cloned from strains 2362, 1593, and 2297. Proteins of a variety of other masses have been demonstrated in strains 1593, 1691, and 2297 following SDS-PAGE of solubilized crystals and protein detection by staining or immunoblotting (Baumann et al. 1985). However, the significance of these proteins is unclear since proteins detected may be influenced by the age of the cells when the crystals are harvested (Broadwell and Baumann 1986), the method used to dissolve the crystals (Baumann et al. 1985), and the purity of the crystal preparation.

Sgarrella and Szulmajster (1987) have reported that a toxic protein of 38 kDa isolated from strain 1593M underwent aggregation into higher molecular weight forms as the pH was shifted from 7.5 to 8.5. The opportunity clearly exists for the creation of considerable confusion if analytical techniques are not consistent. Using a probe prepared from a toxin determinant cloned from strain 1593M, Louis, Jayaraman, and Szulmajster (1984) examined some highly toxic, moderately toxic, and nontoxic strains for the presence of homology to the probe DNA. In general, the highly toxic strains showed homologous DNA sequences not found in the less-toxic strains. D. Wilcox (pers. comm.) used probes prepared from the sequence of the 41.9-kDa protein to demonstrate the presence of this gene in highly toxic strains and its absence in less-toxic or nontoxic strains.

The deduced amino acid sequence of the 41.9-kDa crystal protein from strain 2362 (phage-group 3, serotype 5a,5b) differed in 5 amino acids from that of strain 2297 (serotype 25). The 51.4-kDa protein differed in 4 amino acids between the two strains (Baumann, Broadwell, and Baumann 1988; Hindley and Berry 1988). Antisera to the 41.9- and 51.4-kDa proteins did not react with solubilized proteins from the crystals of *B. thuringiensis* subsp. *kurstaki* or *B.t.i.* (Baumann et al. 1985). Furthermore, no similarity was found between the nucleotide coding sequences of the 51.4-kDa or the 41.9-kDa proteins from *B. sphaericus* 2362 and the published sequences of toxin proteins from Diptera-, Lepidoptera-, or Coleoptera-active *B. thuringiensis* (Baumann, Broadwell, and Baumann 1988; Bowditch and Baumann, pers. comm.).

Ganesan et al. (1983) and Louis, Jayaraman, and Szulmajster (1984) reported having cloned the toxin genetic determinant from strain 1593M onto plasmid pGsp03. The *E. coli* carrying this plasmid produced 4 polypeptides of 21, 19, 15, and 12 kDa. However, the toxic protein isolated from strain 1593M was shown to be 38 kDa, and the four polypeptides produced in *E. coli* may be proteolytic degradation products of a larger toxic protein (Sgarrella and Szulmajster 1987).

Toxicity to *Culex* larvae has been demonstrated for the 41.9-kDa crystal protein (Baumann et al. 1985) and for the 110-kDa (S-layer–derived) protein (Broadwell and Baumann 1986). Against second- and third-instar larvae of *Culex pipiens*, the LC_{50} of the 110-kDa protein was 115 ng/ml. The 51.4-kDa protein lacked toxicity (Baumann et al. 1985). The LC_{50} of whole purified crystals was reported to be 6.1 ng of protein/ml. The 41.9-kDa protein also showed toxicity to tissue culture–grown cells of *Cx. quinquefasciatus*, but the 110-kDa protein lacked toxicity to these cells (Broadwell and Baumann 1986, 1987). In the presence of larval gut proteases from any of three mosquito larval species (including the relatively nonsusceptible *Aedes aegypti*), the 41.9-kDa protein was cleaved to a 40-kDa form (mass determined by SDS-PAGE), which demonstrated a 54-fold increase in toxicity to *Cx. quinquefasciatus* cells. A similar conversion was found when whole crystals were fed to larvae (Aly, Mulla, and Federici 1986; Broadwell and Baumann 1986, 1987; Davidson, Bieber et al. 1987). The cleavage by gut protease removes 6 amino acids from the N-terminal end of the molecule (in addition to the 4 removed in the posttranslational modification within the bacterial cell) and approximately 20 amino acids from the C-terminal end (Baumann, Broadwell, and Baumann 1988). The mechanism by which this increases the toxicity of the molecule for tissue culture–grown cells is unknown. Interestingly, the 41.9-kDa protein produced when the gene from either strain 2362 or strain 2297 was cloned in *E. coli* was not toxic to *Culex* larvae, but toxicity was expressed when it was combined with the 51.4-kDa protein also produced in *E. coli* (Baumann et al. 1987; Broadwell and Baumann, pers. comm.). The reason for the requirement for the nontoxic 51.4-kDa protein to produce larval toxicity from the 41.9-kDa protein (when that protein was obtained from *E. coli*) has yet to be explained.

In an effort to provide the *B. sphaericus* toxin in a form that may persist longer in the aquatic environment, Tandeau de Marsac, de la Torre, and Szulmajster (1987) have cloned and expressed a toxin gene from strain 1593M in the cyanobacterium *Anacystis nidulans*. The nature and the amount of the toxic protein produced by *A. nidulans* was not reported, and the toxicity was about 1,000 times lower than that produced in *B. sphaericus*. Nonetheless, this represents an interesting approach to the creative manipulation of the toxin.

15.3 PATHOLOGY IN THE HOST

When a mosquito larva of a susceptible species such as *Cx. quinquefasciatus* ingests a lethal dose of *B. sphaericus* spores and crystals, symptoms may appear within as little as 30 minutes to 1 hour. Because the larval cuticle is transparent, it is possible to observe the marked expansion of the midgut, which permits the peritrophic membrane to fall into zigzag folds. The larvae undergo tremors, become sluggish, and eventually die, often while hanging from the water surface. Mortality in larvae fed a large dose may begin as early as 4 hours, but generally 48 hours are required for full expression of mortality in bioassays (Davidson 1981a, 1984).

The pathology of *B. sphaericus* infections in susceptible larvae has been studied both by light and electron microscopy. Kellen et al. (1965), who discovered the first mosquito larvicidal strain of *B. sphaericus*, described the pathology of this weakly insecticidal strain in host larvae. Death required up to 7 days and appeared to result from a progressive degeneration of gut cells. In more toxic strains such as SSII-1 and 1593, however, the pathology is much faster and more typical of an intoxication process. Strain SSII-1 vegetative cells were found to be rapidly digested in the larval gut, probably releasing the toxin (Davidson, Singer, and Briggs 1975). The outer cell wall layer of these vegetative cells was digested first, and then the cytoplasm of the cells was rapidly dissolved away. Only midgut cells of the intoxicated larvae showed visible ultrastructural changes. Midgut cells separated from one another at the bases accompanying the swelling of the gut. Large cytolysosomes appeared in the posterior midgut cells, and eventually these cells were

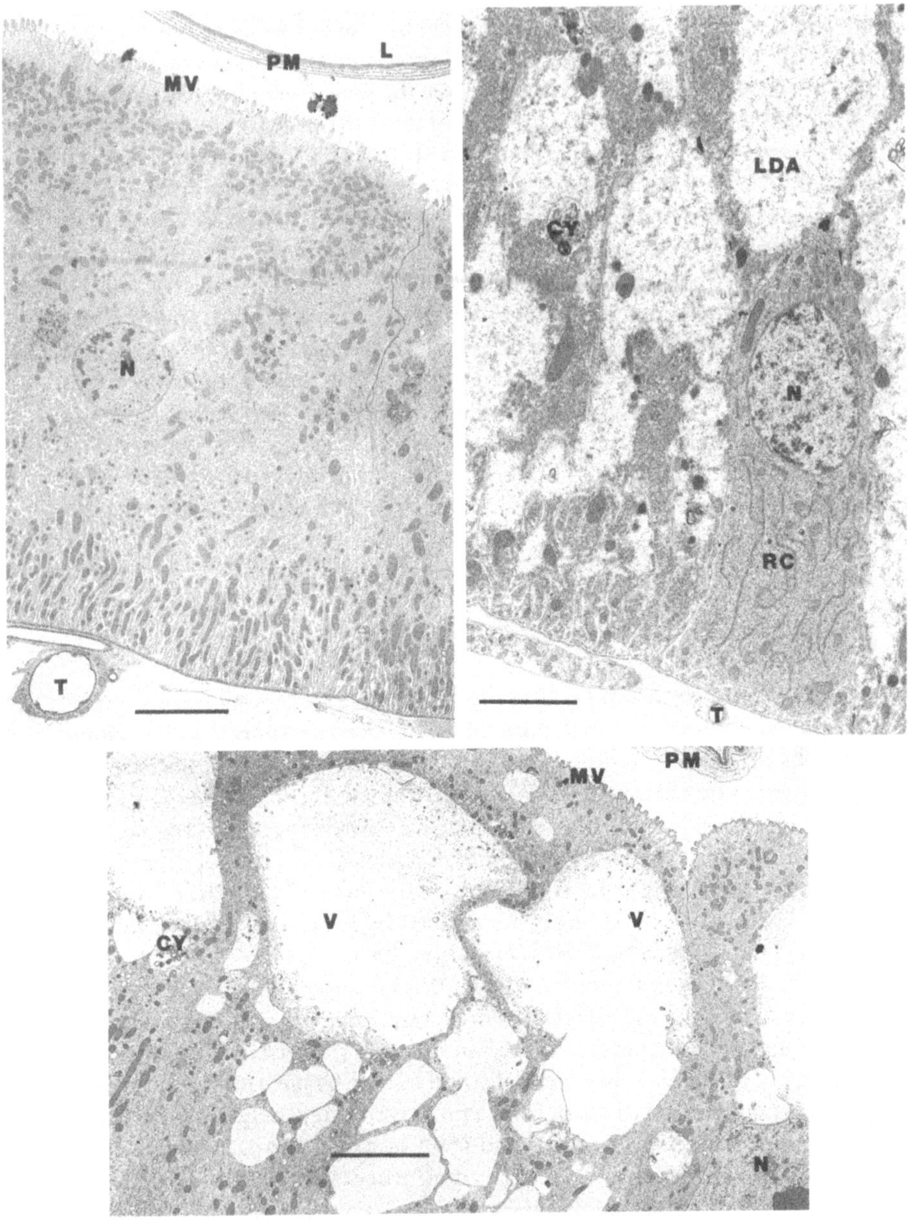

FIGURE 15.4 Sections of fourth-instar mosquito larval midgut cells before (*a*) and after (*b, c*) feeding on *B. sphaericus* 2297 spore/crystal complex: *a*, normal midgut cell of *An. stephensi*, with nucleus (*N*), microvilli (*MV*), peritrophic membrane (*PM*) enclosing the lumen (*L*), and tracheolar cell (*T*); Bar = 5 μm; *b*, midgut cell of *An. stephensi* 6 hours after feeding on *B. sphaericus* spore/crystal complex. Most typical alteration is appearance of large areas of low electron density (*LDA*) in the cytoplasm, particularly in the basal regions of the cell. An unaltered regenerative cell (*RC*) is also shown. Bar = 2.5 μm; *c*, midgut cell of *Cx. pipiens pipiens* 12 hours after feeding on *B. sphaericus* spore/crystal complex. Large vacuoles (*V*) appeared in apical and medial

observed to slough from the basement membrane. Invasion of the larval hemocoel by gut bacteria did not occur until after death of the larva (Davidson 1979).

Pathological changes in larvae fed strain 1593 spores (and crystals) were observed by Davidson (1981b) and by Karch and Coz (1983), in larvae fed 2297 spore/ crystal complexes by Charles (1987), and in larvae fed 2362 spore/crystal complexes by Singh and Gill (1988). Spores of strain 1593 began to germinate at about 4 hours after feeding; the first larval deaths occurred at the same time. The vegetative cells emerging from the spores showed no signs of digestion in the gut of the dying or dead larva. Symptoms of intoxication in the larva were similar following ingestion of either 1593 or 2297 spores (fig. 15.4). Large areas of low electron density in midgut cells of *Anopheles stephensi* and enlarged vacuoles or cytolysosomes in the midgut cells of *Cx. quinquefasciatus* or *Cx. pipiens* were characteristic of *B. sphaericus* intoxication (fig. 15.4 b, c) (Davidson 1981b; Charles 1987; Singh and Gill 1988). Changes in *Culex* species larvae resembled those observed in SSII-1 infections of this species (Davidson 1979). In *Ae. aegypti* larvae fed a very large, lethal dose of 2297 spore/crystal complexes, mitochondria swelled in midgut cells, but other changes characteristic of the intoxication in *Anopheles* or *Culex* species were not observed (Charles 1987). Singh and Gill (1988) continued ultrastructural observations of *Cx. quinquefasciatus* larvae intoxicated with strain 2362 out to 36 hours and detected damaged mitochondria and lamellar bodies in neural tissue, as well as damage to skeletal muscle. These late-appearing lesions may be involved in the general body paralysis observed in *B. sphaericus*–intoxicated larvae.

Even in those larvae that survive an initial exposure to toxin, there are apparently long-lasting effects, which may limit the potential for a normal life span. Lacey, Day, and Heitzman (1987) have shown that fewer survivors than untreated controls can successfully pupate. Also, there are lower amounts of nutritional reserve materials in adults that develop from survivors than in control adults. The reduced vigor that may result from this nutritional deficiency would add to the benefit derived from use of *B. sphaericus* for mosquito control.

Following death of the larva, *B. sphaericus* cells multiplied in the larval cadaver and formed new spores (Silapanuntakul et al. 1983; Davidson et al. 1984; Des Rochers and Garcia 1984; Charles and Nicolas 1986). The potential for recycling in the larval host is one of the most attractive features of this organism as a microbial control agent, as each larva generally contains around 10^5–10^6 spores at completion of sporulation (Davidson et al. 1984;

FIGURE 15.4 *(Continued)*

parts of the cell, which may be coming from cytolysosomes (*CY*). Bar = 5 μm. (From Charles 1987. Reprinted by permission of the Institut Pasteur. Kindly provided by Dr. J.-F. Charles, Institut Pasteur, Paris.)

Charles and Nicolas 1986). The cadavers are then toxic for other larvae (Des Rochers and Garcia 1984).

15.4 HOST RANGE OF THE TOXIN

Because a variety of bacterial preparations, bioassay procedures, and data calculation methods have been used in published assays, it is difficult to compare sensitivity of mosquito species to the *B. sphaericus* toxin. However, if assays involving the highly sensitive species *Cx. quinquefasciatus* were included, comparison of the sensitivity of this species with other species could be used as an estimate of sensitivity. In general, larvae of *Culex* species are all very susceptible to the toxin (LC_{50} = 0.002–0.005 ppm dry primary powder). An interesting exception is the non-human-biting *Culex cinereus,* which competes for the same environmental breeding sites with the susceptible *Cx. quinquefasciatus* (Nicolas and Dossou-Yovo 1987). The susceptibility of larvae of *Aedes* species varies from species to species. For example, *Ae. aegypti* larvae are at least 4,000-fold less susceptible than *Cx. quinquefasciatus* larvae to strain 2362 spores (Mulla *et al.* 1984). However, *Aedes triseriatus* larvae are only about 63-fold less susceptible than *Cx. quinquefasciatus* to a closely related strain, 2013-4 (Lacey and Singer 1982). In comparative assays, the susceptibility of *Anopheles quadrimaculatus, albimanus, freeborni, gambiae,* and *stephensi* larvae to *B. sphaericus* was found to be about 10- to 50-fold less than *Cx. quinquefasciatus* (Lacey and Singer 1982; Mulla et al. 1984; Obeta and Okafor 1983; Bourgouin, Larget-Thiery, and de Barjac 1984). Black fly larvae are not killed by *B. sphaericus* (Guillet 1984), even though one of the most pathogenic strains for mosquito larvae, 2362, was isolated from these insects (Weiser 1984).

The reasons for the insensitivity of certain mosquito larvae to the *B. sphaericus* toxin are not fully understood. Destruction of the toxin in the gut of resistant species was suggested by Mian and Mulla (1983), who found that the midgut enzymes of *Ae. aegypti* degraded the *B. sphaericus* toxin, whereas the enzymes from *Cx. quinquefasciatus* did not. However, Aly, Mulla, and Stritzke (1985) found that a spore preparation was reduced in toxicity following passage through the larvae of either species. Davidson, Bieber, et al. (1987) found that toxin treated with gut extracts of *Ae. aegypti* was not reduced in insecticidal activity, and toxin activation (as indicated by effect upon *Cx. quinquefasciatus* cell cultures) was accomplished by both *Cx. quinquefasciatus* and *Ae. aegypti* gut extracts. Similarly, Broadwell and Baumann (1987) and Aly, Mulla, and Federici (1986) found that *Cx. quinquefasciatus, An. gambiae,* and *Ae. aegypti* larvae or gut extracts were all able to reduce the amount of high molecular weight protein present in the strain 2362 crystals and to convert the 41.9-kDa form to 40 kDa. Therefore, the ability to activate the toxin is probably not a major factor in determining differences in susceptibility. Cultured cells of *Ae. aegypti* and *Aedes albopictus* were found to be only slightly sensitive or insensitive to dissolved *B. sphaericus* toxin, which was highly cytotoxic to *Cx. quinquefasciatus* cells (Davidson 1986). Lepidopteran and mouse cells were insensitive to the activated toxin, and fluorescent-labeled toxin was bound poorly or not at all to insensitive cells (Davidson, Shellabarger, et al. 1987). Similarly, cultured cells of *An. gambiae, An. stephensi* and *Aedes dorsalis* were less sensitive to the toxin than cells of *Cx. quinquefasciatus* (Broadwell and Baumann 1987). These data suggest that resistance of larvae of certain mosquito species such as *Ae. aegypti* resides in a fundamental cellular characteristic, for example, the lack or reduced population of appropriate receptors on the cell membrane.

Binding of fluorescent-labeled toxin to larval midgut cells was recently demonstrated. Toxin bound strongly to cells of the posterior midgut and to the lumen of the gastric caecum in *Cx. quinquefasciatus,* but did not bind to midgut cells of *Ae. aegypti* larvae, although the toxin was ingested by the latter species. Internalization of the toxin in small vesicles was observed in *Cx. quinquefasciatus.* The lectin, wheat germ agglutinin, reduced both binding of toxin and bioassay activity in these larvae (fig. 15.5) (Davidson 1988). In *Anopheles* species larvae, toxin bound to the midgut in less clearly defined regions, was not internalized, and leaked rapidly from the cells (Davidson 1988a). These results suggest that binding in *Anopheles* species may be of weak affinity or nonspecific, whereas in *Culex* binding appears to be specific and of high affinity. The greatly reduced susceptibility of *Ae. aegypti* larvae is probably due to failure of the toxin to bind to midgut cells in these larvae.

Recently, two unusual activities of the *B. sphaericus* toxin have been reported. Alkali-solubilized toxin was lethal to adult *Cx. quinquefasciatus* mosquitoes when introduced by enema into the midgut, but was not toxic when fed to these insects. The toxin was not lethal to adult *Ae. aegypti* by either route of administration (Stray, Klowden, and Hurlbert 1988). These results indicate that receptors for the toxin are present on the adult *Cx. quinquefasciatus* gut as well as on larval cells. Crude soluble toxin was lethal to the eggs of the nematode *Trichostrongylus colubriformis,* and this toxicity was eliminated by filtration through 1.2 μm or smaller sterilizing filters (Bone and Tinelli 1987).

15.5 PATHOLOGY OF THE TOXIN IN CULTURED CELLS

The mode of action of the *B. sphaericus* toxin is presently unknown. Evidence from histopathology studies suggests that the midgut is the first organ affected by this toxin (Davidson 1979, 1981b; Karch and Coz 1983; Charles 1987; Singh and Gill 1988). Because the mosquito larva is small and aquatic, studies involving feeding a specific amount of bacteria or

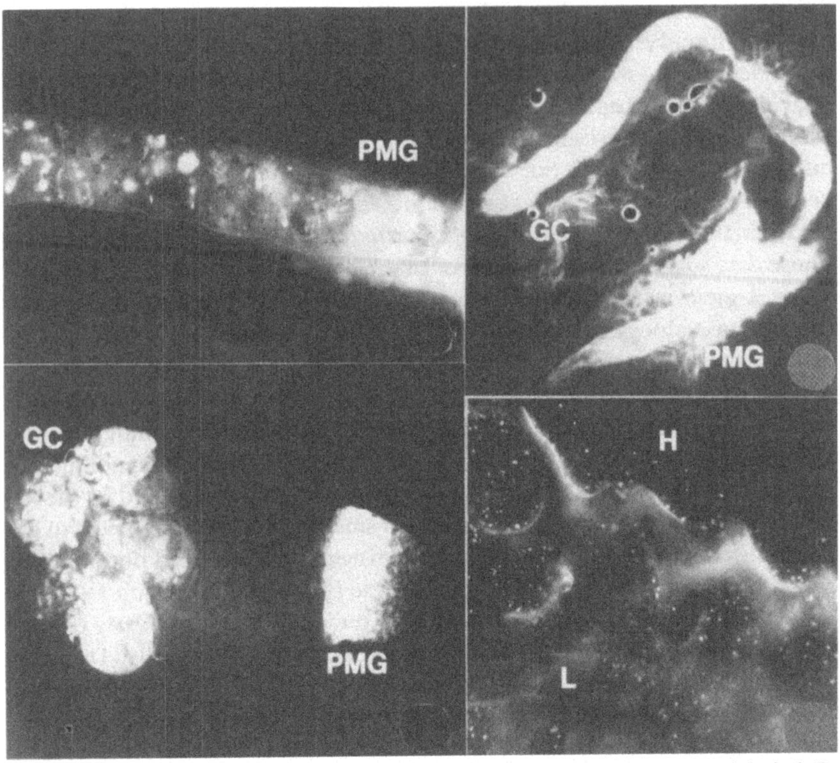

FIGURE 15.5 Midgut of *Cx. quinquefasciatus* larvae fed fluorescent-labeled *B. sphaericus* toxin: (*a*) fed for 15 minutes. Fluorescence restricted to midgut lumen and especially bright on ingested food particles and peritrophic membrane in the posterior midgut (*PMG*). 150×; (*b*), fed for 1 hour. Peritrophic membrane contents brightly fluorescent; fluorescent toxin binding to cells is restricted to the lumen of the gastric caecum (*GC*) and the posterior midgut. 75×; (*c*), fed for 24 hours. Lumen of gastric caecum and posterior midgut are brightly fluorescent; central and anterior midgut did not bind toxin. Peritrophic membrane has been removed. 75×; (*d*), midgut cells of larva fed fluorescent toxin for 2 hours. Fluorescence concentrated on microvillar surface of cells and in small vesicles within cells. *H* = portion of cells bounding the hemocoel; *L* = gut lumen. 150×. (From Davidson 1988a. Reprinted by permission of the Entomological Society of America.)

injection into the hemocoel are difficult, if not impossible. An in vitro assay system would provide a mechanism for studying the intimate associations of this toxin with the larval host, including toxin activation, binding, and molecular mode of action. Such an in vitro system has been devised using cultured *Cx. quinquefasciatus* cells (Davidson 1986). The toxin, extracted from strain 1593 or 2362 by alkali with or without a reducing agent, was activated by midgut homogenates of *Cx. quinquefasciatus* larvae. Toxin can also be activated by mammalian trypsin or α-chymotrypsin (Davidson, Bieber, et al. 1987; Broadwell and Baumann 1987). The activated toxin produced cytopathology in cultured *Cx. quinquefasciatus* cells, including rounding up, release from support, and eventual lysis. Preparations not treated with gut homogenates caused cytopathology much more slowly and affected fewer cells in the population. Similar cytopathology was produced by purified 41.9-kDa toxin (Baumann et al. 1985), and pathology caused by crude toxin extracts was neutralized both by antibody to the crude extracts and antibody to purified 41.9-kDa toxin. As stated above, *Aedes* species cells were much less sensitive or insensitive to activated toxin (Davidson 1986).

The interaction of the toxin with cultured mosquito cells has been studied by Broadwell and Baumann (1987) and by Davidson, Shellabarger, et al. (1987). Both laboratories have evidence that a glycoprotein or other carbohydrate-containing compound functions as a receptor on susceptible cells. Preincubation of toxin with muramic acid, N-acetylglucosamine, N-acetylgalactosamine, N-acetylneuraminic acid, chitobiose, or chitotriose all reduced the effectiveness of the toxin preparation. Preincubation of cells with wheat germ agglutinin and, to a lesser extent, concanavalin A also reduced the susceptibility of the cells to the toxin. Davidson, Shellabarger, et al. (1987) used fluorescent (FITC)-labeled toxin and unlabeled toxin followed by specific antibody and fluorescent-labeled secondary antibody to demonstrate binding to the cells. Toxin appeared to be internalized within 1 hour by the process of receptor-mediated endocytosis. Binding of fluorescent-labeled toxin was inhibited by N-acetylglucosamine and by unlabeled toxin, and was eliminated by protease pretreatment of the cells. Blockage of cell swelling by sugars that did not interfere with binding suggested that this toxin may act by forming pores in the cell membrane.

Ultrastructural effects of the activated *B. sphaericus* toxin on cultured *Cx. quinquefasciatus* cells included very rapid swelling of the mitochondrial cristae and endoplasmic reticulum, which occurred within 5 minutes of treatment. Mitochondria became highly condensed, vacuoles enlarged, and the cytoplasmic density was reduced as the cells became swollen. Changes in cell membrane ultrastructure were observed, which led to the suggestion that this toxin may act upon the cell membrane (Davidson and Titus 1987). It has also been reported (Lakshmi and Gopinathan 1988) that the toxin inhibited the activity of choline acetyl transferase and oxygen uptake by mitochondria.

15.6 COMPARISON OF THE *B. SPHAERICUS* TOXIN WITH THE *B.t.i.* TOXIN

The mosquito larvicidal toxin of *B. sphaericus* differs in several important ways from that of *B.t.i.* Although both toxins are produced during sporulation and concentrated in a parasporal inclusion, the principal toxin of *B. sphaericus* is a protein of 41.9 kDa, while the toxins of *B.t.i.* are larger. The onset of pathological symptoms in susceptible hosts fed *B. sphaericus* is considerably slower than for those fed *B.t.i.,* requiring several hours for onset of symptoms and up to 48 hours for full expression of mortality. The host range of *B. sphaericus* is more restricted; the insensitivity of black fly larvae and low sensitivity of *Ae. aegypti* larvae, both of which are very sensitive to *B.t.i.,* are particularly noteworthy differences. The low molecular weight toxin (26–28 kDa) present in the *B.t.i.* parasporal body lyses a variety of insect and mammalian cells. It is hemolytic and is lethal to mice upon injection (Thomas and Ellar 1983). The toxin of *B. sphaericus,* even when activated and highly cytotoxic to *Cx. quinquefasciatus* cells, does not lyse cells other than mosquito cells, is not hemolytic (Davidson 1986), and is not toxic to mice (Shadduck, Singer, and Lause 1980; de Barjac et al. 1987). The hemolytic toxin of *B.t.i.* binds to a lipid receptor on cells, whereas evidence suggests that the *B. sphaericus* toxin binds to a glycoprotein receptor (Davidson, Shellabarger, et al. 1987; Broadwell and Baumann 1987). Finally, these toxins are not related immunologically, as antiserum raised to either does not cross-react by production of precipitate or by neutralization of activity of the other (Davidson 1983; Baumann et al. 1985). This body of evidence strongly suggests that these two toxins act by different modes of action.

Acknowledgments

Research was supported by U.S. NIH Grant No. R01 AI 22702 to E.W.D. and by grants from the UNDP/World Health Organization/World Bank Special Programme for Research and Training in Tropical Diseases (E.W.D. and A.A.Y.).

References

Aly, C.; Mulla, M. S.; and Federici, B. A. 1986. Ingestion, dissolution, and proteolysis of the *Bacillus sphaericus* toxin by mosquito larvae. In *Fundamental and applied aspects of invertebrate pathology,* ed. R. A. Samson, J. M. Vlak, and D. Peters, 549. Foundation of the Fourth Internat. Colloq. Invertebr. Pathol., Wageningen, Netherlands.

Aly, C.; Mulla, M. S.; and Stritzke, K. 1985. Fate of *Bacillus sphaericus* strain 2362 toxin in the gut of mosquito larvae. Mosquito Control Research, Annual Report, University of California, 59-62. Cited by permission.

Angus, T. A. 1956. Extraction, purification, and properties of *Bacillus sotto* toxin. *Can. J. Microbiol.* 2: 416–426.

Arapinis, C.; de la Torre, F.; and Szulmajster, J. 1988, Nucleotide and deduced amino acid sequence of the *Bacillus sphaericus* 1593M gene encoding a 51.4 kD polypeptide which acts synergistically with the 42 kD protein for expression of the larvicidal toxin. *Nucleic Acids Res.* 16: 7731.

Barjac, H. de, and Charles, J.-F. 1983, Une nouvelle toxine active sur les moustiques presente dans des inclusions crystallines produites par *Bacillus sphaericus. C. R. Acad. Sci.* (Paris) 296D: 905–910.

Barjac, H. de; Cosmao Dumanoir, V.; Hamon, S.; and Thiery, I. 1987. Safety tests on mice with *Bacillus sphaericus* serotype H-5a5b, strain 2362. WHO Document WHO/VBC/87.948.

Barjac, H. de; Larget-Thiery, I.; Cosmao Dumanoir, V.; and Ripouteau, H. 1985. Serological classification of *Bacillus sphaericus* strains on the basis of toxicity to mosquito larvae. *Appl. Microbiol. Biotechnol.* 21: 85–90.

Barjac, H. de; Thiery, I.; Cosmao Dumanoir, V.; Frachon, E.; Laurent, P.; Charles, J.-F.; Hamon, S.; and Ofori, J. 1988. Another *Bacillus sphaericus* serotype harbouring strains very toxic to mosquito larvae: Serotype H6. *Ann. Microbiol.* (Inst. Pasteur) 139: 363–377.

Baumann, P.; Baumann, L.; Bowditch, R. D.; and Broadwell, A. H. 1987. Cloning of the gene for the larvicidal toxin of *Bacillus sphaericus* 2362: Evidence for a family of related sequences. *J. Bacteriol.* 169: 4061–4067.

Baumann, L.; Broadwell, A. H.; and Baumann, P. 1988. Sequence analysis of the mosquitocidal toxin genes encoding 51.4- and 41.9-kilodalton proteins from *Bacillus sphaericus* 2362 and 2297. *J. Bacteriol.* 170: 2045–2050.

Baumann, P.; Unterman, B. M.; Baumann, L.; Broadwell, A. H.; Abbene, S. J.; and Bowditch, R. D. 1985. Purification of the larvicidal toxin of *Bacillus sphaericus* and evidence for high-molecular-weight precursors. *J. Bacteriol.* 163: 738–747.

Berry, C., and Hindley, J. 1987. *Bacillus sphaericus* strain 2362: Identification and nucleotide sequence of the 41.9kDa toxin gene. *Nucleic Acids Res.* 15: 5891.

Bone, L. W., and Tinelli, R. 1987. *Trichostrongylus colubriformis:* Larvicidal activity of toxic extracts from *Bacillus sphaericus* (strain 1593) spores. *Exper. Parasitol.* 64:514–516.

Bourgouin, C.; Larget-Thiery, I.; and Barjac, H. de 1984. Efficacy of dry powders from *Bacillus sphaericus:* RB80, a potent reference preparation for biological titration. *J. Invertebr. Pathol.* 44: 146–150.

Bourgouin, C.; Tinelli, R., Bouvet, J.-P., and Pires, R. 1984. *Bacillus sphaericus:* 1593-4 purification of fractions toxic for mosquito larvae. In *Bacterial protein toxins,* ed. J. Alouf et al., 387–388. London: Academic Press.

Bowditch, R.; Baumann, P.; and Yousten, A. 1989. Cloning and sequencing the gene encoding a 125-kilodalton surface-layer protein from *Bacillus sphaericus* 2362 and a related cryptic gene. *J. Bacteriol.* 171: 4178–4188.

Broadwell, A. H.; and Baumann, P. 1986. Sporulation-associated activation of *Bacillus sphaericus* larvicide. *Appl. Environ. Microbiol.* 52: 758–764.

———. 1987. Proteolysis in the gut of mosquito larvae results in further activation of the *Bacillus sphaericus* toxin. *Appl. Environ. Microbiol.* 53: 1333–1337.

Bulla, L. A., Jr.; Bechtel, D.; Kramer, K.; Shetna, Y.; Aronson, A.; and Fitz-James, P. C. 1980. Ultrastructure, physiology, and biochemistry of *Bacillus thuringiensis. Crit. Rev. Microbiol.* 8: 147–204.

Charles, J.-F. 1987. Ultrastructural midgut events in culicidae larvae fed with *Bacillus sphaericus* 2297 spore/crystal complex. *Ann. Microbiol.* (Inst. Pasteur) 138: 471–484.

Charles, J.-F.; Kalfon, A.; Bourgouin, C.; and Barjac, H. de 1988. *Bacillus sphaericus* asporogenous mutants: Morphology, protein pattern, and larvicidal activity. *Ann. Microbiol.* (Inst. Pasteur) 139: 243–259.

Charles, J.-F.; and Nicolas, L. 1986. Recycling of *Bacillus sphaericus* 2362 in mosquito larvae: A laboratory study. *Ann. Microbiol.* (Inst. Pasteur) 137B: 101–111.

Davidson, E. W. 1979. Ultrastructure of midgut events in the pathogenesis of *Bacillus sphaericus* strain SSII-1 infections of *Culex pipiens quinquefasciatus* larvae. *Can. J. Microbiol.* 25: 178–184.

———. 1981a. Bacterial diseases of insects caused by toxin-producing bacilli other than *Bacillus thuringiensis. Pathogenesis of invertebrate microbial diseases,* ed. E. W. Davidson, 269–291. Totowa, N.J.: Allanheld, Osmun.

———. 1981b. A review of the pathology of bacilli infecting mosquitoes, including an ultrastructural study of larvae fed *Bacillus sphaericus* 1593 spores. *Dev. Indust. Microbiol.* 22: 69–81.

———. 1982. Purification and properties of soluble cytoplasmic toxin from the mosquito pathogen *Bacillus sphaericus* strain 1593. *J. Invertebr. Pathol.* 39: 6–9.

———. 1983. Alkaline extraction of toxin from spores of the mosquito pathogen, *Bacillus sphaericus* strain 1593. *Can. J. Microbiol.* 29: 271–275.

———. 1984. Microbiology, pathology, and genetics of *Bacillus sphaericus:* Biological aspects which are important to field use. *Mosq. News* 44: 147–152.

———. 1986. Effects of *Bacillus sphaericus* 1593 and 2362 spore/crystal toxin on cultured mosquito cells. *J. Invertebr. Pathol.* 47: 21–31.

———. 1988. Binding of the *Bacillus sphaericus* toxin to midgut cells of mosquito larvae: Relationship to host range. *J. Med. Entomol.* 25 (3): 151–157.

———. 1989. Variation in binding of the *Bacillus sphaericus* toxin and wheat germ agglutinin to larval midgut cells of six species of mosquitoes. *J. Invertebr. Pathol.* 53: 251–259.

Davidson, E. W.; Bieber, A. L.; Meyer, M.; and Shellabarger, C. 1987. Enzymatic activation of the *Bacillus sphaericus* mosquito larvicidal toxin. *J. Invertebr. Pathol.* 50: 40–44.

Davidson, E. W.; and Myers, P. 1981. Parasporal inclusions in *Bacillus sphaericus*. *FEMS Microbiol. Lett* 10: 261–265.

Davidson, E. W.; Shellabarger, C.; Meyer, M.; and Bieber, A. L. 1987. Binding of the *Bacillus sphaericus* mosquito larvicidal toxin to cultured insect cells. *Can. J. Microbiol.* 33: 982–989.

Davidson, E. W.; Singer, S.; and Briggs, J. D. 1975. Pathogenesis of *Bacillus sphaericus* strain SSII-1 infections in *Culex pipiens quinquefasciatus* larvae. *J. Invertebr. Pathol.* 25: 179–184.

Davidson, E. W.; and Titus, M. 1987. Ultrastructural effects of the *Bacillus sphaericus* mosquito larvicidal toxin on cultured mosquito cells. *J. Invertebr. Pathol.* 50: 213–220.

Davidson, E. W.; Urbina, M.; Payne, J.; Mulla, M. S.; Darwazeh, H.; Dulmage, H. T.; and Correa, J. A. 1984. Fate of *Bacillus sphaericus* 1593 and 2362 spores used as larvicides in the aquatic environment. *Appl. Environ. Microbiol.* 47: 125–129.

Des Rochers, B., and Garcia, R. 1984. Evidence for persistence and recycling of *Bacillus sphaericus*. *Mosq. News* 44: 160–165.

Ganesan, S.; Kamdar, H.; Jayaraman, K.; and Szulmajster, J. 1983. Cloning and expression in *Escherichia coli* of a DNA fragment from *Bacillus sphaericus* coding for biocidal activity against mosquito larvae. *Mol. Gen. Genet.* 189: 181–183.

Guillet, P. 1984. La lutte contre l'Onchocerose humaine et les perspectives d'integration de la lutte biologique. *Entomophaga* 29: 121–132.

Hindley, J., and Berry, C. 1987. Identification, cloning, and sequence analysis of the *Bacillus sphaericus* 1593 41.9 kD larvicidal toxin gene. *Mol. Microbiol.* 1: 187–194.

———. 1988. *Bacillus sphaericus* 2297: Nucleotide sequence of 41.9 kDa toxin gene. *Nucleic Acids Res.* 16: 4168.

Holt, S. C.; Gauthier, J.; and Tipper, D. 1975. Ultrastructural studies of sporulation in *Bacillus sphaericus*. *J. Bacteriol.* 122: 1322–1338.

Kalfon, A.; Charles, J.-F.; Bourgouin, C.; and Barjac, H. de. 1984. Sporulation of *Bacillus sphaericus* 2297: An electron microscope study of crystal-like inclusion biogenesis and toxicity to mosquito larvae. *J. Gen. Microbiol.* 130: 893–900.

Karch, S., and Coz, J. 1983. Histopathologie de *Culex pipiens* Linne soumis a l'activite larvicide de *Bacillus sphaericus* 1593-4. *Cah. ORSTOM, sér. Ent. méd et Parasitol.* 21:225–230.

Kellen, W. R.; Clark, T. B., Lindegren, J. E.; Ho, B. C.; Rogoff, M. H.; and Singer, S. 1965. *Bacillus sphaericus* Neide as a pathogen of mosquitoes. *J. Invertebr. Pathol.* 7: 442–448.

Lacey, L. A.; Day, J.; and Heitzman, C. 1987. Long-term effects of *Bacillus sphaericus* on *Culex quinquefasciatus*. *J. Invertebr. Pathol.* 49: 116–123.

Lacey, L. A., and Singer, S. 1982. The larvicidal activity of new isolates of *Bacillus sphaericus* and *Bacillus thuringiensis* (H-14) against anopheline and culicine mosquitoes. *Mosq. News* 42: 537–543.

Lakshmi, N., and Gopinathan, K. P. 1986. Purification of larvicidal protein from *Bacillus sphaericus* 1593. *Biochem. Biophys. Res. Commun.* 141: 756–761.

———. 1988. Effect of *Bacillus sphaericus* 1593 toxin on choline acetyl transferase and mitochondrial oxidative activities of the mosquito larvae. *Indian J. Biochem. Biophys.* 25: 253–256.

Lewis, L.; Yousten, A. A.; and Murray, R. G. E. 1987. Characterization of the surface protein layers of the mosquito pathogenic strains of *Bacillus sphaericus*. *J. Bacteriol.* 169:72–79.

Louis, J.; Jayaraman, K.; and Szulmajster, J. 1984. Biocide gene(s) and biocidal activity in different strains of *Bacillus sphaericus*: Expression of the gene(s) in *E. coli* maxicells. *Mol. Gen. Genet.* 195: 23–28.

Mian, L. S., and Mulla, M. S. 1983. Effect of proteolytic enzymes on the activity of *Bacillus sphaericus* against *Aedes aegypti* and *Culex quinquefasciatus*. *Bull. Soc. Vector Ecol.* 8: 122–127.

Mulla, M. S.; Darwazeh, H. A.; Davidson, E. W.; Dulmage, H. T.; and Singer, S. 1984. Larvicidal activity and field efficacy of *Bacillus sphaericus* strains against mosquito larvae and their safety to nontarget organisms. *Mosq. News* 44: 336–342.

Myers, P., and Yousten, A. A. 1978. Toxic activity of *Bacillus sphaericus* SSII-1 for mosquito larvae. *Infect. Immun.* 19: 1047–1053.

———. 1980. Localization of a mosquito-larval toxin of *Bacillus sphaericus* 1593. *Appl. Environ. Microbiol.* 39: 1205–1211.

———. 1981. Toxic activity of *Bacillus sphaericus* for mosquito larvae. *Dev. Indust. Microbiol.* 22: 41–52.

Myers, P.; Yousten, A. A.; and Davidson, E. 1979. Comparative studies of the mosquito-larval toxin of *Bacillus sphaericus* SSII-1 and 1593. *Can. J. Microbiol.* 25: 1227–1231.

Nicolas, L., and Dossou-Yovo, J. 1987. Differential effects of *Bacillus sphaericus* strain 2362 on *Culex quinquefasciatus* and its competitor *Culex cinereus* in West Africa. *Med. Vet. Entomol.* 1: 23–27.

Obeta, J. A. N., and Okafor, N. 1983. Production of *Bacillus sphaericus* strain 1593 primary powder on media made from locally obtainable Nigerian agricultural products. *Can. J. Microbiol.* 29: 704–709.

Payne, J. M., and Davidson, E. W. 1984. Insecticidal activity of the crystalline parasporal inclusions and other components of the *Bacillus sphaericus* 1593 spore complex. *J. Invertebr. Pathol.* 43: 383–388.

Sgarrella, F., and Szulmajster, J. 1987. Purification and characterization of the larvicidal toxin of *Bacillus sphaericus* 1593M. *Biochem. Biophys. Res. Comm.* 143: 901–907.

Shadduck, J.; Singer, S.; and Lause, S. 1980. Lack of mammalian pathogenicity of entomocidal isolates of *Bacillus sphaericus. Environ. Entomol.* 9: 403–407.

Shively, J. M. 1974. Inclusion bodies of procaryotes. *Ann. Rev. Microbiol.* 28: 167–187.

Silapanuntakul, S.; Pantuwatana, S.; Bhumiratana, A.; and Charoensiri, K. 1983. The comparative persistence of toxicity of *Bacillus sphaericus* strain 1593 and *Bacillus thuringiensis* serotype H-14 against mosquito larvae in different kinds of environments. *J. Invertebr. Pathol.* 42:387–392.

Singer, S. 1973. Insecticidal activity of recent bacterial isolates and their toxins against mosquito larvae. *Nature* 244: 110–111.

———. 1974. Entomogenous bacilli against mosquito larvae. *Dev. Industr. Microbiol.* 15: 187–194.

———. 1980. *Bacillus sphaericus* for the control of mosquitoes. *Biotechnol. Bioeng.* 22: 1335–1355.

Singh, G. J. P., and Gill, S. S. 1988. An electron microscopy study of the toxic action of *Bacillus sphaericus* in *Culex quinquefasciatus* larvae. *J. Invertebr. Pathol.* 52: 237–247.

Stray, J. E.; Klowden, M. J.; and Hurlbert, R. E. 1988. Toxicity of *Bacillus sphaericus* crystal toxin to adult mosquitoes. *Appl. Environ. Microbiol.* 54: 2320–2321.

Tandeau de Marsac, N.; de la Torre, F.; and Szulmajster, J. 1987. Expression of the larvicidal gene of *Bacillus sphaericus* 1593M in the cyanobacterium *Anacystis nidulans* R2. *Mol. Gen. Genet.* 209: 396–398.

Thiery, I., and Barjac, H. de. 1989. Selection of the most potent *Bacillus sphaericus* strains based on activity ratios determined on three mosquito species. *Appl. Microbiol. Biotechnol.* 31: 577–581.

Thomas, W. E., and Ellar, D. J. 1983. *Bacillus thuringiensis* var. *israelensis* crystal δ-endotoxin: Effects on insect and mammalian cells in vitro and in vivo. *J. Cell Sci.* 60: 181–197.

Tinelli, R.; Barjac, H. de; and Bourgouin, C. 1980. Isolement d'une fraction de spore de *Bacillus sphaericus* toxique pour les larves d'Anopheles. *C. R. Acad. Sci.* (Paris) 291D: 537–539.

Tinelli, R., and Bourgouin, C. 1982. Larvicidal toxin from *Bacillus sphaericus* spores. Isolation of toxic components. *FEBS Lett.* 142: 155–158.

Weiser, J. 1984. A mosquito-virulent *Bacillus sphaericus* in adult *Simulium damnosum* from Northern Nigeria. *Zbl. Mikrobiol.* 139: 57–60.

Wickremesinghe, R. S. B., and Mendis, C. L. 1980. *Bacillus sphaericus* spore from Sri Lanka demonstrating rapid larvicidal activity on *Culex quinquefasciatus. Mosq. New* 40: 387–389.

Yousten, A. A. 1984. Bacteriophage typing of mosquito pathogenic strains of *Bacillus sphaericus. J. Invertebr. Pathol.* 43: 124–125.

Yousten, A. A., and Davidson, E. 1982. Ultrastructural analysis of spores and parasporal crystals formed by *Bacillus sphaericus* 2297. *Appl. Environ. Microbiol.* 44: 1449–1455.

16

Genetics of *Bacillus sphaericus*

WILLIAM F. BURKE, JR.
KAREN A. ORZECH

16.1. INTRODUCTION

Since it was first described by Neide in 1904, *Bacillus sphaericus* has been the subject of a number of studies to refine the taxonomy of this species. However, for the most part the species still remains an enigma. Early taxonomic investigations placed considerable emphasis on spore morphology and position within the sporangium and on nutritional properties of the vegetative cells. However, because isolates assigned to this species are generally unreactive toward the substrates routinely used in taxonomic studies of *Bacillus* (Gordon, Haynes, and Pana 1973), *B. sphaericus* is considered somewhat of a catchall species. This conclusion has been supported by a variety of genetic and biochemical techniques.

16.2. EVIDENCE OF GENETIC HETEROGENEITY WITHIN THE SPECIES *B. SPHAERICUS*

Relationships among strains of *B. sphaericus* and similar species were examined by Krych, Johnson, and Yousten (1980) by the measurement of the guanine plus cytosine (GC) content of the DNA. Phenotypic characteristics of the 62 strains were examined. In this study, 5 homology groups were identified. The type-strain ATCC 14577 was included in homology group 1. Only 3 strains of 62 examined fell within this group, and none of these are pathogenic to mosquito larvae. Significantly, all 7 strains pathogenic to mosquito larvae were placed in group 2a. All strains within this subgroup had 79% or greater homology to strain 1593. Again, the heterogeneity within the species was indicated by DNA homologies, outside of group 1, ranging from 14 to 26% to the type-strain. No single phenotypic trait studied allowed the separation of homology groups, with the exception of mosquito pathogenicity of group 2a.

In another study, pathogenic and nonpathogenic strains of *B. sphaericus* were studied using both serology and numerical analysis of 160 substrates (de Barjac, Véron, and Cosmao Dumanoir 1980). Both the flagellar agglutination technique and numerical analysis can be used to clearly distinguish those strains of *B. sphaericus* that are entomocidal from those that are not. Three different flagellar serotypes were assigned to the entomopathogens, H1a, H2, and H5. A more recent study (de Barjac et al. 1985) analyzed 54 strains isolated from vectors or mosquito larval habitats. The mosquito pathogens fell into five H-serotypes, and each serotype displayed a different level of toxicity. Toxicity was greatest in all 21 strains of serotype H5a,5b, which includes strains 1593, 1881, and 1691. Strain 2297 displayed the same high level of toxicity, yet belonged to H-serotype 25. Less-toxic strains, SSII-1 and 1404, fell under H-serotype 2.

In addition to flagellar serotyping, bacteriophage typing was found to identify comparable subgroups among the mosquito pathogens of DNA homology group 2a (Yousten et al. 1980). Furthermore, it was recognized that phage-typing and serotyping were related to the level of toxicity displayed by the strains. The most toxic strains bioassayed exhibited a group 3 phage-typing pattern like that of strain 1593. Strain 2362 belongs to this same phage group (Yousten and Hedrick 1982). However, strain 2297, also very toxic, belongs to a distant group, phage group 4 (Yousten 1984).

These DNA homology studies, serotyping, and phage-typing further define the taxonomic position of both entomocidal and nonentomocidal strains of *B. sphaericus*.

The genetic heterogeneity of strains assigned to this species was also demonstrated by investigations of the minimal nutritional requirements of numerous strains of *B. sphaericus*. White and Lotay (1980) determined the nutritional requirements for growth and for sporulation of 27 strains of *B. sphaericus*. These researchers developed a chemically defined minimal medium containing inorganic salts, ammonium sulfate, and sodium acetate (as the sole carbon source). Of the 27 strains examined, 3 (including the type-strain) grew and sporulated in this minimal medium; and 13 strains grew and sporulated when biotin and thiamine were added. Two more strains grew in the minimal medium with these vitamins and glutamic acid; 4 of the remaining strains needed several amino acids, but not acetate; and 4 other strains required a purine as well as amino acids.

Recently, two chemically defined minimal media have been developed for selective isolation of *B. sphaericus* from soil and aquatic habitats. One medium, which contains acetate as the sole carbon source, is designed for isolation of both pathogenic and nonpathogenic strains of *B. sphaericus* (Massie, Roberts, and White 1985); the other medium, which uses arginine as the major carbon source and streptomycin as a selective agent, allows good recovery of *B. sphaericus* pathogenic strains and selects against certain other spore formers, including 68% of the nonpathogenic strains tested (Yousten, Fretz, and Jelley 1985).

Massie, Roberts, and White (1985), using their selective medium, found that all of the strains that they isolated from the soil and identified as *B. sphaericus* based on morphology were capable of growth on acetate alone. Nevertheless, several of these strains had one or more atypical features, most commonly the ability to metabolize glucose, which sometimes led to a positive result in the methyl red or Voges-Proskauer test. The type-strain of this species (ATCC 14577) and three other strains grow with acetate as the sole source of carbon and do not require vitamins. These four strains make up group 1 as defined by Krych, Johnson, and Yousten (1980) based on DNA homology. If all of the new strains isolated by Massie, Roberts, and White (1985) belong to group 1 (because all grow with acetate alone), then the diversity of other properties becomes more surprising and again points to the genetic heterogeneity within this species.

Yousten, Fretz, and Jelley (1985) found that their selective medium for mosquito-pathogenic strains of *B. sphaericus* completely selected against nonpathogenic strains in DNA homology groups 1, 3, and 5. However, five of five strains in homology group 2b and two of five strains in homology group 4 grew weakly on this medium. These results are further indicators of the genetic diversity within this species.

16.3 NATURALLY OCCURRING ANTIBIOTIC RESISTANCES IN *B. SPHAERICUS*

In the course of examining various phenotypic characteristics of *B. sphaericus* in preparation for genetic exchange experiments, the minimal inhibitory concentrations (MICs) of several antibiotics were determined for insecticidal and noninsecticidal strains of this organism. Prior to this work, one isolate of *B. sphaericus*, strain 1593, was reported to be resistant to streptomycin at concentrations of 10 μg/ml (Hertlein, Levy, and Miller 1979) and to 100 μg/ml (Yousten, Fretz, and Jelley 1985). In our study, all of the *B. sphaericus* strains examined were found to be resistant to streptomycin at concentrations of 4 mg/ml or greater (table 16.1; Burke and McDonald 1983). All of the *B. sphaericus* strains examined, with the exception of the type-strain, were found to be resistant to chloramphenicol at concentrations of 8 μg/ml or greater after 24 hours of incubation (table 16.2). Values presented in table 16.2 represent the uninduced level of chloramphenicol resistance.

Work by S. Mongkolsuk and P. Lovett (pers. comm.) and by Harwood, Williams, and Lovett (1983) on inducible chloramphenicol resistance in *Bacillus pumilus* suggested that many of the naturally occurring chloramphenicol resistance in *Bacillus* species are inducible. In order to examine this, one strain of *Bacillus subtilis* and one strain of *B. sphaericus* were streaked repeatedly at 24-hour intervals on two types of media, BHI (Brain Heart Infusion) agar supplemented with thymine and the same agar medium

TABLE 16.1.
Minimal Inhibitory Concentrations of Streptomycin

Strain	Streptomycin Concentration (mg/ml)		
	NAYE[a]	BHI[a]	MH[a]
ATCC 14577	5	>20	>20
ATCC 7054	4	>20	6
CCEB 740	5	8	4
CCEB 921	12	>20	>20
CCEB 922	12	>20	>20
1593[b]	14	16	>20
2362[b]	18	>20	>20
SSII-1[b]	18	>20	>20
2297[b]	16	20	>20

[a]NAYE, nutrient agar-yeast extract; BHI, brain heart extract; MH, Mueller-Hinton.
[b]Strains with insecticidal activity.

containing increasing concentrations of chloramphenicol. Induced strains were maintained at chloramphenicol concentrations that permitted only minimal growth over a 24-hour period at 30° C as determined by visual examination. As controls, strains were carried on the same medium without antibiotic and were considered to be uninduced. Also, *B. subtilis* 168 was used as an internal control because it is not naturally resistant to chloramphenicol (Duvall et al. 1983). MICs were determined for both induced and uninduced cells and are presented in table 16.3. These results reaffirm that *B. subtilis* 168 is uninducible under these conditions, while *B. sphaericus* 1593 is definitely inducible.

TABLE 16.2.
Minimal Inhibitory Concentrations of Chloramphenicol

Strain	Chloramphenicol Concentration (μg/ml)		
	NAYE[a]	BHI[a]	MH[a]
ATCC 14577	2	2	2
ATCC 7054	8	10	8
CCEB 740	8	8	8
CCEB 921	10	8	12
CCEB 922	10	10	10
1593[b]	12	12	16
2362[b]	12	12	16
SSII-1[b]	12	12	16
2297[b]	10	12	14

[a]NAYE, nutrient agar-yeast extract; BHI, brain heart extract; MH, Mueller-Hinton.
[b]Strains with insecticidal activity.

TABLE 16.3.
Inducible Chloramphenicol Resistance in *B. sphaericus* 1593 and *B. subtilis* 168

Bacterial strain	Plasmid	MIC (μg/ml) of Uninduced Cells	MIC (μg/ml) of Induced Cells
B. subtilis			
168	None	5	5
168	pNN101	25	75
B. sphaericus			
1593	None	25	50
1593	pNN101	30	75

Subsequent to these studies, *B. sphaericus* strains 1593 and ATCC 14577 (the type-strain) were determined to be resistant to D-cycloserine at concentrations of > 100 μg/ml in Nutrient Agar-Yeast Extract Sporulation medium (NYSM). This compound may be useful in media used to isolate organisms from natural environments. Preliminary studies examining this possibility have shown that D-cycloserine may be a beneficial adjunct in developing more efficient media for recovery of organisms from the environment during field tests (table 16.4). The increase in the number of germinating cells in medium containing between 50 and 100 μg/ml was consistently observed.

Gould (1966) has shown that the spores of certain strains of *Bacillus cereus* are stimulated to germinate by D-cycloserine. However, unlike *B.*

TABLE 16.4.
Comparison of Recovery of *B. sphaericus* Spores on NYSM Containing Either D-cycloserine Alone or D-cycloserine and Streptomycin

NYSM Containing		
Cycloserine (μg/ml)	Streptomycin (μg/ml)	Spore Recovery (% viable spores)[a]
0	0	100
50	0	131
50	100	143
100	0	143
100	100	134
150	0	148
150	100	36
200	0	56

[a]% viable spores calculated by dividing the number of colony-forming units (cfu) on plates containing antimicrobial agents by the number of cfu on control plates without antimicrobials.

sphaericus, these *B. cereus* strains were sensitive to the antimicrobial action of D-cycloserine during vegetative growth. A possible explanation for the stimulation of germination may be that D-cycloserine, which is an analogue of D-alanine, may interfere with enzymes such as alanine racemase, which are located in the spore coat and function to convert L-alanine to D-alanine as a mechanism of inhibiting germination. The affinity of many enzymes that catalyze reactions involving D-alanine as a substrate or product have a much greater affinity for D-cycloserine than the natural substrate. This hypothesis remains to be investigated more fully.

16.4 AUXOTROPHIC MUTANTS OF *B. SPHAERICUS*

Walker and Maddus (1979) used ultraviolet irradiation to isolate mutants of *B. sphaericus* 1593. These authors reported that the mutants resulting from this procedure included cysteine, glycine, histidine, and adenosine auxotrophs as well as mutants tolerant to sodium azide. The strains carrying these mutations retained their pathogenicity toward mosquito larvae.

B. sphaericus strains are nearly indistinguishable from each other except by phage-typing (Yousten et al 1980 and serotyping (de Barjac, Véron, and Cosmao Dumanoir; de Barjac 1985). Even these techniques cannot distinguish strains within a phage-typing or serotyping class. The isolation of auxotrophic mutants provides a valuable means of differentiating one strain from another in laboratory experiments. For these reasons, efforts in our laboratory were directed toward the induction of genetic lesions in *B. sphaericus* 1593 so that derivatives of this strain could be readily identified when used in genetic exchange studies. To this end, 1593 spores were mutagenized with ethyl methanesulfonate (EMS); cells were selected for thymine-auxotrophy on medium containing trimethoprim and thymine. Using this procedure, several thymine-requiring mutants were isolated. ASB 13092, an extremely stable, thymine-requiring mutant of 1593, was mutagenized with EMS and screened for auxotrophic mutants by the method described by Carlton and Brown (1981). Most of the mutants isolated in this manner carried one or more auxotrophic mutations for an amino acid as well as the thymine requirement. Arginine, histidine, methionine, isoleucine-valine, and threonine auxotrophs of this thymine-requiring mutant were isolated. All of these auxotrophs retained their ability to sporulate and their pathogenicity.

16.5 TRANSFORMATION OF *B. SPHAERICUS* WITH PLASMID DNA

The study of the genetic organization of *B. sphaericus* and the progress in strain improvement using recombinant DNA techniques have been hampered for a long time by the lack of an in vitro system for the introduction of genetic information. Conventional transformation regimens employed for *B. subtilis* proved completely unsuccessful in our laboratory. Therefore, we examined the possibility of adapting the protoplast transformation system developed for *B. subtilis* by Chang and Cohen (1979). Adaptations of this technique proved successful for introduction of small plasmid molecules into *B. sphaericus* (McDonald and Burke 1984). The plasmids first used to successfully transform this organism were pUB110 (a *Staphylococcus* plasmid), which codes for resistance to kanamycin and neomycin, and pBC16 (a *Bacillus* plasmid), which codes for resistance to tetracycline.

It should be noted that, although the plasmid pBC16 was used in these genetic exchange experiments, Polak and Novick (1982) demonstrated that this plasmid can be found in *B. sphaericus* strains in nature. They examined four 4.5-kb Tc^r plasmids from *B. sphaericus* strains, originally isolated from a hog feedlot in the United States, which were subjected to restriction endonuclease analysis and found to be indistinguishable from pBC16, a Tc^r plasmid of the same size isolated from *B. cereus* in Germany several years previously (Berhard, Schrempf, and Goebel 1978), and from pJP3656 obtained from *B. subtilis*. Polak and Novick also reported that pBC16 is highly homologous to and incompatible with pUB110, the other plasmid used in these genetic experiments.

The frequency of the first *B. sphaericus* protoplast transformation using pUB110 DNA isolated from *B. subtilis* was less than one transformant per μg of DNA. However, when pUB110 DNA, isolated from a *B. sphaericus* 1593 transformant obtained in the first transformation, was used as the donor DNA, a frequency of 1.0×10^4 transformants per μg of DNA was obtained. Although the frequencies obtained with pBC16 using DNA isolated from *B. sphaericus* as compared with DNA isolated from *B. subtilis* were not as dramatic as those obtained with pUB110 (only a three-fold increase in transformation frequency with pBC16 from *B. sphaericus*), the same general trend was observed. These data suggested the operation of a restriction-modification system in *B. sphaericus* 1593 that was effective at inactivating pUB110 isolated from *B. subtilis*.

In order to confirm the presence of a restriction-modification system, 1593 cells were converted to protoplasts with lysozyme and lysed osmotically. This cell lysate was incubated with plasmid DNA isolated from *B. subtilis* and from *B. sphaericus* as well as λ DNA. The lysate was shown to contain a restriction system that digested λ, pBR322, pUB110, pBC16, and pBD64 (a hybrid plasmid consisting of the chloramphenicol resistance gene from pC194 ligated to pUB110; Gryczan, Shivakumar, and Dubnau 1980) to varying degrees.

This newly discovered restriction-modification system from strain 1593 was compared to well-characterized restriction-modification systems found in two other strains of *B. sphaericus* (Koncz, Kiss, and Venetianer 1978; Shibata et al. 1976). Using pBD64 and pBR322 as substrates, the three restriction systems of *B. sphaericus* 1593, R, and IAM 1286 were found to be identical. These results indicate that the restriction system found in 1593 is an isoschizomer of *Hae*III.

16.6 *B. SPHAERICUS* MUTANTS WITH LIMITED RESTRICTION ENDONUCLEASE ACTIVITY

In order to improve the introduction of heterologous genes into *B. sphaericus*, it was necessary to isolate and characterize mutants deficient in restriction endonuclease activity. To circumvent the restriction system, it was necessary to isolate mutants with limited restriction endonuclease activity. This was accomplished by mutagenizing *B. sphaericus* 1593 cells with EMS and then immediately subjecting these cells to protoplast transformation with pUB110 DNA isolated from *B. subtilis*. This plasmid was selected for these experiments because it has four sites recognized by the *Bsp* 1593 restriction system. The rationale for this experiment was that cells that survive the mutagenesis with an intact restriction system should degrade almost all of the pUB110 molecules immediately after entering these cells. As a result, these cells remain sensitive to neomycin, the antibiotic resistance marker on pUB110. However, cells that have had the restriction system inactivated, as a consequence of mutagenesis of the gene encoding the restriction enzyme, are unable to degrade the pUB110 DNA. These cells become resistant to neomycin and can be selected directly. Neomycin-resistant transformants obtained in this manner are potential restrictionless mutants.

Using this technique a number of transformants were obtained. Cell lysates of transformants obtained in this manner were incubated with indicator DNA such as pBD64 and pBR322, which contain multiple sites for the *Bsp* 1593 restriction enzyme. By analyzing the resulting digestion products, isolates were evaluated for the presence of the restriction system. Those isolates found to lack the restriction system were cured of the pUB110 plasmid and examined for their ability to sporulate and for their toxicity toward mosquito larvae. Only those mutants that maintained both toxicity and the ability to sporulate were retained. Mutants obtained in this manner have proved to be exceedingly useful in subsequent investigations involving the transfer and expression of heterologous genes in this mosquiticidal organism.

16.7 EVIDENCE FOR THE EXISTENCE OF A *Bsp* 1593 MODIFICATION SYSTEM

To demonstrate the presence of a modification system in *B. sphaericus* 1593 and its restrictionless derivatives, lysates of both wild-type and restrictionless strains were incubated with pUB110 isolated from *B. sub-*

tilis and *B. sphaericus.* The results of these experiments clearly showed that the 1593 lysate was able to restrict pUB110 isolated from *B. subtilis,* but not from *B. sphaericus,* as was expected. Also, lysates of the restrictionless mutants did not restrict pUB110 DNA isolated from either *B. sphaericus* or *B. subtilis.*

16.8 CONJUGAL TRANSFER OF pAMβ1 IN *B. SPHAERICUS* 1593

Plasmids that mediate resistance to macrolides, lincosamides, and streptogramin B (MLS) (Weisblum 1975) have been found to be widely disseminated in nature. Among these plasmids is pAMβ1, a 26-kb plasmid (Clewell et al. 1974; LeBlanc and Lee 1984), which has been shown to be transmissible to numerous species of streptococci (LeBlanc et al. 1978) as well as other gram-positive bacteria (LeBlanc and Lee 1984). Although the technique of protoplast transformation has been successfully modified and optimized for *B. sphaericus* (McDonald and Burke 1984), there are certain drawbacks to this method of genetic exchange. The protoplast technique is tedious, time-consuming, and expensive. Therefore, other methods of genetic transfer were investigated. Because of the promiscuous nature of this plasmid in gram-positive bacteria, a study of the pAMβ1-directed conjugal transfer system in *B. sphaericus* was initiated.

Forced cell-to-cell contact has been found to facilitate the transfer of pAMβ1 from donor to recipient (LeBlanc et al. 1978); and, by employing the method of filter mating, the plasmid pAMβ1 has been successfully transferred from *Streptococcus faecalis* to *B. sphaericus* (Orzech and Burke 1984). Transfer of pAMβ1 from one *B. sphaericus* strain to another has also been demonstrated. All Em^r transcipients analyzed for the presence of plasmid DNA were shown to harbor a plasmid of the same electrophoretic mobility as pAMβ1.

The transfer frequencies of pAMβ1 from *S. faecalis* or *B. sphaericus* to several isogenic strains of *B. sphaericus* 1593 are shown in table 16.5. These frequencies represent ratios of the number of transcipients per ml of mating mixture to the number of recipients per ml. DNase appears to have had no effect on the frequency of transfer, suggesting a conjugative-type mechanism rather than transformation.

Strain ASB 13052 is a mutant with limited restriction endonuclease activity that was isolated as described previously. The frequencies of transfer of pAMβ1 from JH2.2 (pAMβ1) to either strain ASB 13052 or strain 1593, a strain with restriction endonuclease activity, are essentially identical. The frequencies for both transfers are approximately 1.5×10^{-3}, indicating that *Bsp* 1593 restriction system has little if any effect on incoming pAMβ1 DNA. In order to attempt to explain this observation, pAMβ1 was digested with a

TABLE 16.5.
Frequency of Transfer of pAMβ1 from *Streptococcus faecalis* JH2.2 to *B. sphaericus* and from *B. sphaericus* to *B. sphaericus*

Donor Strain	*B. sphaericus* Recipient Strain	Number of Recipients[a]	Number of Transcipients[a]	Transfer Frequency[b]
S. faecalis				
JH2.2	1593	2.65×10^8	1.75×10^4	1.4×10^{-3}
JH2.2	ASB 13052	1.25×10^8	1.98×10^5	1.6×10^{-3}
JH2.2	1593	2.40×10^8	1.01×10^5	4.2×10^{-4}[c]
JH2.2	1593	2.80×10^8	1.06×10^5	3.8×10^{-4}[d]
B. sphaericus				
ASB 13119[e]	1593	1.80×10^8	2.80×10^4	1.6×10^{-4}

[a]Per ml of mating mixture after incubation for 18–24 hours.
[b]Calculated by dividing the number of transcipients per ml by the number of recipients per ml after resuspension from the filter.
[c]No DNase treatment.
[d]DNase treatment.
[e]Strain ASB 13119 was constructed by mating *S. faecalis* JH2.2 (pAMβ1) with *B. sphaericus* 1593.

lysate containing *Bsp* 1593 restriction enzyme and with the commercially available isoschizomer of *Bsp* 1593, *Hae*III. No digestion of pAMβ1 by either of these enzymes was observed. The lack of a *Bsp* 1593 (or *Hae*III) restriction site in pAMβ1 has been confirmed independently by D. B. Clewell and S. D. Ehrlich (pers. comm.).

Experiments were performed to examine the effect of the species of the donor on the transfer frequency of pAMβ1. Specific matings were run in parallel using either *S. faecalis* JH2.2 (pAMβ1) or *B. sphaericus* ASB 13096 (pAMβ1) as donors to 1593. Transfer of pAMβ1 from JH2.2 to 1593 was consistently 10 times greater than that of pAMβ1 transfer from *B. sphaericus* to *B. sphaericus*. There are several possible explanations for these results. Proteins involved in the mobilization and transfer of pAMβ1 may not be as efficiently transcribed and/or translated in *B. sphaericus* as they are in *S. faecalis*, or the differences in composition and structure of the cell walls may greatly affect the operation of the plasmid-transfer mechanism.

16.9 CONSTRUCTION OF PLASMID VECTORS FOR CLONING IN *B. SPHAERICUS* 1593

Two plasmid vectors for cloning in *B. sphaericus* 1593 were constructed in *B. subtilis* from two parent plasmids, pBC16 and pBD64 (Norton, Orzech, and Burke 1985). The plasmids were developed in *B. subtilis* because the genetics of this organism has been more extensively studied and techniques for genetic exchange in this bacterium are better developed than in *B. sphaericus.* The development of these plasmids was motivated by the limited usefulness of the available *Bacillus* cloning vectors when used in the protoplast transformation system developed for *B. sphaericus.* When characterized, the 5.9-kb chimeric plasmid pNN101 was found to consist of the *Msp* I fragment containing the chloramphenicol acetyltransferase (CAT) gene from pBD64 inserted into an *Msp* I fragment site in pBC16. Then pNN101 was shown to replicate, express, and be stably maintained in *B. sphaericus* 1593 without affecting the mosquito larvicidal activity of this organism (see tables 16.3 and 16.6). A derivative of this plasmid, pNN302, was constructed in which a unique *Hin*dIII site was introduced into the CAT gene without loss of chloramphenicol resistance.

These plasmids represent improvements over the naturally occurring Staphylococcal plasmids, such as pUB110, and the naturally occurring *Bacillus* plasmid, pBC16, which are commonly used as *B. subtilis* cloning vectors. The improvement results from the presence of two drug-resistance markers on these new constructs, making possible the use of insertional inactivation of one of the markers coupled with continued selection for the other marker. Each of the antibiotic-resistance genes present on pNN302 contains unique restriction endonuclease sites, which upon digestion generate cohesive 5′ termini. However, the usefulness of the chloramphenicol acetyltransferase gene encoded by pNN101 and pNN302 may be somewhat limited in *B. sphaericus* due to the discovery, after the vector had been constructed, that this organism has an endogenous, inducible chloramphenicol-resistance gene. Fortunately, this limitation does not apply to the *B. subtilis* system.

An additional advantage of both pNN101 and pNN302 is that both of these plasmids encode resistance to two antibiotics whose inhibitory action is not suppressed by the DM3 medium required for regeneration of transformed protoplasts (Chang and Cohen 1979). The suppression of the inhibitory action of certain antibiotics can greatly diminish the usefulness of protoplast transformation. The incorporation into pNN302 of resistances to antibiotics, which are not suppressed by DM3, facilitates selection of transformants following protoplast transformation. Further, the presence of this new cloning vector in *B. sphaericus* 1593 does not interfere with the larvicidal activity of this organism (table 16.6).

When the insertional inactivation of the chloramphenicol-resistance gene in pNN302 was tested by insertion of *Hin*dIII-digested λ DNA, it was found that only large fragments of λ DNA possessed the ability to inactivate the CAT gene. No plasmids with inserts of < 9.5 kb were found in any of the colonies screened for insertional inactivation of the CAT gene, despite the presence of multiple λ-fragments in the size range of 0.125 kb to 6.557 kb in the initial ligation mixture. It is possible that insertion of these large fragments of DNA near the distal end of the CAT gene resulted in the production of a protein with a drastically altered tertiary structure. That portion of the fusion protein encoded by the inserted DNA may sterically block access to

TABLE 16.6.
Minimal Inhibitory Concentrations of Antibiotics, Plasmid Stabilities, and Larvicidal Activity of pNN101 and pNN302 Transformants

Strain	Plasmid	MIC (μg/ml)		Plasmid Stability[a]	LC_{50}[b]
		Cm[c]	Tc	$Cm^S + Tc^S$	
B. subtilis					
168	None	<5	<5	—	—
168	pNN101	45	>100	0/360	—
168	pNN302	45	>100	0/210	—
B. sphaericus					
1593	None	15	<1	—	4.1×10^2
1593	pNN101	20	70	0/358	8.5×10
1593	pNN302	20	70	0/120	9.5×10^1

[a] Ratio of antibiotic-sensitive isolates to the total number of isolates examined.
[b] Lethal concentration of cells expressed as colony-forming units per ml at which 50% of the treated *Culex quinquefasciatus* larvae died.
[c] Determined under uninduced conditions.

the active site or actually interfere with the physical integrity of the active site of this enzyme. The insertion of a linker or small DNA fragment may simply produce a fusion protein in which there is minimal change in the essential tertiary structure of the chloramphenicol acetyltransferase. Thus, the plasmid pNN302 selects for the cloning of large fragments of DNA. Since *B. subtilis* has no developed system for the cloning of large fragments of DNA, such as cosmid- and λ-cloning in *Escherichia coli*, this method may prove useful in the development of gene libraries in *B. subtilis*.

16.10 CLONING OF *B. SPHAERICUS* LARVICIDAL TOXIN GENES

As part of the effort to elucidate the nature of the mosquito-larvicidal activity, genes coding for toxin proteins have been cloned and analyzed with respect to DNA sequence and deduced amino acid sequences. Information obtained in this manner has been exceptionally useful in relating specific genes to proteins believed to function as components of the larvicidal activity.

Ganesan et al. (1983) were the first to report the cloning of a gene from an entomocidal *B. sphaericus* that encoded biocidal activity toward mosquito larvae. The clone consisted of a 3.7-kb *Sau*3A insert from *B. sphaericus* 1593. Maxicell analysis of the expressed gene products revealed at least four peptides ranging in size from 12 to 21 kilodaltons (kDa) (Louis, Jayaraman, and Szulmajster 1984). Since the predominant proteins in crystals obtained from 48-hour cultures of *B. sphaericus* strains 1593, 2362, and 2297 have molecular masses of 125, 110, 63, and 43 kDa as determined by SDS electrophoresis (Baumann, Broadwell, and Baumann 1988), it would appear that the nature of the toxin reported by Ganesan et al. (1983) and by Louis, Jayaraman, and Szulmajster (1984) is quite distinct from the toxins considered to be the major *B. sphaericus* larvicidal proteins.

Tandeau de Marsac, de la Torre, and Szulmajster (1987) reported that the original clone used in the studies of Ganesan et al. (1983) and Louis, Jayaraman, and Szulmajster (1984) gradually lost larvicidal activity, forcing them to "reclone" the *B. sphaericus* 1593M toxin gene. The gene they successfully cloned was quite different from that examined previously. The significance of the original clone described by Ganesan et al. (1983) remains in question.

The toxin gene cloned by Tandeau de Marsac, de la Torre, and Szulmajster (1987) encodes a peptide with a molecular mass of 42–43 kDa. This is quite similar to the 43-kDa peptide found in the *B. sphaericus* crystal (Baumann, Broadwell, and Baumann 1988). These workers successfully demonstrated the expression of this toxin gene in the cyanobacterium *Anacystis nidulans* R2. The rationale for this experiment was to examine the potential for introducing functional mosquiticidal toxin genes into cyanobacteria indigenous to the larval feeding zones.

The gene coding for the 43-kDa biocidal toxin has been cloned and sequenced from *B. sphaericus* strains 1593 (Hindley and Berry 1987; Lee et al. 1987), 2362 (Baumann et al. 1987; Baumann, Broadwell, and Baumann 1988; Berry and Hindley 1987), and 2297 (Baumann, Broadwell, and Baumann; Hindley and Berry 1988). From DNA sequence data, the deduced size of the protein toxin was determined to be 41.9 kDa. The nucleotide sequences encoding the 41.9-kDa protein from all three of these strains were found to contain two open reading frames. One of the reading frames codes for the 41.9-kDa protein and the other for a 51.4-kDa protein (Baumann, Broadwell, and Baumann 1988; Arapinis, de la Torre, and Szulmajster 1988). The 51.4-kDa protein was previously designated the 63-kDa protein.

Interestingly, recombinants expressing the 41.9-kDa toxin are not toxic to *Culex* larvae. However, when this recombinant is combined with a clone expressing the 51.4-kDa protein, larval toxicity is regained (Baumann et al. 1987). Recently Arapinis, de la Torre, and Szulmajster (1988) and Baumann, Broadwell, and Baumann (1988) have suggested that the 51.4-kDa protein acts synergistically with the 41.9-kDa protein to produce toxicity. In all three of the toxic *B. sphaericus* strains examined, the genes encoding these two proteins are physically adjacent to one another on the *B. sphaericus* chromosome. Sequencing data further suggest that these crystal protein genes reside within a single transcriptional operon (Baumann, Broadwell, and Baumann 1988).

The gene encoding the 125-kDa precursor of the 122-kDa *B. sphaericus*

2362 surface-layer protein has been cloned, sequenced, and examined for mosquiticidal activity (Bowditch, Baumann, and Yousten 1989). The 122-kDa surface protein and its 125-kDa precursor, observed in crystal preparations of *B. sphaericus* strains 1593, 2362, and 2297 (Broadwell and Baumann 1986), were found to lack insecticidal activity. However, the 122-kDa protein is believed to be the precursor of the 110-kDa larvicide (Broadwell and Baumann 1986).

Acknowledgments

Previously unpublished research by the authors was supported by the following sources: Biomedical Sciences component of the UNDP/World Bank/WHO Special Programme for Research and Training in Tropical Diseases; Public Health Services Grant AI20918 from the National Institute of Allergy and Infectious Diseases; and Biomedical Research Grant SO7 RR07112 awarded to Arizona State University by the Biomedical Research Grant Program, Division of Research Resources, National Institutes of Health.

References

Arapinis, C.; de la Torre, F.; and Szulmajster, J. 1988. Nucleotide and deduced amino acid sequence of the *Bacillus sphaericus* 1593M gene encoding a 51.4 kD polypeptide which acts synergistically with the 42 kD protein for expression of the larvicidal toxin. *Nucl. Acids Res.* 16: 7731.

Barjac, H. de; Larget-Thiery, I.; Cosmao Dumanoir, V. C.; and Ripouteau, H. 1985. Serological classification of *Bacillus sphaericus* strains on the basis of toxicity to mosquito larvae. *Appl. Microbiol. Biotechnol.* 21: 85–90.

Barjac, H. de; Véron, M.; and Cosmao-Dumanoir, V. C. 1980. Caractérisation biochimique et sérological de souches de *Bacillus sphaericus* pathogènes ou non pour les moustiques. *Ann. Microbiol.* (Inst. Pasteur) 131B: 191–201.

Baumann, P.; Baumann, L.; Bowditch, R. D.; and Broadwell, A. H. 1987. Cloning of the gene for the larvicidal toxin of *Bacillus sphaericus* 2362: Evidence for a family of related sequences. *J. Bacteriol.* 169: 4061–4067.

Baumann, L.; Broadwell, A. H.; and Baumann, P. 1988. Sequence analysis of the mosquitocidal toxin genes encoding 51.4- and 41.9-kilodalton proteins from *Bacillus sphaericus* 2362 and 2297. *J. Bacteriol.* 170: 2045–2050.

Bernhard, K.; Schrempf, H.; and Goebel, W. 1978. Bacteriocin and antibiotic resistance plasmids in *Bacillus cereus* and *Bacillus subtilis*. *J. Bacteriol.* 133: 897–903.

Berry, C., and Hindley, J. 1987. *Bacillus sphaericus* 2362: Identification and nucleotide sequence of the 41.9kDa toxin gene. *Nucleic Acids Res.* 15: 5891.

Bowditch, R. D.; Baumann, P.; and Yousten, A. A. 1989. Cloning of the gene encoding the 125-kilodalton surface-layer protein from *Bacillus sphaericus* 2362 and a related cryptic gene. *J. Bacteriol.* 171: 4178–4188.

Broadwell, A. H.; and Baumann, P. 1986. Sporulation-associated activation of *Bacillus sphaericus* larvicide. *Appl. Environ. Microbiol.* 52: 758–764.

Burke, W. F., Jr.; and McDonald, K. O. 1983. Naturally occurring antibiotic resistance in *Bacillus sphaericus* and *Bacillus licheniformis*. *Curr. Microbiol.* 9: 69–72.

Carlton, B. C., and Brown, B. J. 1981. Gene mutation. In *Manual of Methods of General Bacteriology,* ed. P. Gerhardt, R. G. E. Murray, R. N. Costilow, E. W. Nester, W. A. Wood, N. R. Krieg, and G. B. Phillips, 222–242. Washington, D.C.: American Society of Microbiology.

Chang, S., and Cohen, S. N. 1979. High frequency transformation of *Bacillus subtilis* protoplasts by plasmid DNA. *Mol. Gen. Genet.* 168: 111–115.

Clewell, D. B.; Yagi, Y.; Dunny, G. M.; and Schultz, S. K. 1974. Characterization of three plasmid deoxyribonucleic acid molecules in a strain of *Streptococcus faecalis:* Identification of a plasmid determining erythromycin resistance. *J. Bacteriol.* 117: 283–289.

Duvall, E. J.; Williams, D. M.; Lovett, P. S.; Rudolph, C.; Vasantha, N.; and Guyer, M. 1983. Chloramphenicol-inducible gene expression in *Bacillus sphaericus. Gene* 24: 171–177.

Ganesan, S.; Kamdar, H.; Jayaraman, K.; and Szulmajster, J. 1983. Cloning and expression in *Escherichia coli* of a DNA fragment from *Bacillus sphaericus* coding for biocidal activity against mosquito larvae. *Mol. Gen. Genet.* 189: 181–183.

Gordon, R. E.; Haynes, W. C.; and Pang, C. H.-N. 1973. The Genus *Bacillus.* Agric. Handbk. no. 427, U.S. Dept. Agric. Research Serv., Washington, D.C.

Gould, G. W. 1966. Stimulation of L-alanine–induced germination of *Bacillus cereus* spores by D-cycloserine and o-carbamyl-D-serine. *J. Bacteriol.* 92: 1261–1262.

Gryczan, T. J.; Shivakumar, A. G.; and Dubnau, D. 1980. Characterization of chimeric cloning vehicles in *Bacillus subtilis. J. Bacteriol.* 141: 246–253.

Harwood, C. R.; Williams, D. M.; and Lovett, P. S. 1983. Nucleotide sequence of a *Bacillus pumilus* gene specifying chloramphenicol acetyltransferase. *Gene* 24: 163–169.

Hertlein, B. C.; Levy, R.; and Miller, T. W., Jr. 1979. Recycling potential and selective retrieval of *Bacillus sphaericus* from soil in a mosquito habitat. *J. Invertebr. Pathol.* 33: 217–221.

Hindley, J., and Berry, C. 1987. Identification, cloning, and sequence analysis of the *Bacillus sphaericus* 1593 41.9 kd larvicidal toxin gene. *Mol. Microbiol.* 1: 187–194.

———. 1988. *Bacillus sphaericus* strain 2297: Nucleotide sequence of the 41.9 kDa toxin gene. *Nucl. Acids Res.* 16: 4168.

Koncz, C.; Kiss, A.; and Venetianer, P. 1978. A biochemical characterization of the restriction-modification system of *Bacillus sphaericus. Eur. J. Biochem.* 89: 523–529.

Krych, V. K.; Johnson, J. L.; and Yousten, A. A. 1980. Deoxyribonucleic acid homologies among strains of *Bacillus sphaericus. Int. J. Syst. Bacteriol.* 30: 476–484.

LeBlanc, D. J.; Hawley, R. J.; Lee, L. N., and St. Martin, E. J. 1978. "Conjugal" transfer of plasmid DNA among oral streptococci. *Proc. Natl. Acad. Sci.* (U.S.A.) 75: 3484–3487.

LeBlanc, D. J., and Lee, L. N. 1984. Physical and genetic analyses of Streptococcal plasmid pAMβ1 and cloning of its replication region. *J. Bacteriol.* 157: 445–453.

Louis, J.; Jayaraman, K.; and Szulmajster, J. 1984. Biocide gene(s) and biocidal activity in different strains of *Bacillus sphaericus:* Expression of the gene(s) in *E. coli* maxicells. *Mol. Gen. Genet.* 195: 23–28.

McDonald, K. O., and Burke, W. F., Jr. 1984. Plasmid transformation of *Bacillus sphaericus* 1593. *J. Gen. Microbiol.* 130: 203–208.

Massie, J.; Roberts, G.; and White, P. J. 1985. Selective isolation of *Bacillus sphaericus* from the soil by use of acetate as the only major source of carbon. *Appl. Environ. Microbiol.* 49: 1478–1481.

Neide, E. 1904. Botanische Beschreibung einiger sporenbildenden Bacterien. *Zentralbl. Bacteriol. Parasitenk, Infektionskr. Hyg. Abt. 2,* 12: 1–32.

Norton, N. B.; Orzech, K. A.; and Burke, W. F., Jr. 1985. Construction and characterization of plasmid vectors for cloning in the entomocidal organism *Bacillus sphaericus* 1593. *Plasmid* 13: 211–214.

Orzech, K. A., and Burke, W. F., Jr. 1984. Conjugal transfer of pAMβ1 in *Bacillus sphaericus* 1593. *FEMS Microbiol. Lett.* 25: 91–95.

Polak, J., and Novick, R. P. 1982. Closely related plasmids from *Staphylococcus aureus* and soil bacilli. *Plasmid* 7: 152–162.

Shibata, T.; Ikawa, S.; Kim, C.; and Ando, T. 1976. Site-specific deoxyribonucleases in *Bacillus subtilis* and other *Bacillus* strains. *J. Bacteriol.* 128: 473–476.

Tandeau de Marsac, N.; de la Torre, F.; and Szulmajster, J. 1987. Expression of the larvicidal gene of *Bacillus sphaericus* 1593M in the cyanobacterium *Anacystis nidulans* R2. *Mol. Gen. Genet.* 209: 396–398.

Walker, A. C., and Maddux, N. L. 1979. Isolation of mutant strains and new growth media for

Bacillus sphaericus. Abstracts of the Annual Meeting of the American Society for Microbiology, 52.

Weisblum, B. 1975. Altered methylation of ribosomal ribonucleic acid in erythromycin-resistant *Staphylococcus aureus.* In *Microbiology-1974,* ed. D. Schlessinger, 199–206. American Society for Microbiology, Washington, D.C.

White, P. J., and Lotay, H. K. 1980. Minimal nutritional requirements of *Bacillus sphaericus* NCTC9602 and 26 other strains of this species: The majority grow and sporulate with acetate as the major source of carbon. *J. Gen. Microbiol.* 118: 13–19.

Yousten, A. A. 1984. Bacteriophage typing of mosquito pathogenic strains of *Bacillus sphaericus. J. Invertebr. Pathol.* 43: 124–125.

Yousten, A. A.; Barjac, H. de; Hedrick, J.; Cosmao-Dumanoir, V.; and Myers, P. 1980. Comparison between bacteriophage typing and serotyping for the differentiation of *Bacillus sphaericus* strains. *Ann. Microbiol.* (Inst. Pasteur) 131B: 297–308.

Yousten, A. A.; Fretz, S. B.; and Jelley, S. A. 1985. Selective medium for mosquito-pathogenic strains of *Bacillus sphaericus. Appl. Environ. Microbiol.* 49 (6): 1532–1533.

Yousten, A. A., and Hedrick, J. 1982. Bacteriophage typing of mosquito pathogenic strains of *Bacillus sphaericus.* In *Proceedings of the Third International Colloquium on Invertebrate Pathology,* 476–482. Brighton, England.

Yurks, M. A., and Singer, S. 1986. Isolation and description of antimicrobic-resistant (marker) strains of select insecticidal and noninsecticidal varieties of *Bacillus sphaericus. J. Indust. Microbiol.* 1: 245–250.

17

Local Production of *Bacillus sphaericus*

AMARET BHUMIRATANA

17.1 INTRODUCTION

Recent reports (Aronson, Beckman, and Dunn 1986; Khachatourians 1986; Kristiansen 1986; Burgess 1981) have described in detail various measures to reduce the population of insect vectors, particularly mosquitoes, through biological means. These methods are thought to hold considerable advantages over chemical controls, which often have long residual action and may be toxic to nontarget organisms (Burgess 1981). Also, insect populations have developed resistance to a number of these chemicals. Several investigators (Davidson, Singer, and Briggs 1975; Kellen et al. 1965; Panbangred, Bhumiratana, and Pantuwatana 1979; Reeves 1970; Reeves and Garcia 1971; Samasanti, Pantuwatana, and Bhumiratana 1982; Singer 1974) have reported larvicidal effects on various mosquito species by *Bacillus* species as well as by other microbial agents. *Bacillus sphaericus* and *B. t. i.* are two species that have demonstrated very high toxicity toward mosquito larvae (de Barjac 1978a, 1978b; Goldberg and Margalit 1977; Singer 1977). In particular, a number of strains of *B. sphaericus* have been classified on the basis of toxicity to mosquito larvae (de Barjac et al. 1985) and are considered to be the most promising ones for possible use as mosquito control agents. Furthermore, the persistence and the recycling ability of *B. sphaericus* reported in a number of field trial experiments (Cheong and Yap 1985; Mulligan, Schaeffer, and Miura 1978; Silapanuntakul et al. 1983) have made this microorganism a prime candidate as a future biological control agent either alone or as an important ingredient in an integrated mosquito control program. New genetic improvements (Norton, Orzech, and Burke 1985) and new methods for formulating microbial insecticides may create an even better potential for this microorganism.

Recent findings (Yap 1985) on the effectiveness of *B. sphaericus* in controlling the population of *Anopheles* mosquitoes (the major vector for transmitting malaria) will be very significant in further development of *B. sphaericus* as a major microbial insecticide in mosquito control programs. Malaria is still a very common tropical disease in many developing countries, and recent reports have indicated difficulties in treating the parasites. The annual incidence of hundreds of millions of cases is still common in endemic areas. As such, each year considerable effort and financial resources are put into malarial control programs. Future success in controlling malarial disease may depend upon the ability to reduce the population of *Anopheles* vectors.

Many reports have shown different spectra of susceptibilities with regard to the different bacteria and different targeted mosquito larvae; for example, *Aedes aegypti* mosquito larvae seem to be more susceptible to *B.t.i.*, and *Culex* species seem to be more susceptible to *B. sphaericus*. However, it is generally agreed that *Anopheles* mosquito larvae are more susceptible to *B. sphaericus* than they are to other microbial agents. This susceptibility varies according to the mosquito species, the environment, and the feeding habitat of the mosquito larvae. It is perhaps a little early to conclude that *B. sphaericus* can be used to control *Anopheles* species. More large-scale field trials and critical evaluation are needed prior to reaching such a conclusion. Even so, the present available data on toxicity, safety, persistence, and possible recycling capability suggest that suitably formulated *B. sphaericus* has one of the most promising potentials for controlling *Culex* and *Anopheles* larvae.

If *B. sphaericus* is proven to be successful in controlling *Anopheles* mosquito larvae and subsequently in controlling malaria, it will certainly find a place in the commercial microbial insecticide market. The market will be substantial enough to warrant local industrial production.

17.2 THE NEED FOR LOCAL PRODUCTION OF *B. SPHAERICUS*

The term *local production* of microbial insecticides sometimes has different meanings to different people. To some, this term refers to production at the very site of application, that is, to the so-called cottage-level or cottage-industry type of production. It appears that such a process for *B. sphaericus* will not be possible due to the requirement for liquid fermentation and strict sterility control during the fermentation process. This would be impossible to attain in a cottage-type industry. Even though the fermentation process for production of *B. sphaericus* is relatively simple, its proper operation still requires the attention of persons with knowledge in microbiology and/or fermentation technology.

This author prefers to use the term *local production* of microbial insecticides to refer to the development of production (fermentation and formulation) facilities in developing countries, in contrast to the existing production facilities in more industrialized nations. This concept of local production is similar to the one described by Vandekar and Dulmage (1983). It seems that the present state of the art in the use of *B. sphaericus* as a mosquito control agent definitely warrants the setting up of local facilities for its production.

There are a number of advantages in promoting development of local production facilities for *B. sphaericus* in developing countries.

The most important advantage of local production concerns stability. One of the disadvantages of using microbial agents to control agricultural pests has been the instability of the microbial agents and the variation in the toxicity of the formulations. This lack of stability was most likely the result of the lengthy shipping periods and long and variable storage temperatures before the product reached the consumer. Such conditions have led to the unpopularity of *B. thuringiensis*–based microbial insecticides as the choice for controlling lepidopteran pests. Similar disadvantages might be encountered in the mosquito control program if the *B. sphaericus* were to be produced overseas and subsequently imported for use in developing countries. To avoid these stability and variability problems with *B. sphaericus* agents, local production should be encouraged.

The second advantage of local production concerns appropriate formulations. As has been stated by many authors, the requirement for appropriate formulations of microbial insecticides will be crucial for success in using *B. sphaericus* as an agent to control mosquito larvae. The development of these formulations will depend largely on the result of actual field-trial data. Presently, there are very few commercial formulations of *B. sphaericus*. However, examining the many formulations of *B. t. i.* available at present, one finds that all have been developed by companies in industrialized nations with, perhaps, very limited knowledge and few tests appropriate to the actual field conditions in tropical countries where the product is most needed. Local production of *B. sphaericus* would certainly be highly beneficial in providing material for conducting appropriate field studies and for developing formulations suitable for local environmental conditions.

Preliminary field trial studies in developing countries (Cheong and Yap 1985; Silapanontakul et al. 1983) have indicated some potential for the use of *B. sphaericus* as a microbial agent to control mosquitoes. Unfortunately, all of these field trials concentrated on the control of *Culex* larvae. Not much attention has been directed to the control of *Anopheles* larvae. It is also unfortunate that most of the field trials undertaken to date in developing countries have been conducted using imported products. There is an urgent need for local production of *B. sphaericus* for field trial studies, in particular, for field trials on the control of *Anopheles* mosquito larvae.

In summary, local production of *B. sphaericus* would not only allow the development of a general fermentation capability in developing countries, but would also provide more stable and more suitable product formulations. Due to the very different environmental conditions between tropical countries and temperate industrialized countries, there probably will be no single formulation that will be effective for all field conditions.

As will be discussed in a later section, local production of *B. sphaericus* can employ inexpensive, locally available raw materials (Dharmsthiti, Puntuwatana, and Bhumiratana 1985; Obeta and Okafor 1983) as fermentation media, and this can substantially reduce the cost of microbial insecticides. Therefore, local production and formulation of *B. sphaericus* may lead to substantial savings on the cost of microbial insecticides when they eventually reach the stage of industrial production and/or commercialization.

17.3 GROWTH AND CULTIVATING CONDITIONS FOR *B. SPHAERICUS*

The method for cultivation of *B. sphaericus* is very simple, and the organism does not require a complex growth-medium composition. Selective media (Yousten, Fretz, and Jelley 1985; Massie, Roberts, and White 1985), as well as routine media for optimal growth, sporulation, and toxin formation in conventional laboratory conditions, have been worked out (Kalfon et al. 1983). Since *B. sphaericus* is an aerobic bacterium, sufficient aeration must be provided for all media; with *B. sphaericus* 1593, the effect of oxygen on growth, sporulation, and toxin formation for mosquito larvae has been examined (Yousten, Wallis, and Singer 1984; Yousten, Madehekar, and Wallis 1984). These studies indicate that spurting with ordinary air is sufficient for cultivation of *B. sphaericus* and that there is no advantage in increasing aeration or even in continuing the expense of aeration after forespore development has been completed.

The relationship between the presence of inclusion bodies and toxicity of *B. sphaericus* toward mosquito larvae is still somewhat confusing. Although Kalfon et al. (1984) have shown increase in toxicity to correlate with the formation of crystalline inclusions, some strains of *B. sphaericus* may possess toxicity against mosquito larvae in the absence of inclusion bodies. Good growth of *B. sphaericus* can be obtained by using continuous production techniques, and it has been shown to produce at least 15 to 25% prespore levels. The final product using this preparation has shown excellent efficacy in operative field spraying (Hertlein et al. 1981).

The ease in propagating *B. sphaericus* strain 2262 to obtain high level of toxicity toward mosquito larvae will undoubtedly lead to rapid development of the capability for industrial-scale production of *B. sphaericus*. If a large demand for *B. sphaericus* is established, there will be no difficulty for companies that already have expertise in production of other microbial insecticides, such as *B. thuringiensis* HD1 subsp. *kurstaki* and *B.t.i.*, to begin producing *B. sphaericus*. Of course, the market demand for *B. sphaericus* must be sufficiently high to warrant industrial-scale production. As far as the capability for local production in developing countries is concerned, however, the story probably will be much different, since there are as yet no fermentation industries with experience in production of any type of microbial insecticide. Most of the major effort in local production of *B. sphaericus* at present is confined to

TABLE 17.1.
List of Potential Raw Materials for Use in Formulating Fermentation Media for Local Production of *B. sphaericus* and *B.t.i.*

Raw Material	Comment; Reference
Agricultural products	
Bambara bean, cowpea, groundnut cake, soya bean	These materials are extracted in boiling water and added to basal medium containing dried cow blood and mineral salts; Obeta and Okafor 1983.
Ground corn, cottonseed meal, cassava, yams	Vandekar and Dulmage 1983.
Animal products	
Blood, beef bones, chicken parts, fishmeal, animal dung	Vandekar and Dulmage 1983. Hertlein et al. 1981.
Industrial byproducts	
Byproduct from monosodium glutamate factory	Dharmsthiti, Puntuwatana, and Bhumiratana 1985.
Other	
Coconut milk, corn-steep liquor, molasses, whey	Vandekar and Dulmage 1982.

the search for locally available raw materials for fermentation, and a number of reports have indicated the success in this endeavor. The various local raw materials that have been tried are listed in table 17.1. Most of the studies cited were aimed at obtaining microbial growth similar to or higher than that attained by routine laboratory media. The raw materials used included agricultural products, as reported by Obeta and Okafor (1983) and Vandekar and Dulmage (1983). It should be noted that by-products with high sugar content, such as molasses, are not suitable for *B. sphaericus* culture, since this *Bacillus* does not use sugars. Arcas et al. (1984) reported the use of malt sprout extract, and Obeta and Okafor (1983) reported a formulation for *B. sphaericus* using dried bovine blood and extracts of leguminous seeds. Hertelein et al. (1981) have reported the use of animal protein, animal dung, sewage, and agricultural waste for production of *B. sphaericus* and *B. thuringiensis*. In all these reports, high cell yields could be demonstrated. However, the preparation of media required extensive pretreatments such as extraction and/or hydrolysis. In contrast, the H-media formulation introduced by Dharmsthiti, Puntuwatana, and Bhumiratana (1985) possessed several advantages: (1) the media were very easy to prepare and required no pretreatment for any of the constituents; (2) the media supported high growth and toxicity of *B. sphaericus*; and (3) the cost of the major raw material for fermentation was extremely low since it was a by-product from an existing fermentation industry.

The use of media formulated from industrial by-products (Dharmsthiti,

Puntuwatana, and Bhumiratana 1985) may result in nutrient fluctuations depending on the quality of the raw material from factory lots. We suggested that this problem could be easily overcome by determining the amount of L-tyrosine in each batch of factory material (called HDL) and making an appropriate dilution. The testing interval for L-tyrosine need not be frequent since each batch of HDL is produced in large quantity and can be stored for very extended periods. Thus, small variations in various factory lots should not create undue difficulties should fermentation process be scaled up.

Based on the present available data concerning raw materials for production of *B. sphaericus,* it can be concluded that abundant, suitable alternatives to laboratory media can be located readily. Thus, the availability of cheap raw materials should not be a major obstacle in the attempt to stimulate local production of the bacterium. Indeed, the substitution of locally available raw materials may even result in substantial savings in production costs. For example, in one case (Dharmsthiti Puntuwatana, and Bhumiratana 1985), the cost of media was reduced from approximately $1.20 per liter to approximately $0.003 per liter. As will become clearer in a later section, the cost of the raw materials can be reduced so much that it contributes very little to the total production cost of *B. sphaericus.*

Although a number of reports have described the production of *B. sphaericus,* most of them involved laboratory-scale experiment only. There has as yet been no report on a cost analysis for production of the bacterium on a pilot scale. In the example cited in the preceding paragraph, the cost of raw materials for laboratory production could be reduced to a minimum, but there is still a need for data at the pilot scale in order to assess the economic feasibility of full-scale *B. sphaericus* production.

Recent studies in our laboratory at pilot-scale level (100-liter fermentation) indicated that the cost for raw materials was less than 2% of the total cost for producing the 100-l culture of *B. sphaericus.* In this case, the operating cost (shaking, agitation, aeration, cooling, sterilization) reached 36% and the labor cost contributed another 62%. The impetus for promoting a local production capability is not only to make the product available in a relatively short time, but also to reduce the transportation costs, to increase the shelf life, and to reduce the cost so that the locally produced material will be able to compete in both quality and price with an imported product. For *B. sphaericus,* this may be a premature concern since there is as yet no commercially available product for comparison.

Because the fermentation technology involved in production of *B. sphaericus* is relatively simple, there appears to be sufficient existing technical capability in developing countries to undertake the task of producing it locally. It also appears, if *B. thuringiensis* is a measure, that the production cost for *B. sphaericus* may be competitive with imported products. This assumes, of course, that the quality of the product can be strictly controlled and that the market demand for *B. sphaericus* will be sufficiently large enough to warrant industrial-scale production.

17.4 FORMULATION OF *B. SPHAERICUS*

In the development of local production of *B. sphaericus*, techniques for formulation are probably the most difficult problem to solve. Of course, for *B.t.i.*, a number of formulations are presently available. These include flowable concentrates, wettable powders, granules, briquettes, and the recently developed encapsulated forms (Margalit, Markus, and Pelah 1984). Unfortunately, most of the know-how for these formulations is in the hands of private industries. There are few published reports on formulations (Dulmage, Correa, and Martinez 1970; Vandekar and Dulmage 1983), and none of the relevant studies have been undertaken in developing countries. At the same time, in spite of the different types of formulations available, there is no conclusive proof that any one is better than any other. Limited field trials with laboratory-prepared formulations by Silapanantakul et al. (1983) and Yap (1985) seemed to indicate that liquid formulations would work well. The commercial success of *B. sphaericus* as a microbial larvicide will rely heavily on its use in malarial control programs, and so the most appropriate formulation developed must be suitable for controlling *Anopheles* mosquitoes.

At the present state of development, the most suitable formulation may be concentrated cultures of *B. sphaericus*. Such concentrates have a number of advantages over other types of formulations. First, they are the simplest one to prepare, requiring only centrifugation of the fermentation broth and the addition of certain additives and preservatives (Vandekar and Dulmage 1983). There is no problem with developing specialized technology. Also, the cost of such formulations would be considerably lower than that of other types, which require drying of the product and/or attachment to specific carriers. Another advantage of concentrated preparations is that they are dispersible, and, therefore, the toxin would rapidly spread during application and give quick exposure to mosquito larvae. Of course, rapid dispersibility of the concentrated preparation could be considered disadvantageous in situations where it led to excessive dilution of the toxin. However, recent data have indicated that residual action due to either persistence or recycling may help to off-set this dilution disadvantage.

Another disadvantage of liquid formulations compared to dried powder formulations is their bulkiness. This may lead to difficulties in transportation and handling with distance, but local production facilities are relatively close to the site of field application. A further disadvantage of liquid concentrate preparations is the relatively short shelf life of the product when compared to formulations such as wettable powder or briquettes. The hot climate that prevails for most of the year in tropical countries will aggravate the problem and reduce the shelf life even further. Maintaining the product in cold storage would be almost impossible given the present practice and economic situation in developing nations except at the production plant itself. This limitation in the shelf life of the liquid-based product will be the most difficult problem to tackle. Some benefit can be obtained if the concentrates contain preservatives. An appropriate preservative must satisfy the following criteria: it must be safe and cheap and must not interfere with the toxicity of the formulation. Preliminary results from our laboratory indicate that a mixture of *B. sphaericus* or *B. t. i.* and MSF (monomolecular surface film) agent may be advantageous in this regard. Perhaps a similar formulation with *B. sphaericus* would prove useful and highly efficacious. Other types of preservatives should be sought.

Limitations resulting from the use of liquid formulation have led to development of other types of formulations. A variety of formulations of primary powder or granular formulations are used in operational control programs to yield effective control of *Aedes* and *Culex* larvae (Mulla 1985). However, these formulations may be less effective against *Anopheles* larvae due mainly to the differential feeding behavior among mosquito species. Larvae of *Culex* and *Aedes* species are water filterers, whereas *Anopheles* larvae are adapted to collect particles from the air-water interface (Aly et al. 1987). Other differences in feeding behavior such as the size of the particles, the filtration rates, and the rate of ingestion can contribute greatly toward the effectiveness of each formulation. In a recent study by Aly et al. (1987) the combination of two effects, retention of toxin at the primary feeding zone of larvae and addition of a gustatory carrier particle, has resulted in a substantial increase in the effectiveness of microbial insecticides against larvae of three *Anopheles* species. The use of phagostimulation by components associated with food has also been reported for *Culex pipiens* (Dadd 1970) and *Aedes vexans*. There have also been reports indicating that mosquito larvae ingest food particulates up to eight times faster than inert materials. However, larvicidal activity of *B. sphaericus* against larvae of *Culex* species had been found to decrease substantially in organically enriched and unshaded habitats (Lacey et al. 1988). The fate of microbial insecticides in a natural habitat was well studied in laboratory-simulated field waters (Ohana, Margalit, and Barak 1987).

In 1985 Novak, Gubler, and Underwood reported the use of a number of slow-release formulation of temephos and *B. t. i.* on corncobs carriers and coconut husks; although good control was reported by using these carriers with temephos, only limited effectiveness was found with formulations containing *B. t. i.* Nonetheless, the idea of using residues such as corncobs, coconut husks, and similar materials should be pursued further. Preliminary results from the author's group with the use of *B. sphaericus* mixed with coconut husks indicated that the husk-based formulations provide good control of *Culex quinquefaciatas* in laboratory conditions.

17.5 ECONOMICS OF LOCAL PRODUCTION OF *B. SPHAERICUS*

The economics of industrial production of *B. thuringiensis* were briefly discussed in a recent review by Rowe and Margaritis (1987). It is very difficult at this stage of development and preliminary field trial study to predict the cost of *B. sphaericus* as a microbial insecticide. But it will be shown that if the price of *B. sphaericus* is based on that of equivalent formulations of *B.t.i.,* the use of *B. sphaericus* will appear to be cost-effective in controlling *Cx. quinquefasciatus.* However, in regard to local production, one must compare *B. sphaericus* preparation produced locally in a developing country at or near the site of field application with the preparation prepared from industrial production in industrialized countries. Comparisons will be difficult since presently no extensive data are available on the cost of *B. sphaericus* produced at pilot-scale or industrial-scale levels. However, preliminary economic evaluation appears to favor local production of *B. sphaericus* based on data from pilot-scale experiments, the relatively low cost of locally available raw materials, and the level of productivity in terms of toxicity. And, since the formulation most likely to be suitable for local production is the liquid flowable formulation, there will be no technical difficulties involved in the downstream processing required for formulating the product. As such, the cost of the downstream process will probably not contribute greatly to the final cost of the product. The technology for production and recovery of *B. sphaericus* is relatively simple; the equipment required is not sophisticated, and pilot and industrial-scale equipment can be easily fabricated locally. Thus, the only major problem that may hinder local production is the size of fermentors and/or production capacity.

Economies of scale do affect the unit cost of a fermentation product, particularly for production of *B. sphaericus.* From our pilot-scale fermentation data on 100-liter fermentation capacity, one may extrapolate to 45-kiloliter (kl) fermentation level. It can be roughly calculated that if the capacity of a fermentor is increased from 100 l to 45 kl. The production cost, which includes material cost, operating cost, and labor cost, for production of *B. sphaericus* could be lowered as much as 20-fold. This is mainly due to reduction in the cost of operating (agitation, aeration, and cooling) and labor cost. Of course, the investment cost, which is not taken into consideration in this case, will be much higher for the 45-kl fermentor. As such, for local production the main question to be addressed is, What is the minimum production capacity possible for the process to be economically feasible?

The size of the market for the product will also be a major deciding factor. In most situations where the volume of product being demanded by the market is relatively small, the size of the production capacity will also be small, and will subsequently lead to an increase in unit price of the final product. New innovations in fermentor design will have to be found in order to achieve the goal of building economically feasible local production. A new fermentor design would have to meet the requirement of especially high productivity in order to off-set the reduction in price resulting from increased production capacity. More research on creating special models of small, high-productivity fermentors will ultimately help to achieve the aim in local production. Recently, the author's group obtained very interesting results in experimenting with a cell recycle system using attachment of membrane unit to a 10-liter fermentor. Preliminary results indicated that productivity of *B. sphaericus* and *B.t.i.* could be increased more than 10-fold over the conventional batch fermentation process. If these data hold true for larger-scale fermentation, the production cost for local production of microbial insecticides will be substantially reduced.

17.6 CONCLUSIONS

B. sphaericus has a number of potential advantages for future use as a microbial insecticide. However, the future impact of this microbial insecticide will depend largely on its ability to be more persistent, its recycling ability in field trials, and its possibility for use as a microbial control agent in malarial control programs. The future for development of local production of *B. sphaericus* will depend largely on the capability of producing this bacterial agent at a price cheap enough to be competitive with products that will be available on the world market in the near future. *B. sphaericus* preparations from local production facilities will have certain advantages: cheaper raw materials, shorter shipment period, shorter shelf-life requirement, and better quality of low-cost flowable formulation. Nonetheless, it appears that to achieve the aim of local production capability, the final price of the microbial agent has to be competitive with the commercially available products. One way to achieve this aim is to develop a system that will lead to production of high bacterial mass with a small-size fermentor so that the commercial advantage of economies of scale can be minimized.

Acknowledgments

This research was supported in part by the UNDP/World BANK/WHO Special Programme for Research and Training in Tropical Diseases. We thank Dr. T. W. Flegel for critical reading and Ms. Prapaiporn Praisupa for typing this manuscript.

References

Aly, C.; Mulla, M. S.; Schnetter, W.; and Xu, B. 1987. Floating bait formulations increase effectiveness of *Bacillus thuringiensis* var. *israelensis* against *Anopheles* larvae. *J. Amer. Mosq. Control Assoc.* 3: 583–588.

Arcas, J.; Yantorno, O.; Arraras, E.; and Ertola, R. 1984. A new medium for growth and delta-endotoxin production by *Bacillus thuringiensis* var. *kurstaki. Biotech. Lett.* 6: 495–500.

Aronson, A. I.; Beckman, W.; and Dunn, P. 1986. *Bacillus thuringiensis* and related insect pathogens. *Microbiol. Rev.* 50: 1–24.

Barjac, H. de. 1978a. Toxicité de *Bacillus thuringiensis* var. *israelensis* pour les larves d'*Aedes aegypti* et d'*Anopheles stephensi, C. R. Acad. Sci.* (Paris) 286b: 1175–1178.

———. 1978b. Une nouvelle variété de *Bacillus thuringiensis* très toxique pour les moustiques: *Bacillus thuringiensis* var. *israelensis* sérotype 14. *C. R. Acad. Sci.* (Paris) 286b: 797–800.

Barjac, H.de; Larget-Thiery, I.; Cosmao Dumanoir, V. C.; and Ripouteau, H. 1985. Serological classification of *Bacillus sphaericus* strains on the basis of toxicity to mosquito larvae. *Appl. Microbiol. Biotechnol.* 21: 85–90.

Burgess, H. D., ed. 1981. *Microbial control of pests and plant diseases, 1970–1980.* London: Academic Press.

Cheong, W. C., and Yap, H. H. 1985. Bioassays of *Bacillus sphaericus* (strain 1593) against mosquitoes of public health importance in Malaysia. *Southeast Asian J. Trop. Med. Pub. Hlth.* 16: 54–58.

Dadd, R. H. 1970. Comparison of rates of ingestion of particulated solids by *Culex pipiens* larvae: Phagostimulant effect of water-soluble yeast extract. *Entomol. Exp. Appl.* 13: 407–419.

Davidson, E. W.; Singer, S.; and Briggs, J. D. 1975. Pathogenesis of *Bacillus sphaericus* strain SSII-1 infections in *Culex pipiens quinquefasciatus larvae. J. Invertebr. Pathol.* 25: 179–184.

Dharmsthiti, S. C.; Puntuwatana, S.; and Bhumiratana, A. 1985. Production of *Bacillus thuringiensis* subsp. *israelensis* and *Bacillus sphaericus* 1593 on media using a byproduct from a monosodium glutamate factory. *J. Invertr. Pathol.* 46: 231–238.

Dulmage, H. T.; Correa, J. A.; and Martinez, A. J. 1970. Coprecipitation with lactose as a means of recovering the spore-crystal complex of *Bacillus thuringiensis. J. Invertebr. Pathol.* 15 (1): 15–20.

Goldberg, L. J., and Margalit, J. 1977. A bacterial spore demonstrating rapid larvicidal activity against *Anopheles sergentii, Uranotaenia unguiculata, Culex univitattus, Aedes aegypti,* and *Culex pipiens. Mosq. News* 37: 355–358.

Hertlein, B. C.; Hornby, J.; Levy, R.; and Miller, T. W., Jr. 1981. Prospects of spore-forming bacteria for vector control with special emphasis on their local production potential. *Dev. Industr. Microbiol.* 22: 53–60.

Kalfon, A.; Charles, J. F.; Bourgouin, C.; and Barjac, H. de. 1984. Sporulation of *Bacillus sphaericus* 2297: An electron microscope study of crystal-like inclusion biogenesis and toxicity to mosquito larvae. *J. Gen. Microbiol.* 130: 893–900.

Kalfon, A.; Larget-Thiery, I.; Charles, J.-F.; and Barjac, H. de. 1983. Growth, sporulation, and larvicidal activity of *Bacillus sphaericus. Eur. J. Appl. Microbiol. Biotechnol.* 18: 168–173.

Kellen, W. R.; Clark, T. B.; Lindegren, J. E.; Ho, B. C.; Rogoff, M. H.; and Singer, S. 1965. *Bacillus sphaericus* Neide as a pathogen of mosquitoes. *J. Invertebr. Pathol.* 7: 442–448.

Khachatourians, G. G. 1986. Production and use of biological pest control agents. *Trends in Biotechnol.* 415: 120–124.

Kristiansen, B., and Lewis, C. 1986. Bacterial insecticides: Recent developments. *Trends in Biotechnol.* 4 (3): 56–58.

Lacey, L. A.; Lacey, C. M.; Peacock, B.; and Thiery, I. 1988. Mosquito host range and field activity of *Bacillus sphaericus* isolate 2297 (serotype 25). *J. Amer. Mosq. Control Assoc.* 3: 51–56.

Margalit, J.; Markus, A.; and Pelah, Z. 1984. Effect of encapsulation on the persistence of *Bacillus thuringiensis* var. *israelensis* (H-14). *Appl. Microbiol. Biotechnol.* 19: 382–383.

Massie, J.; Roberts, G.; and White, P. J. 1985. Selective isolation of *Bacillus sphaericus* from the soil by use of acetate as the only major source of carbon. *Appl. Environ. Microbiol.* 49 (6): 1478–1481.

Mulla, M. S. 1985. Field evaluation and efficacy of bacterial agents and their formulations against mosquito larvae. In Integrated mosquito control methodologies, ed. M. Laird and J. W. Miles, 2: 227–250. New York: Academic Press.

Mulligan, F. S., III; Shaeffer, C. H.; and Miura, T. 1978. Laboratory and field evaluation of *Bacillus sphaericus* as a mosquito control agent. *J. Econ. Entomol.* 71: 774–777.

Norton, N. B.; Orzech, K. A.; and Burke, W. F., Jr. 1985. Construction and characterization of plasmid vectors for cloning in the entomocidal organism *Bacillus sphaericus* 1593. *Plasmid* 13: 211–214.

Novak, R. J.; Gubler, D. J.; and Underwood, D. 1985. Evaluation of slow-release formulations of temephos (Abate®) and *Bacillus thuringiensis* var *israelensis* for the control of *Aedes aegypti* in Puerto Rico. *J. Amer. Mosq. Control Assoc.* 1: 449–453.

Obeta, J. A. N., and Okafor, N. 1983. Production of *Bacillus sphaericus* strain 1593 primary powder on media made from locally obtainable Nigerian agricultural products. *Can. J. Microbiol.* 29: 704–709.

Ohana, B.; Margalit, J.; and Barak, Z. 1987. Fate of *Bacillus thuringiensis* subsp. *israelensis* under simulated field conditions. *Appl. Environ. Microbiol.* 53: 828–831.

Panbangred, W.; Bhumiratana, A.; and Pantuwatana, S. 1979. Toxicity of *Bacillus thuringiensis* toward *Aedes aegypti* larvae. *J. Invertebr. Pathol.* 33: 340–347.

Reeves, E. L. 1970. Pathogens of mosquitoes. *Proc. Calif. Mosq. Control Assoc.* 38: 20–22.

Reeves, E. J., and Garcia, C. 1971. Pathogenicity of biocrystalliferous *Bacillus* isolate for *Aedes aegypti* and other aedine mosquito larvae. In *Proceedings Fourth International Coloquium Insect Pathol.,* College Park, Md. August 1970, 219–228.

Rowe, G. E., and Margaritis, A. 1987. Bioprocess developments in the production of bioinsecticides by *Bacillus thuringiensis. CRC Crit. Rev. Biotechnol.* 6: 87–127.

Samasanti, W.; Pantuwatana, S.; and Bhumiratana, A. 1982. Role of the parasporal body in causing toxicity of *Bacillus thuringiensis* toward *Aedes aegypti* larvae. *J. Invertebr. Pathol.* 39: 41–48.

Silapanuntakul, S.; Pantuwatana, S.; Bhumiratana, A.; and Charoensiri, K. 1983. The comparative persistence of toxicity of *Bacillus sphaericus* strain 1593 and *Bacillus thuringiensis* serotype H-14 against mosquito larvae in different kinds of environments. *J. Invertebr. Pathol.* 42: 387–392.

Singer, S. 1974. Entomogeneous bacilli against mosquito larvae. *Dev. Industr. Microbiol.* 15: 187–194.

———. 1977. Bacterial pathogens of Culicidae (Mosquitos). *Bull. WHO* (Suppl.) 55: 47–62.

Vandekar, M., and Dulmage, H. T. 1983, Guidelines for the production of *Bacillus thuringiensis* H-14: Proceedings of a consultation held in Geneva, Switzerland, 25–28 October 1982. Geneva: UNDP/WORLD BANK/WHO Special Programme for Research and Training in Tropical Disease.

Yap, H. H. 1985. Biological control of mosquitoes, especially malaria vectors, *Anopheles* species. *Southeast Asian J. Trop. Med. Pub. Hlth.* 16 (1): 163–172.

Yousten, A. A.; Fretz, S. B.; and Jelley, S. A. 1985. Selective medium for mosquito-pathogenic strains of *Bacillus sphaericus. Appl. Environ. Microbiol.* 49 (6): 1532–1533.

Yousten, A. A.; Madehekar, N.; and Wallis, D. 1984. Fermentation conditions affecting growth, sporulation, and mosquito larval toxin formation by *Bacillus sphaericus. Dev. Industr. Microbiol.* 25: 757–762.

Yousten, A. A.; Wallis, D. A.; and Singer, S. 1984. Effect of oxygen on growth, sporulation, and mosquito larval toxin formation by *Bacillus sphaericus* 1593. *Curr. Microbiol.* 11: 175–178.

18

Persistence and Formulation of *Bacillus sphaericus*

LAWRENCE A. LACEY

18.1 INTRODUCTION

The development of *Bacillus sphaericus* as a microbial control agent of mosquitoes has been somewhat hampered by the success and broader spectrum of *Bacillus thuringiensis israelensis* (*B.t.i.*). Despite its narrower host range, *B. sphaericus* offers some distinct advantages over its commercially produced counterpart. Most notable is the increased duration of larvicidal activity against certain species of mosquitoes, especially in organically enriched habitats. Several studies attest to prolonged control of mosquito larvae under natural and simulated natural conditions (Hornby, Hertlein, and Miller 1984; Des Rochers and Garcia 1984; Lacey, Urbina, and Heitzman 1984; Mulla et al. 1984a; Silapanuntakul et al. 1983; Vankova 1984; Mulligan, Schaefer, and Wilder 1980). Under certain conditions, the bacterium may actually recycle within mosquito cadavers (Charles and Nicolas 1986; Karch and Coz 1986; Nicolas, Dossou-Yovo, and Hougard 1987).

18.2 FACTORS AFFECTING RESIDUAL ACTIVITY

18.2.1 Definitions

A number of factors can influence the level and duration of larvicidal activity in mosquito habitats: water quality and depth, solar radiation, target species, and larval density are some of the best studied. Residual larvicidal activity may simply be due to persistence of accessible toxin in the environment, to recycling of the bacterium in larval cadavers with subsequent amplification of spores and toxins, or to both phenomena.

18.2.2 Water Quality, Depth, and Feeding Behavior

The toxin of *B. sphaericus* can apparently persist in a variety of habitats, including those that are polluted. In shallow containers, in clear water, larvicidal activity has been reported as persisting for nine months (Silapanuntakul et al. 1983). In some extremely enriched and/or deep breeding sites, however, curtailment of larvicidal activity has been reported (Mulligan, Schaefer, and Miura 1978; Mulligan, Schaefer, and Wilder 1980; Silpanuntakul et al. 1983; Des Rochers and Garcia 1984; Hoti and Balaraman 1984; Mulla et al. 1984a; 1988; Karch and Charles 1987). Ostensibly, toxin either settled out of the feeding zone of the larvae or ingestion of sufficient quantities was inhibited by excess particulate matter. Studies by Mulligan, Schaefer, and Wilder (1980), Aly (1983), and Ramoska and Pacey (1979) support this hypothesis. Even though larvicidal activity was relatively short-lived in the surface water of catch basins that had been treated with extremely high concentrations of *B. sphaericus*, Mulligan, Schaefer, and Wilder (1980) found that larvae were still killed by exposure to water taken from the bottom of the same basins for up to one month after treatment. Karch and Charles (1987), however, found that spores that had settled to the bottom of cesspools rapidly lost their toxicity against *Culex pipiens* larvae. Ramoska and Pacey (1979) demonstrated an inverse relationship between the efficacy of *B. sphaericus* and the amount of food available to larvae of *Culex quinquefasciatus* Say and *Anopheles albimanus* Wiedemann. The effect of increased food availability and decreased diving rate in *Aedes vexans* (Meigen) was presented by Aly (1983). He indicated that settled toxin of *B.t.i.* would be less available to *Ae. vexans* in habitats with abundant food. Microbial degradation of toxin is another factor that may limit residual larvicidal activity in organically enriched habitats in which there is no recycling of *B. sphaericus*. Nicolas, Dossou-Yovo, and Hougard (1987), however, attributed persistence of activity in polluted habitats to slow sedimentation of spores.

18.2.3 Solar Radiation

Sunlight can apparently also decrease residual larvicidal activity. Although ultraviolet (UV) radiation inactivates spores of *B. sphaericus* without denaturing the larvicidal toxin (Burke, McDonald, and Davidson 1983), exposure to solar radiation in shallow water can denature the toxin (Mulligan, Schaefer, and Miura 1978; Des Rochers and Garcia 1984). Bacterial spores and toxin that are contained within larval cadavers are apparently protected from solar degradation (Des Rochers and Garcia 1984).

18.2.4 Species Susceptibility

The factors that play the greatest roles in terms of determining level of larvicidal activity are the isolate of *B. sphaericus* and species of mosquito. Under field and laboratory conditions, members of the genera *Culex* and *Psorophora* are extremely susceptible to *B. sphaericus* toxin, whereas *Aedes aegypti* is nearly totally refractive and *Anopheles* species are moderately susceptible (Ramoska, Burgess, and Singer 1978; Lacey and Singer 1982; Silapanuntakul et al. 1983; Mulla et al. 1984b, 1985; Lacey et al. 1986, 1988). Differences in feeding behavior of the various genera alone do not account for the variations in susceptibility. Although Ramoska and Hopkins (1981) observed faster initial ingestion rates of ^{14}C-labeled *B. sphaericus* by *Cx. quinquefasciatus* larvae than by *An. albimanus* larvae, the anophelines eventually accumulated the same number of bacterial cells but responded with lower mortality. *Ae. aegypti* larvae fed at the same rate as the *Cx. quinquefasciatus* larvae, but were not affected by the bacterium. Nicolas and Dossou-Yovo (1987) observed differential activity of *B. sphaericus* against *Cx. quinquefasciatus* and a cohabiting competitor, *Culex cinereus*. The decreased larvicidal activity on *Cx. cinereus* could enable it to outcompete reinvading *Cx. quinquefasciatus*, resulting in increased duration of control of the latter. *Cx. cinereus* is not a pest of humans.

Little information is available on the effect of mosquito species on length of residual activity. Larval density appears to be an important factor in determining the occurrence of recycling of *B. sphaericus* and hence in duration of activity. Larval numbers, in turn, may be a function of species and/or environmental conditions. Although susceptibility may not be necessarily correlated with larval feeding behavior, residual activity of *B. sphaericus* in certain habitats might be. Susceptible bottom-feeding species would appear to have the longest possible contact time with settled spores and toxin. In contrast, exposure time of surface-feeding species to nonfloating formulations would only be evanescent.

Yousten (1984) and Davidson and Yousten (chapter 15) have elucidated the influence of isolate and serotype on degree of larvicidal activity. Most research on residual activity and recycling of *B. sphaericus* has been conducted with the serotype H5a,5b isolates (e.g., 1593 and 2362), but there is a paucity of information on the effect of serotype and isolate on persistence.

18.2.5 Recycling

In laboratory studies with the SSII-1 isolate, Davidson, Singer, and Briggs (1975) demonstrated germination, growth, and sporulation of *B. sphaericus* in cadavers of *Cx. quinquefasciatus* with a subsequent log increase in spores over that ingested by the larvae. The retrieval of viable and infective spores of the 1593 isolate several months after introduction into larval habitats led Hertlein, Levy, and Miller (1979) to conclude that *B. sphaericus* is capable of recycling in nature.

Laboratory and field studies by Silapanuntakul et al. (1983); Des Rochers and Garcia (1984); Hornby, Hertlein, and Miller (1984); Charles and Nicolas (1986); Karch and Coz (1986); and Nicolas, Dossou-Yovo, and Hougard

(1987) indicate that, under certain conditions, larva-to-larva recycling occurs and may perpetuate effective control for several months. Amplification of *B. sphaericus* in larvae may provide significant recycling under field conditions when large numbers of susceptible larvae are present at the time of treatment (Davidson et al. 1984; Charles and Nicolas 1986; Karch and Coz 1986). Although toxin may be less effective or short-lived in habitats with high organic content, the larval density that these situations engender may facilitate larva-to-larva recycling as in the study of Hornby, Hertlein, and Miller (1984). Results obtained by Davidson et al. (1984) and by Karch and Coz (1984) indicate that *B. sphaericus*–killed larvae are highly insecticidal to healthy larvae. In many situations, however, recycling of *B. sphaericus* is not observed despite repeated application (Hoti and Balaraman 1984). Additional biological and physical factors that affect efficacy and that may also influence persistence of toxin are presented by Davidson (1985) and Lacey and Undeen (1986) and in section 18.3.4 of this chapter.

18.3 FORMULATION: FIELD EFFICACY

Until lately, most field trials of *B. sphaericus* have been conducted with primary powders and whole cultures. Although these preparations have proved to be fairly efficacious, their activity could be further enhanced through appropriate formulation. Proper formulation can improve control performance by enabling greater contact with target larvae, ensuring stability under storage and field conditions, providing a variety of application options, and increasing the ease of handling (Lacey 1984). Improper formulation may result in lowered efficacy of an otherwise effective entomopathogen.

The need to control mosquito larvae in a variety of habitats has resulted in the development of a number of formulations of the commercially produced *B.t.i.* Most of these have also been experimentally produced using *B. sphaericus* (table 18.1). Recent interest in *B. sphaericus* by industry and the World Health Organization has facilitated production of sufficient quantities of experimental formulations to enable large-scale field trials (Lacey et al. 1986, 1988).

18.3.1 Granules

Dense canopy in habitats such as salt marsh and maturing rice fields may impede conventional spray application of wettable powder and flowable-concentrate formulations from reaching target larvae. Granular formulations in which bacteria are coated upon or incorporated within a carrier allow penetration of canopy and provide even distribution within the larval

TABLE 18.1.
Larvicidal Activity of Experimental Formulations of *Bacillus sphaericus* (serotype 5a,5b) under Natural and Simulated Natural Conditions

Formulation	Target Species	Habitat	Dosage	Degree of Control	Reference
Wettable powder[a]	*Anopheles annulipes, Culex annulirostris, Cx. quinquefasciatus*	sod-lined ponds	0.5–2 kg/ha	20–100% 2 days	Davidson, Sweeney, and Cooper 1981
[b]	*Cx. tarsalis*	sod-lined ponds	0.8 kg/ha	98–100% 1 day	Mulligan, Schaefer, and Miura 1978
[c]	*Cx. quinquefasciatus*	catch basins	500 mg/l	95–100% 1–4 weeks	Mulligan, Schaefer, and Wilder 1980
Sustained release pellet[d]	*Cx. quinquefasciatus*	container (15 1)	1–2 pel/cont	97–100% 8+ weeks	Lacey, Urbina, and Heitzman 1984
	Cx. restuans	tires	1 pel/tire	75–100% 7+ weeks	Lacey and Reiter, unpub. data
	Cx. restuans	woodland pools	4 pel/pool	92–99% 8+ days	Lacey et al. 1988
Briquette[e]	*Cx. quinquefasciatus*	sod-lined ponds	1/2 Br/pond	79–95% 2+ weeks	Lacey et al. 1988
Granule[f]	*Cx. nigripalpus, Cx. quinquefasciatus*	sod-lined ponds	2.5–5 kg/ha	63–100% 1.5 weeks	Lacey et al. 1988
	An. quadrimaculatus, An. crucians	mature rice fields	2.5–5 kg/ha	17–68% 1 week	Lacey et al. 1988

	Psorophora columbiae	reflooded rice fields	1–5 kg/ha	65–100% 2 days	Lacey et al. 1988
[g]	*Cx. peus*	wastewater lagoons	2.8–22.4 kg/ha	80+% 14–21 days	Mulla et al. 1988
	Cx. pipiens	lagoons and pools	2.8–8.4 kg/ha	100% 4 days	Berry et al. 1987
	Cx. pipiens, Aedes trivittatus	retention pond	2.8–5.6 kg	84–98% 3 days	Berry et al. 1987
Flowable concentrate[b]	*An. quadrimaculatus*	mature rice field	0.6–1.2 l/ha	71–82% 2 days	Lacey et al. 1986
	Ps. columbiae	reflooded rice field	0.3–0.6 l/ha	50–98% 2 days	Lacey et al. 1986
	Cx. peus	wastewater lagoon	2.2–5.6 kg/ha	100% 14–21 days	Mulla et al. 1988

[a]Abbott Laboratories, lot 11-115-BD, isolate 1593.
[b]Stauffer Chemical Co., lot no. 4920-34-3 (1593).
[c]Stauffer Chemical Co., lot no. MV-716 (1593).
[d]Made with 30% primary powder of the 1593 isolate (Lacey, Urbina, and Heitzman 1984) or the 2362 isolate (Lacey et al. 1988) produced by H. T. Dulmage, USDA, Brownsville, Tex.; pellets formulated by Lacey et al. 1988, USDA, Gainesville, Fla.
[e]Made with 5% primary powder (2362) produced by H. T. Dulmage; briquettes formulated by Summit Chemical Co. for Biochem Products (Solvay Labs).
[f]Made with 5% primary powder (2362) produced by H. T. Dulmage; formulated by D. Ross for Biochem Products.
[g]Abbott Laboratories experimental granules (ABG-6185 and 6185-A and B) 2362 isolate.
[h]Made with 12.3% primary powder (2362); produced and formulated by Biochem Products.

habitat. Modifying the carrier to increase flotation will extend contact time with surface-feeding species such as the anophelines, whereas sinking pellets may be more effective against bottom-feeding species. An experimental granular formulation, identical to the Bactimos® formulation of *B.t.i.* in terms of physical characteristics and inert constituents, facilitated penetration of vegetation and provided excellent control of *Culex* and *Psorophora* larvae. Considerably reduced activity against *Anopheles* species, however, was observed (Lacey et al. 1988).

18.3.2 Flowable Concentrates

Flowable-concentrate formulations have provided the greatest number of options in terms of application and ease of handling for *B.t.i.* (Lacey and Undeen 1986). A flowable concentrate of the 2362 isolate of *B. sphaericus* produced by Biochem Products provided effective control of *Psorophora columbiae* (Dyar and Knab) at fairly low dosage rates and moderate control of *Anopheles quadrimaculatus* Say breeding in rice fields using a Beecomist® ULV spray head (Lacey et al. 1986).

18.3.3 Sustained-release Formulations

Incorporating primary powder of *B. sphaericus* within a matrix that enables both sustained release of inoculum and flotation may extend residual control in habitats where conditions are not favorable for larva-to-larva recycling and/or where the toxin settles from the larval feeding zone. The briquette and sustained-release pellets listed in table 18.1 provided extended control under these conditions and in full sunlight. In each study it was not possible to assess the full potential of the formulation due to either a crash in control populations or drying of the habitat. Laboratory studies on the use of matrices to enhance the residual activity of *B. sphaericus* are reported by Kuppusamy, Hoti, and Balaraman (1989).

18.3.4 Storage Considerations

In addition to addressing formulation requirements for use of *B. sphaericus* in specific habitats, the effects of formulation ingredients (surfactants, diluents, etc.) as well as temperature and moisture on stability of the toxin must also be considered (Couch and Ignoffo 1981; Sawicka and Couch 1983). *Bacillus sphaericus* toxin remains extremely stable under optimal storage conditions of neutral pH and 4° C (Lacey 1985) and fairly stable at room temperature (Mian and Mulla 1983; Hertlein, Hornby, and Miller

1980). Wraight and Molloy (pers. comm.), however, observed a steady decline in larvicidal activity of a primary powder stored out of a desiccator and subjected to temperature fluctuations of 13–30° C. Toxin that is exposed to high pH (10.8) will become denatured immediately (Lacey 1985).

Prolonged storage under even slightly basic conditions at room temperature results in an accelerated decline in larvicidal activity, ostensibly due to denaturing of the toxin. A rapid decline in spore viability, on the other hand, is observed during storage under acidic conditions (Lacey 1985). Although viable spores are not required for larvicidal activity (Burke, McDonald, and Davidson 1983; Lacey and Smittle 1985), they are required for persistence of activity due to recycling. Formulation additives that prevent secondary fermentation during storage should not affect spore viability, longevity, or sensitivity to various environmental stimuli.

Finally, in addition, to the impact of environmental parameters on *B. sphaericus*, the impact of formulation constituents on the environment must be given strong consideration. These components should be as innocuous to nontarget organisms as is the active moiety.

18.4 RECOMMENDATIONS FOR FUTURE FORMULATION RESEARCH

The prototypes listed in table 18.1 are the results of initial attempts at formulation of *B. sphaericus*. Further enhancement of efficacy will be commensurate with the amount of research invested.

A variety of improvements in handling, efficacy, and persistence of activity are possible through formulation modification and increase in concentration of active moiety. The production of sustained-release floating granules that provide both high initial release of toxin and long-term lower release rates will enable residual control in a number of habitats, including those with dense canopy, high populations of older larvae, and continuous oviposition. Further increase in efficacy might also be obtained by including feeding attractants/stimulants within the matrix of the granules. The effect of such substances and baits on larval feeding rates has been presented by Dadd (1970); Dadd, Kleinjan, and Merrill (1982); Dadd and Kleinjan (1985); and Aly (1983, 1985). The effectiveness of floating formulations of *B. t. i.* has been improved with the addition of bait (Aly et al. 1987). Greater field stability may result from providing protection to spores and toxin from detrimental environmental factors such as sunlight. Couch and Ignoffo (1981) present information on the formulation of other entomopathogens with UV protectants. Recent developments in the formulation of *B. sphaericus* and recommendations for future avenues of research are summarized by WHO (1989).

In certain habitats it may be desirable or even required to use microbial control agents that do not contain viable spores (e.g., potable water supplies).

In these situations it is possible to use gamma-irradiated formulations in which viable spores have been almost totally reduced yet still retain full larvicidal capability (Lacey and Smittle 1985). It may also be possible to employ asporogenous mutants in this capacity if a suitable strain is developed.

In addition to traditional methods of increasing toxin production (optimal fermentation conditions, search for and selection of more active strains) and broadening host range, genetic engineering might someday be employed (Davidson 1984).

Development of economically feasible methods for increased refinement of fermentation residues will enable higher concentration of toxin. Improved refinement coupled with genetically engineered strains that yield greater amounts of toxic will permit production of toxin-based, rather than primary powder-based formulations. Emulsifiable concentrate and microencapsulated toxin-based formulations will provide a broad range of application options that are not currently available. Cheung and Hammock (1985) have formulated purified toxin of *B.t.i.* as a microlipid-droplet encapsulation and observed improved efficacy against *Anopheles freeborni* Aitken. Techniques for the refinement and purification of *B. sphaericus* toxin for research purposes have already been developed by Davidson (1983); de Barjac and Charles (1983); Payne and Davidson (1984); and Baumann et al. (1985). Considering the relatively low percentage of active moiety in the existing formulations of *B. sphaericus*, a substantial increase in larvicidal toxin would permit control of susceptible species using application rates even lower than those currently used for conventional chemical larvicides.

References

Aly, C. 1983. Feeding behavior of *Aedes vexans* larvae influencing efficacy of *Bacillus thuringiensis* var. *israelensis* toxin. *Bull. Soc. Vector Ecol.* 8: 94–100.

______. 1985. Feeding rate of larval *Aedes vexans* stimulated by food substances. *J. Amer. Mosq. Control Assoc.* 1: 506–510.

Aly, C.; Mulla, M. S.; Schnetter, W.; and B.-Z. Xu. 1987. Floating bait formulations increase effectiveness of *Bacillus thuringiensis* var. *israelensis* against *Anopheles* larvae. *J. Amer. Mosq. Control Assoc.* 3: 583–588.

Barjac, H. de, and Charles, J.-F. 1983. Une nouvelle toxin active sur les moustiques presente dans des inclusions crystallines produites par *Bacillus sphaericus*. *C. R. Acad. Sci.* (Paris) 296D: 905–910.

Baumann, P.; Unterman, B. M.; Baumann, L.; Broadwell, A. H.; Abbene, S. J.; and Bowditch, R. D. 1985. Purification of the larvicidal toxin of *Bacillus sphaericus* and evidence for high-molecular-weight precursors. *J. Bacteriol.* 163: 738–747.

Berry, W. J.; Novak, M. G.; Khounlo, S.; Rowley, W. A.; and Melchior, G. L. 1987. Efficacy of *Bacillus sphaericus* and *Bacillus thuringiensis* var. *israelensis* for control of *Culex pipiens* and floodwater *Aedes* larvae in Iowa. *J. Amer. Mosq. Control Assoc.* 3: 579–582.

Burke, W. F.; McDonald, K. O.; and Davidson, E. W. 1983. Effect of UV light on spore viability and mosquito larvicidal activity of *Bacillus sphaericus* 1593. *Appl. Environ. Microbiol.* 46: 954–956.

Charles, J.-F., and Nicolas, L. 1986. Recycling of *Bacillus sphaericus* 2362 in mosquito larvae: A laboratory study. *Ann. Microbiol.* (Inst. Pasteur) 137B: 101–111.

Cheung, P. Y. K., and Hammock, B. D. 1985. Micro-lipid-droplet encapsulation of *Bacillus thuringiensis* subsp. *israelensis* δ-endotoxin for control of mosquito larvae. *Appl. Environ. Microbiol.* 50 (4): 984–988.

Couch, T. L., and Ignoffo, C. M. 1981. Formulation of insect pathogens. In *Microbial control of pests and plant diseases, 1970–1980,* ed. D. Burges, 621–634. London: Academic Press.

Dadd, R. H. 1970. Comparison of rates of ingestion of particulate solids by *Culex pipiens* larvae: Phagostimulant effect of water-soluble yeast extract. *Ent. Exp. Appl.* 13: 407–419.

Dadd, R. H., and Kleinjan, J. E. 1985. Phagostimulation of larval *Culex pipiens* L. by nucleic acid nucleotides, nucleosides, and bases. *Physiol. Entomol.* 10: 37–44.

Dadd, R. H.; Kleinjan, J. E.; and Merrill, L. D. 1982. Phagostimulant effects of simple nutrients on larval *Culex pipiens* (Diptera: Culicidae). *Ann. Entomol. Soc. Am. J.* 75: 605–612.

Davidson, E. W. 1983. Alkaline extraction of toxin from spores of the mosquito pathogen, *Bacillus sphaericus* strain 1593. *Can. J. Microbiol.* 29: 271–275.

———. 1984. Microbiology, pathology, and genetics of *Bacillus sphaericus:* Biological aspects which are important to field use. *Mosq. News* 44: 147–152.

———. 1985. *Bacillus sphaericus* as a microbial control agent for mosquito larvae. In *Integrated mosquito control methodologies,* ed. M. Laird and J. W. Miles, 2: 213–226. London: Academic Press.

Davidson, E. W.; Singer, S.; and Briggs, J. D. 1975. Pathogenesis of *Bacillus sphaericus* strain SSII-1 infections in *Culex pipiens quinquefasciatus* larvae. *J. Invertebr. Pathol.* 25: 179–184.

Davidson, E. W.; Sweeney, A. W.; and Cooper, R. 1981. Comparative field trials of *Bacillus sphaericus* strain 1593 and *B. thuringiensis* var. *israelensis* commercial powder formulations. *J. Econ. Entomol.* 74: 350–354.

Davidson, E. W.; Urbina, M.; Payne, J.; Mulla, M. S.; Darwazeh, H.; Dulmage, H. T.; and Correa, J. A. 1984. Fate of *Bacillus sphaericus* 1593 and 2362 spores used as larvicides in the aquatic environment. *Appl. Environ. Microbiol.* 47: 125–129.

Des Rochers, B., and Garcia, R. 1984. Evidence for persistence and recycling of *Bacillus sphaericus. Mosq. News* 44: 160–165.

Hertlein, B. C.; Hornby, J.; Levy, R.; and Miller, T. W., Jr. 1980. Shelf-life of larvicidal preparations based on the strain 1593 of *Bacillus sphaericus.* WHO/VBC/80.790. Mimeo.

Hertlein, B. C.; Levy, R.; and Miller, T. W., Jr. 1979. Recycling potential and selective retrieval of *Bacillus sphaericus* from soil in a mosquito habitat. *J. Invertebr. Pathol.* 33: 217–221.

Hornby, J. A.; Hertlein, B. C.; and Miller, T. W., Jr. 1984. Persistent spores and mosquito larvicidal activity of *Bacillus sphaericus* 1593 in well-water and sewage. *J. Ga. Entomol. Soc.* 19: 165–167.

Hoti, S. L., and Balaraman, K. 1984. Recycling potential of *Bacillus sphaericus* in natural mosquito breeding habitats. *Indian J. Med. Res.* 80 (July): 90–94.

Karch, S., and Charles, J.-F. 1987. Toxicity, viability, and ultrastructure of *Bacillus sphaericus* 2362 spore/crystal complex used in the field. *Ann. Microbiol.* (Inst. Pasteur) 138: 485–492.

Karch, S., and Coz, J. 1984. Acceleration de l'activite larvicide de *Bacillus sphaericus* sur *Culex pipiens* par l'ingestion de cadavres de larves de moustiques intoxiques par ce bacile. *Cah. ORSTOM, sér. Ent. méd. Parasitol.* 22: 175–177.

———. 1986. Recycling of *Bacillus sphaericus* in dead larvae of *Culex pipiens* (Diptera: Culicidae). *Cah. ORSTOM, sér. Ent. méd. et Parsitol.* 24: 41–43.

Kuppusamy, M.; Hoti, S. L., and Balaraman, K. 1989. Residual activity of briquette and alginate formulations of *Bacillus sphaericus* against mosquito larvae. WHO/VBC/89.977. Mimeo.

Lacey, L. A. 1984. Production and formulation of *Bacillus sphaericus. Mosq. News* 44: 153–159.

———. 1985. Effects of pH and storage temperature on spore viability and mosquito larvicidal activity of *Bacillus sphaericus. Bull. Soc. Vector Ecol.* 10: 102–106.

Lacey, L. A.; Heitzman, C. M.; Meisch, M. V.; and Billodeaux, J. 1986. Beecomist®-applied *Bacillus sphaericus* for the control of riceland mosquitoes. *J. Amer. Mosq. Control Assoc.* 2: 548–551.

Lacey, L. A.; Ross, D. A.; Lacey, C. M.; Inman, A.; and Dulmage, H. T. 1988. Experimental formulations of *Bacillus sphaericus* for the control of anopheline and culicine larvae. *J. Indust. Microbiol.* 3: 39–47.

Lacey, L. A., and Singer, S. 1982. The larvicidal activity of new isolates of *Bacillus sphaericus* and *Bacillus thuringiensis* (H-14) against anopheline and culicine mosquitoes. *Mosq. News* 42: 537–543.

Lacey, L. A., and Smittle, B. J. 1985. The effects of gamma radiation on spore viability and mosquito larvicidal activity of *Bacillus sphaericus* and *Bacillus thuringiensis* var. *israelensis*. *Bull. Soc. Vector Ecol.* 10: 98–101.

Lacey, L. A., and Undeen, A. H. 1986. Microbial control of black flies and mosquitoes. *Ann. Rev. Entomol.* 31: 265–296.

Lacey, L. A.; Urbina, M. J.; and Heitzman, C. M. 1984. Sustained release formulations of *Bacillus sphaericus* and *Bacillus thuringiensis* (H-14) for control of container breeding *Culex quinquefasciatus*. *Mosq. News* 44: 26–32.

Mian, L. S., and Mulla, M. S. 1983. Factors influencing activity of the microbial agent *Bacillus sphaericus* against mosquito larvae. *Bull. Soc. Vector Ecol.* 8 (2): 128–134.

Mulla, M. S.; Axelrod, H.; Darwazeh, H. A., and Matanmi, B. A. 1988. Efficacy and longevity of *Bacillus sphaericus* 2362 formulations for control of mosquito larvae in dairy wastewater lagoons. *J. Amer. Mosq. Control Assoc.* 4: 448–452.

Mulla, M. S.; Darwazeh, H. A.; Davidson, E. W., and Dulmage, H. T. 1984a. Efficacy and persistence of the microbial agent *Bacillus sphaericus* for the control of mosquito larvae in organically enriched habitats. *Mosq. News* 44: 166–173.

Mulla, M. S.; Darwazeh, H. A.; Davidson, E. W., Dulmage, H. T.; and Singer, S. 1984b. Larvicidal activity and field efficacy of *Bacillus sphaericus* strains against mosquito larvae and their safety to nontarget organisms. *Mosq. News* 44: 336–342.

Mulla, M. S.; Darwazeh, H. A.; Ede, L.; Kennedy, B.; and Dulmage, H. T. 1985. Efficacy and field evaluation of *Bacillus thuringiensis* (H-14) and *B. sphaericus* against floodwater mosquitoes in California. *J. Amer. Mosq. Control Assoc.* 1: 310–315.

Mulligan, F. S., III; Schaefer, C. H.; and Miura, T. 1978. Laboratory and field evaluation of *Bacillus sphaericus* as a mosquito control agent. *J. Econ. Entomol.* 71: 774–777.

Mulligan, F. S., III; Schaefer, C. H.; and Wilder, W. H. 1980. Efficacy and persistence of *Bacillus sphaericus* and *Bacillus thuringiensis* H-14 against mosquitoes under laboratory and field conditions. *J. Econ. Entomol.* 73: 684–688.

Nicolas, L., and Dossou-Yovo, J. 1987. Differential effects of *Bacillus sphaericus* strain 2362 on *Culex quinquefasciatus* and its competitor *Culex cinereus* in West Africa. *Med. Vet. Entomol.* 1: 23–27.

Nicolas, L.; Dossou-Yovo, J.; and Hougard, J.-M. 1987. Persistence and recycling of *Bacillus sphaericus* 2362 spores in *Culex quinquefasciatus* breeding sites in West Africa. *App. Microbiol. Biotechnol.* 25: 341–345.

Payne, J. M., and Davidson, E. W. 1984. Insecticidal activity of the crystalline parasporal inclusions and other components of the *Bacillus sphaericus* 1593 spore complex. *J. Invertebr. Pathol.* 43: 383–388.

Ramoska, W. A.; Burgess, J.; and Singer, S. 1978. Field application of a bacterial insecticide. *Mosq. News* 38: 57–60.

Ramoska, W. A., and Hopkins, T. L. 1981. Effects of mosquito larval feeding behavior on *Bacillus sphaericus* efficacy. *J. Invertebr. Pathol.* 37: 269–272.

Ramoska, W. A., and Pacey, C. 1979. Food availability and period of exposure as factors of *Bacillus sphaericus* efficacy on mosquito larvae. *J. Econ. Entomol.* 72: 523–525.

Sawicka, E. M., and Couch, T. L. 1983. Formulations of entomopathogens. In *Pesticide formulations and application systems: Third symposium, ASTM STP 828,* ed. T. M. Kaneko and N. B. Akesson, 5–11. Philadelphia.

Silapanuntakul, S.; Pantuwatana, S.; Bhumiratana, A.; and Charoensiri, K. 1983. The comparative persistence of toxicity of *Bacillus sphaericus* strain 1593 and *Bacillus thuringiensis* serotype H-14 against mosquito larvae in different kinds of environments. *J. Invertebr. Pathol.* 42: 387–392.

Vankova, J. 1984. Persistence and efficacy of *Bacillus sphaericus* strain 1593 and 2362 against *Culex pipiens* larvae under field conditions. *Z. Ang. Ent.* 98: 185–189.

WHO. 1989. Informal consultation on bacterial formulations for cost-effective vector control in endemic areas. WHO/VBC/89.979. Mimeo.

Yousten, A. A. 1984. *Bacillus sphaericus:* Microbiological factors related to its potential as a mosquito larvicide. *Adv. Biotechnol. Processes* 3: 315–343.

19

Formulations and Persistence of *Bacillus sphaericus* in *Culex quinquefasciatus* Larval Sites in Tropical Africa

JEAN-MARC HOUGARD

19.1 INTRODUCTION

Although chemical insecticides are commonly used in public health programs (Mouchet 1980), they present a number of difficulties, such as the resistance of vectors to these compounds, the environmental pollution, or even the cost of the application. Therefore, it becomes necessary to conduct constant research on new substitute insecticides and to select the best formulations suited to the various constraints imposed by the vector control program.

Various strains of *Bacillus sphaericus* are particularly effective against larvae of some Culicidae. Due to the discovery of increasingly toxic strains, their recycling potential, and their innocuity to mammals, this biological insecticide seems very promising for the control of mosquitoes. Among the latter, *Culex quinquefasciatus* plays a significant role in the transmission of Bancroftian filariasis in East Africa (Nelson, Heisch, and Furlong 1962). Although this mosquito is not of great medical importance in West Africa (Brengues 1975), experimental transmissions have been obtained under laboratory conditions (Ogunba 1971). In East Africa and in various islands of the Indian Ocean (Madagascar, the Comoro Islands, and Sechelles), *Cx. quinquefasciatus* can be found in rural zones in polluted water and also in domestic containers (rainwater tanks), peridomestic containers (various unused vessels), and natural sites (tree holes, temporary ponds, rock holes). However, in tropical Africa, the *Cx. quinquefasciatus* larval habitats are composed of cesspools, pit latrines, septic tanks, or drains—mainly reservoirs of sewage water due to human activity (Subra 1980).

The *B. sphaericus* formulations that were tested in the field belong to the serotype H5a,5b: a wettable powder of strain 1593 and a flowable concentrate of strain 2362. In order to better understand the results obtained in the larval habitats, additional experiments based on bacteriological analyses were carried out. Most of the experiments were made in West Africa in Ivory

Coast (Bouaké), where a particularly large number of *Cx. quinquefasciatus* larval habitats exist.

19.2 FIELD EVALUATION

In *Cx. quinquefasciatus* habitats treated with strain 1593, the residual effect obtained by Hornby et al. (1981) was observed to be more than 3 months, while that obtained by other authors has been rather low and even nil (Indian Council of Medical Research 1977). In *Culex pipiens* habitats treated at about 0.2 mg/m^2, the residual effect obtained by Karch (1984) with this same strain was observed over about 20 days, but that obtained by Sinègre et al. (1986) with a flowable concentrate of strain 2362 did not exceed 2 days. With two other strains belonging to the serotype H5a,5b in polluted water, no residual effect was observed by Hoti and Balaraman (1984) in *Culex sitiens* habitats treated at 0.3 g/m^2.

19.2.1 Wettable Powder 1593

Using the wettable powder 1593, we selected 10 positive cesspits (containing larval populations), which were treated at 1, 10, and 50 g/m^2 (3 pits for each concentration and 1 untreated). At 24 hours posttreatment, larval mortality in treated cesspits was complete, even in those cesspits receiving the lowest dosage. Subsequently, the larval populations gradually reappeared (became positive) according to the dosage used (table 19.1); third- and fourth-instar larvae were observed between the eighth and sixteenth day following the treatment at 1 g/m^2, between the twentieth and twenty-sixth day at 10 g/m^2, and between the thirty-second and thirty-sixth day at 50 g/m^2.

In further studies we sampled the larval populations and studied the dynamics of *Cx. quinquefasciatus* for a four-month period following treatment. In the untreated cesspit, the larval and pupal densities were characterized by a series of highly pronounced maxima and minima (fig. 19.1a). Observations made in the same type of habitat by Subra (1971) showed that the aspect of these fluctuations can be generalized for all untreated pits. Once the treated sites have been recolonized, the larval population of the cesspits is characterized, like the untreated cesspit, by a series of minima and maxima. However, in a cesspit treated with 1 g/m^2 (fig. 19.1b), we observed that dead larvae were present during almost the entire observational period. The same phenomenon was temporarily observed in two other cesspits treated at 1 and 50 g/m^2 (fig. 19.1c and d). Generally, the dead larvae were scattered over the entire surface of the cesspit and came into contact with the living larvae, which were found in larger amounts. This larval mortality does not seem to modify the dynamics of the larval population. An analysis of these dead larvae

TABLE 19.1.
Persistence of a Wettable Powder of *B. sphaericus* Strain 1593 in a *Cx. quinquefasciatus* Larval Habitat

Pit Number	Days After Treatment[a]											
(Treatment)	0	1	10	14	18	22	26	28	34	36	38	40
1 (control)	+	+	+	+	+	+	+	+	+	+	+	+
2 (1 g/m²)	+	−	−	+	+	+	+	+	+	+	+	+
3 (1 g/m²)	+	−	+	+	+	+	+	+	+	+	+	+
4 (1 g/m²)	+	−	−	−	+	+	+	+	+	+	+	+
5 (10 g/m²)	+	−	−	−	−	−	+	+	+	+	+	+
6 (10 g/m²)	+	−	−	−	−	+	+	+	+	+	+	+
7 (10 g/m²)	+	−	−	−	−	−	−	+	+	+	+	+
8 (50 g/m²)	+	−	−	−	−	−	−	−	−	−	+	+
9 (50 g/m²)	+	−	−	−	−	−	−	−	+	+	+	+
10 (50 g/m²)	+	−	−	−	−	−	−	−	−	+	+	+

[a]Plus sign indicates presence of 3rd/4th-instar larvae and/or pupae; minus sign indicates their absence.

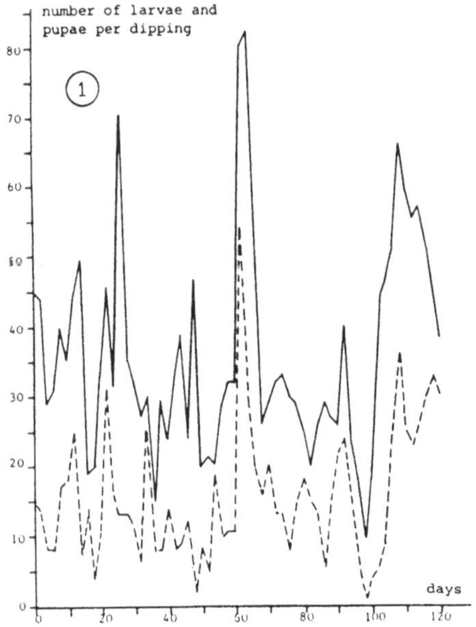

FIGURE 19.1 Dynamics of the larval population of *Cx. quinquefasciatus* in nontreated and *B. sphaericus*–treated larval habitats: graph 1, nontreated; graphs 2 and 3 treated with 1 g/m²; graph 4, 50 g/m². ——— = third- fourth-instar larvae; ------ = pupae; ········ = dead larvae. (Reprinted by courtesy of ORSTOM.)

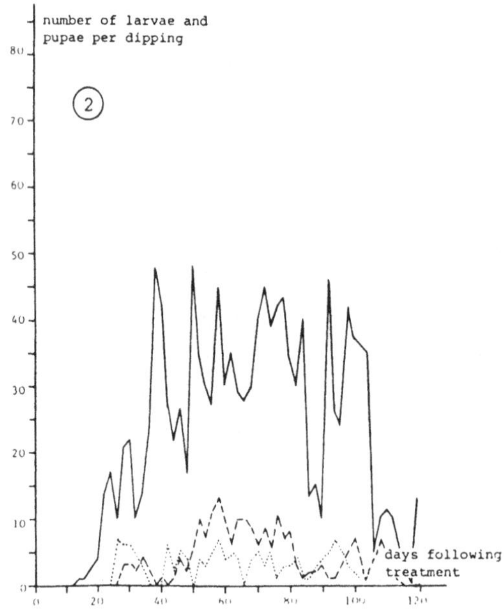

FIGURE 19.1 (*Continued*)

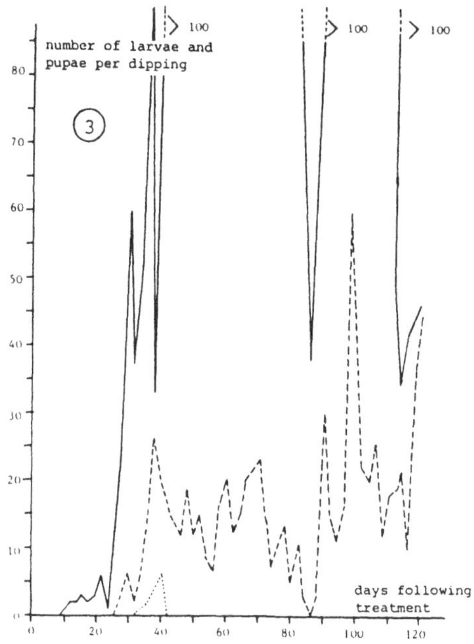

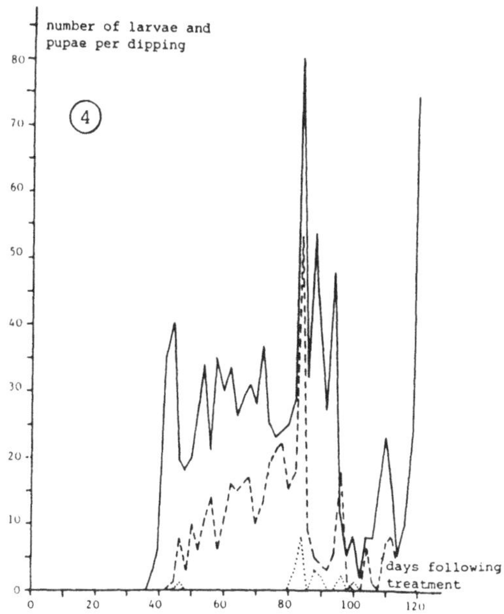

at the Institut Pasteur, Paris, showed that they were infected with *B. sphaericus* spores. More detailed studies specified that these spores belonged to the serotype H5a,5b, which includes strain 1593.

19.2.2 Flowable Concentrate 2362

Another field evaluation, carried out in Ivory Coast by Nicolas, Dossou-Yovo, and Hougard (1987) with a flowable concentrate of strain 2362 has given more promising results. Twenty-four hours after treatment at 10 g/m^2, the mortality ranged from 99 to 100% in the pits and reached 100% 48 hours after treatment. The third- and fourth-instar larvae reappeared between the thirty-fifth and forty-second day depending on the breeding sites, that is, 5–6 weeks after the treatment. Throughout this period, *Cx. quinquefasciatus* eggs and first-instar larvae were found in the sites but never reached the next stage.

Larvae and pupae belonging to psychodid Diptera and occurring in most cesspits were found, as with *B. t. i.* (personal observation), to be insensitive to the *B. sphaericus* strain at the dosage used in this study. Unlike with *B. t. i.* (Hougard, Darriet, and Bakayoko 1983), the *Culex cinereus* larvae are less susceptible to *B. sphaericus* strain 2362 than *Cx. quinquefasciatus* (Nicolas and Dossou-Yovo 1987).

19.3 BACTERIOLOGICAL TESTS

In the cesspits treated at 10 g/m^2 by strain 2362, Nicolas, Dossou-Yovo, and Hougard (1987) conducted bacteriological tests using samples of water taken on the surface, at a 1-meter depth, and from the bottom after two hours, six hours, one day, two days, and three days following the treatment and every four to seven days thereafter (fig. 19.2). Initially the spores were distributed homogeneously within the entire volume of water. Then they settled very slowly both from the surface and from the 1-meter depth; the spore concentration varied from about 10^4 spores/ml just after treatment to a little more than 10^2 after five weeks. The latter concentration coincided with the first new appearance of larvae in the habitat (fig. 19.2).

19.4 DISCUSSION

19.4.1 Persistence of *B. sphaericus* in the Field

It is difficult to compare the different studies carried out under natural conditions given the variety of larval habitats, strains, and formula-

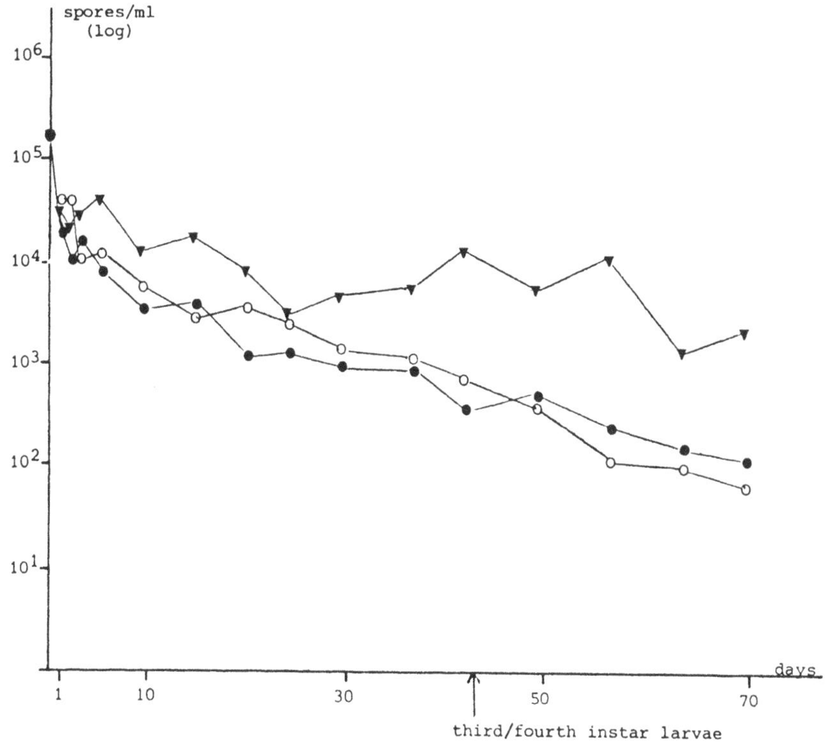

FIGURE 19.2 Settling of spores of a formulated compound of *B. sphaericus* strain 2362 (BSP 1) in a *Cx. quinquefasciatus* larval habitat treated at 10 g/m². ● = sampling at the surface; ○ = sampling at 1-m depth; ▼ = sampling in the mud. (Reprinted by courtesy of Luc Nicolas.)

tions of *B. sphaericus* tested. Although the residual effect varies considerably according to the authors, our results obtained with strain 1593 would seem to indicate that persistence increases with the dosage, thus confirming works by Mulligan, Schaefer, and Wilder (1980) and by Sinègre et al. (1986).

The residual effect observed over three to four weeks, respectively, with strains 1593 and 2362 in cesspools treated at 10 g/m² urged us to raise questions about the fate of the active ingredient in this type of larval habitat. The interpretation given to this phenomenon leads us to three hypotheses, which are based mainly on our laboratory experiments.

First hypothesis. An experiment conducted in the laboratory with wettable powder 1593 has shown that about two weeks are required for the active ingredient to settle completely in distilled water (Hougard 1988), while it takes only a few hours for *B.t.i.* (Guillet, Dempah, and Cox 1980). The bacteriological analysis made in situ with the same formulation of *B. sphaericus* also revealed that toxicity was linked to the presence of spores in the larval feeding zone (Hougard et al. 1985). Nicolas, Dossou-Yovo, and Hougard (1987) have subsequently given quantitative specification of these results using liquid concentrate 2362, thus confirming the results reported by Davidson et al. (1984). The residual effect of *B. sphaericus* depends on the settling rate of the spores; the larvae reappear when the number of spores becomes lower than about 10^2 spores/ml.

The differences in residual effect observed between *B.t.i.* and *B. sphaericus* seem to be linked mainly to the relation between the toxin and the spore. In the case of *B.t.i.,* crystals are not linked to the spores, and they settle more rapidly than the latter; while in the case of *B. sphaericus,* the toxic agents are closely linked to the spore, which would then function as a kind of float (Karch and Hougard 1986).

Second hypothesis. The active ingredient preserves its toxicity. Numerous experiments conducted in the laboratory have shown that the toxic agents of *B. sphaericus* are less stable than those of *B.t.i.* and are degraded mainly by ultraviolet light and sunlight. This phenomenon was also revealed in the field by Hornby et al. (1981) and by Karch (pers. comm.), who pointed out that the residual effect of *B. sphaericus* is greater in habitats characterized by polluted water or in shaded sites; presumably the spores would be, in both cases, protected from sunlight.

These observations can partly account for the variability of the results obtained by the different authors as well as variability of residual effect among the formulations tested. The water observed in the Bouaké cesspools is generally highly polluted, mostly protected from the light, and sometimes in darkness.

Third hypothesis. The active ingredient is still available. We observed along with Mulligan, Schaefer, and Wilder (1980) and Mian and Mulla (1983) that there was, as was shown with *B.t.i.* (Hougard, Darriet, and Bakayoko 1983), a negative correlation between the efficacy of *B. sphaericus* and the amount of suspended matter available for the larvae. However, this is not a major obstacle for using *B. sphaericus* in polluted water insofar as it is sufficient to increase the dosage in order to make up for this phenomenon.

19.4.2 Recycling of *B. sphaericus* in the Field

In the cesspools treated with the wettable powder of *B. sphaericus* 1593, the appearance of dead larvae infected with spores belonging to the same serotype, several weeks after the breeding sites have become positive, leads us to raise the question of the recycling of this bacterium in polluted water. As far as Davidson et al. (1984), Des Rochers and Garcia (1984), and Karch and Coz (1986) are concerned, the larvae are likely to favor spore germination; vegetative cells multiply and again produce viable toxic spores, which are released when the dead larvae disintegrate in the larval habitat. Although this recycling was proven in the laboratory and in the cesspools in Bouaké (Nicolas, Dossou-Yovo, and Hougard 1987), it is not very likely that it plays a significant larvicidal role in the field. The dead larvae settle to the bottom of the cesspits before recycling, thus carrying the spores away from the larval feeding zone.

Concerning our experiment with wettable powder 1593, there is the matter of determining the origin of the spores isolated in the dead larvae, sometimes several months after the treatment was applied. Considering the results obtained by Hertlein, Levy, and Miller (1979) and by Singer (1980), we tend to think that these spores originate from the treatments. According to these authors, the substratum is likely to favor the spores, which would then remain viable and infective for at least nine months. On this assumption, the spores may have settled to the bottom of the treated cesspools and have been resuspended due to water agitation caused by various reasons.

19.5 CONCLUSIONS

When comparing the results obtained with *B. sphaericus* in this type of larval habitat, the chemical insecticides seem to be much more promising because they are more persistent. However, an analysis of the following aspects leads us to reconsider our selection criteria.

Resistance. In the case of the chemical insecticides, the resistance potential leads us to have some reservations concerning the sole use of such compounds. Therefore, we recommend (as do Hamon and Mouchet 1967) alternating insecticides judiciously and making use of quite different products such as the biological control agents, specifically *B. sphaericus.*

Toxicity. The *Cx. quinquefasciatus* larval habitats sometimes may be connected to drinking water points or to phreatic water. For this reason, larvicides whose toxicity is too high or incompletely known for mammals cannot be used. In this case, biological control agents such as *B.t.i.* and *B. sphaericus* are the sole compounds likely to be used currently in such situations.

Economy. Despite the high concentration of *B. sphaericus* recommended for the treatment of polluted water (10 g/m^2), it already seems highly promising, in certain cases, to use a formulation such as the liquid concentrate of strain 2362. For instance, at Dar es Salaam (Tanzania), this formulation can compete costwise with the chlorpyrifos used currently by the sanitary services of the town (table 19.2); the frequency of larviciding at 1 g/m^2 has increased over the last decade from once every 9 to 12 weeks (Bang, Sabuni, and Tonn 1975) to once every 2 weeks (Curtis et al. 1984).

The entomopathogenic bacteria have also the advantage of being able to be produced locally and at a cheap rate by using a culture medium based on

TABLE 19.2.
Comparison of the Approximate Cost (U.S.$) of an Operational Use for a One-year Period of *B. sphaericus* (BSP1) at 10 g/m² and chlorpyrifos (Dursban®) at 1 g/m² in Dar es Salaam (Tanzania) for Control of *Cx. quinquefasciatus*

Parameter	*B. sphaericus* 2362 (flowable conc.)	Chlorpyrifos (E.C. 480 g/l)
Dosage	10 g/m² (0.01 l/m²)	1 g/m² (0.001 l/m²)
Frequency of larviciding	9 cycles a year	24 cycles a year
Quantity of formulation	45 liters	25 liters
Staff[a]	$12 × 9 = $108	$12 × 24 = $288
Equipment[b]	$30 × 9 = $270	$30 × 24 = $720
Insecticide	$8 × 45 = $360	$15 × 25 = $375
TOTAL	$738	$1,383

NOTE: The cost is calculated on the basis of 500 m² of treated sites.
[a]About $12 for two technicians for one cycle of larviciding.
[b]About $30 for one vehicle for one cycle of larviciding.

animal and plant wastes, which are available in tropical countries (Obeta and Okafor 1983).

Considering these points, the flowable concentrate of *B. sphaericus* 2362 (produced in large quantities) seems here and now to be operational, and the amounts used could increase in the near future. Using the same formulation, some trials carried out in the field by Nicolas, Darriet, and Hougard (1987) in Burkina Faso have controlled *Anopheles gambia* larvae for 10–15 days. However, these results have to be carefully considered given the difficulties of larvicidal control efforts in West Africa with its temporary and very numerous larval habitats.

19.6 SUMMARY

In West Africa, two formulations of *B. sphaericus* strains 1593 and 2362 have been selected for field tests in *Cx. quinquefasciatus* larval habitats. The flowable concentrate of the 2362 strain has given the most promising results, with five to six weeks of residual effect when treated at 10 g/m². These results seem to be related, on the one hand, with the low settling of the spores in the polluted water, and on the other hand, with the persistence of toxicity of the active ingredient protected from sunlight or ultraviolet light. Although recycling occurs in situ in larval cadavers, it is doubtful that it plays a major role in larval control in the field. The use of this flowable concentrate, even at a higher dosage, already seems operationally feasible in urban areas where *Cx. quinquefasciatus* is resistant to chemical insecticides.

References

Bang, Y. H., Sabuni, I. B., and Tonn, R. J. 1975. Integrated control of urban mosquitoes in Dar es Salaam using community sanitation supplemented by larviciding. *E. Afr. Med. J.* 52: 578–588.

Brengues, J. 1975. La filariose de Bancroft en Afrique de l'Ouest. *Mém. ORSTOM* (Paris) 79: 1–299.

Curtis, C. J.; Keto, A.; Ramji, B. D.; and Iosson, I. 1984. Assessment of the impact of chlorpyrifos resistance in *Culex quinquefasciatus* on a control scheme. *Insect. Sci. Applic.* 5: 263–267,

Davidson, E. W.; Urbina, M.; Payne, J.; Mulla, M. S.; Darwazeh, H.; Dulmage, H. T., and Correa, J. A. 1984. Fate of *Bacillus sphaericus* 1593 and 2362 spores used as larvicides in the aquatic environment. *Appl. Environ. Microbiol.* 47: 125–129.

Des Rochers, B., and Garcia, R. 1984. Evidence for persistence and recycling of *Bacillus sphaericus*. *Mosq. News* 44: 160–165.

Guillet, P.; Dempah, J.; and Coz, J. 1980. Evaluation de *Bacillus thuringiensis* sérotype 14 de Barjac pour la lutte contre les larves de *Simulium damnosum* s. l. III. Données préliminaires sur la sédimentation de l'endotoxine dans l'eau et sur sa stabilité en zone tropicale. WHO/VBC/80.756. Mimeo.

Hamon, J., and Mouchet, J. 1967. La résistance aux insecticides chez *Culex pipiens fatigans* Wiedemann. *Bull. Org. mond. Santé* 37: 277–286.

Hertlein, B. C.; Levy, R.; and Miller, T. H., Jr. 1979. Recycling potential and selective retrieval of *Bacillus sphaericus* from soil in a mosquito habitat. *J. Invertebr. Pathol.* 33: 217–221.

Hornby, J. A.; Hertlein, B. C.; Levy, R.; and Miller, T. H., Jr. 1981. Persistent activity of mosquito larvicidal *Bacillus sphaericus* 1593 in fresh water and sewage. WHO/VBC/81.830. Mimeo.

Hoti, S. L., and Balaraman, K. 1984. Recycling potential of *Bacillus sphaericus* in natural mosquito breeding habitats. *Indian J. Med. Res.* 80 (July): 90–94.

Hougard, J. M. 1988. Evaluation de l'efficacité de nouveaux larvicides pour la lutte contre les vecteurs d'endémies en Afrique de l'Ouest. Travaux et documents microédités de l'ORSTOM, TDM 38.

Hougard, J. M.; Darriet, F.; and Bakayoko, S. 1983. Evaluation en milieu naturel de l'activité larvicide de *Bacillus thuringiensis* sérotype H-14 sur *Culex quinquefasciatus* Say, 1823 et *Anopheles gambiae* Giles, 1902 s.l. (Diptera: Culicidae) en Afrique de l'Ouest. *Cah. ORSTOM, sér. Ent. Méd. et Parasitol.* 21 (2): 111–117.

Hougard, J. M.; Kohoun, G.; Guillet, P.; Doannio, J.; Duval, J.; and Escaffre, H. 1985. Evaluation en milieu naturel de l'activité larvicide de *Bacillus sphaericus* Neide, 1904 souche 1593-4 dans des gites larvaires à *Culex quinquefasciatus* Say, 1823 en Afrique de l'Ouest. *Cah. ORSTOM, sér. Ent. méd. et Parsitol.* 23 (1): 35–44.

Indian Council of Medical Research. 1977. Vector Control Research Centre, Pondichery. Annual report, 81–83. Mimeo.

Karch, S. 1984. *Bacillus sphaericus*, Agent de lutte biologique contre *Culex pipiens* Linné, 1758 (Culicidae-Diptera) et contre d'autres moustiques. Thèse doct. Ingénieur, Université Paris sud. *ORSTOM.*

Karch, S., and Coz, J. 1986. Recycling of *Bacillus sphaericus* in dead larvae of *Culex pipiens* (Diptera: Culicidae). *Cah. ORSTOM, sér. Ent. Méd. et Parasitol.* 24: 41–43.

Karch, S., and Hougard, J. M. 1986. Etude comparative au laboratoire du devenir de la matière active et des spores de *Bacillus sphaericus* souche 2362 et de *Bacillus thuringiensis* sérotype H-14 en milieu aqueux. *Cah. ORSTOM, sér. Ent. Méd. et Parasitol.* 24 (3): 175–179.

Mian, L. S., and Mulla, M. S. 1983. Factors influencing activity of the microbial agent *Bacillus sphaericus* against mosquito larvae. *Bull. Soc. Vector Ecol.* 8 (2): 128–134.

Mouchet, J. 1980. Lutte contre les vecteurs et nuisances en Santé publique. *Encycl. Méd. Chir.* (Paris). Maladies infectieuses, 8120 B10 (3-1980).

Mulligan, F.S., III; Schaefer, C. H.; and Wilder, W. H. 1980. Efficacy and persistence of *Bacillus sphaericus* and *Bacillus thuringiensis* H-14 against mosquitoes under laboratory and field conditions. *J. Econ. Entomol.* 73: 684–688.

Nelson, G. S.; Heisch, R. B.; and Furlong, M. 1962. Studies in filariasis in East Africa. Part 2, Filarial infection in man, animals, and mosquitoes on the Kenya Coast. *Trans. Royal Soc. Trop. Med. Hyg.* 56: 202–217.

Nicolas, L.; Darriet, F.; and Hougard, J. M. 1987. Efficacy of *Bacillus sphaericus* 2362 against larvae of *Anopheles gambiae* under laboratory and field conditions in West Africa. *Med. Vet. Entomol.* 1: 157–162.

Nicolas, L., and Dossou-Yovo, J. 1987. Differential effects of *Bacillus sphaericus* strain 2362 on *Culex quinquefasciatus* and its competitor *Culex cinereus* in West Africa. *Med. Vet. Entomol.* 1: 23–27.

Nicolas, L.; Dossou-Yovo, J.; and Hougard, J. M. 1987. Persistence and recycling of *Bacillus sphaericus* 2362 spores in *Culex quinquefasciatus* breeding sites in West Africa. *Appl. Microbiol. Biotechnol.* 25: 341–345.

Obeta, J. A. N., and Okafor, N. 1983. Production of *Bacillus sphaericus* strain 1593 primary powder on media made from locally obtainable Nigerian agricultural products. *Can. J. Microbiol.* 29: 704–709.

Ogunba, E. O. 1971. Observations on *Culex pipiens fatigans* in Ibadan, Western Nigeria. *Ann. Trop. Med. Parasit.* 65: 399–402.

Sinègre, G.; Gaven, B.; Jullien, J. L.; Vigo, G.; and Karch, S. 1986. Activité initiale et résiduelle de *Bacillus sphaericus* dans les gites larvaires à *Culex pipiens* du sud de la france. IVè congrès sur la protection de la santé humaine et des cultures en milieu tropical. Marseille, 2–4 juillet 1986.

Singer, S. 1980. *Bacillus sphaericus* for the control of mosquitoes. *Biotechnol. Bioeng.* 22: 1335–1355.

Subra, R. 1971. Etudes écologiques sur *Culex pipiens fatigans* Wiedemann, 1828 (Diptera, Culicidae) dans une zone urbaine de savane soudanienne ouest-africaine: Dynamique des populations préimaginales. *Cah. ORSTOM, sér. Ent. Méd. Parasitol.* 9 (1): 73–102.

———. 1980. Biology and control of *Culex pipiens quinquefasciatus* Say, 1823 (Diptera, Culicidae) with special reference to Africa. WHO/VBC/80.781. Mimeo.

20

Field Trials of *Bacillus sphaericus* for Mosquito Control

HAN-HENG YAP

20.1 INTRODUCTION

Following the successful development of *Bacillus thuringiensis* H14 (*B.t.i.*) as a larvicide for some mosquitoes and black flies (Lacey 1985), another *Bacillus* species, *Bacillus sphaericus* Neide, was developed to complement *B.t.i.* in the control of various mosquito species in diverse habitats (Davidson 1985; Mulla 1985; Singer 1985).

In laboratory tests, the relative efficacy of *B.t.i.* against common mosquitoes of public health importance in decreasing order was *Aedes, Culex, Anopheles,* and *Mansonia* (WHO 1979b; Foo and Yap 1982). In contrast, the relative susceptibility to *B. sphaericus* in decreasing order was *Culex, Anopheles, Mansonia,* and *Aedes* (WHO 1979a; Lacey and Singer 1982; Mulla et al. 1984b; Cheong and Yap 1985; Yap et al. 1988). In general, *B.t.i.* showed very short residual activity, especially in polluted water. There was also no conclusive evidence of recycling of the spores of *B.t.i.* in the environment (Lacey 1985). For *B. sphaericus,* improved techniques in production and formulation (Lacey 1984; Lacey, Urbina, and Heitzman 1984), as well as evidence of persistent and recycling properties in polluted aquatic environments (Davidson et al. 1984; Des Rochers and Garcia 1984; Mulla et al. 1984a), has provided impetus for conducting field trials against susceptible mosquito species, especially those inhabiting polluted environments.

20.2 GENERAL INFORMATION ON FIELD TRIALS

Field trials of *B. sphaericus* as a mosquito larvicide were first reported in 1978 (Ramoska, Burgess, and Singer 1978). To date, there have been more than 20 reports concerning simulated and field trials of several formulations of *B. sphaericus* strains against various species of mosquitoes (tables 20.1 and 20.2). A literature survey indicates that more field trials of *B. sphaericus* were carried out against *Culex* mosquitoes (table 20.1). There

TABLE 20.1.
Field Efficacy of Formulations of *B. sphaericus* Strains against *Culex* Mosquitoes in Diverse Habitats

Species	Locality and Habitat	Strain and Formulation[a]	Application Rate[b]	% Reduction & Residual Effect (RE)	Reference
Cx. nigripalpus	Florida: roadside ditches	C SSII-1, C 1404-9, C 1593-4	$3.1–19 \times 10^4$ cell/ml	89% (48 hours)	Ramoska, Burgess, and Singer 1978
Cx. tarsalis, Cx. quinquefasciatus	California: ponds, pasture plots, dairy drains	C SSII-1, LP 1593-4 WP 1593-4, C 1593-4	LP 0.67 kg/ha WP 0.84–2.24 kg/ha	90–100% (48 hours) 40–60% (48 hours)	Mulligan, Schaefer, and Miura 1978
Cx. quinquefasciatus	California: catch basins	PP 1593-4	10–500 ppm	100% (72 hours); RE for 28 days	Mulligan, Schaefer, and Wilder 1980
Cx. quinquefasciatus	Florida: roadside ditches	C 1593-4	5×10^6 sp/ml (stock solution)	spores viable in soil for 9 months	Hertlein, Levy, and Miller 1979
Cx. nigripalpus	Florida: sewage ponds	C 1593-4	$1 \times 10^4–1 \times 10^5$ sp/ml	RE clear water (30 days); sewage water (60–90 days)	Hertlein, Hornby, and Miller 1980
Cx. quinquefasciatus, Cx. annulirostris	Sydney: experimental ponds	DP 1593	0.5, 1, and 2 kg/ha	95+% (2 days at 1 and 2 kg/ha); no RE after 2d day	Davidson, Sweeney, and Cooper 1981
Cx. pipiens	California: sewage effluent	DP 1593 & 2362	$1 \times 10^3–1 \times 10^6$ sp/ml	81–100% (72 hours)	Des Rochers and Garcia 1983
Cx. quinquefasciatus	Bangkok: experimental jars	C 1593	2.28×10^5 sp/ml	RE 6 months (polluted water); 9 months (tap water)	Silapanuntakul et al. 1983
Cx. tarsalis	California: experimental ponds	DP 1593 & 2362	0.12–0.24 kg/ha	83–100% (2 days); 94–100% (4 days)	Davidson et al. 1984

Cx. pipiens	California: glass jars	PP 2362	—	RE (30 days)	Des Rochers & Garcia 1984
Cx. sitiens	Pondicherry, India: coir pits	WP VCRC B.6 & B.64	3g/1 week for 22 weeks	76–100% (24 hours); no RE	Hoti & Balaraman 1984
Cx. tritaeniorhynchus	Pondicherry, India: rice fields	WP VCRC B.64	0.5 to 2.0 kg/ha	62–99% (24 hours); no RE	Kramer 1984
Cx. quinquefasciatus, Cx. nigripalpus	Florida: sewage tanks	C 1593	—	RE 30–35 days (well water); 60 days (sewage water)	Hornby, Hertlein, and Miller 1984
Cx. quinquefasciatus	Florida: buckets	Pellets 1593	1–2 pellets/bucket	RE 8 weeks	Lacey, Urbina, and Heitzman 1984
Cx. pipiens	New Jersey: brackish; polluted	SC 2362	1–3 pts./ac	RE 12–26 days	Sutherland et al. 1986
Cx. quinquefasciatus, Cx. tarsalis	California: experimental ponds	P 2362	0.22 kg/ha	RE 3 weeks (clear water); nil (polluted water)	Mulla et al. 1984a
Cx. tarsalis/	California: experimental	PP 1593	0.11–0.22 kg/ha	100% (48 hours)	Mulla et al. 1984b
Cx. peus	ponds	PP 2362	0.06–0.11 kg/ha	100% (48 hours)	
Cx. quinquefasciatus	Ivory Coast: cesspools	P 1593-4	10 g/m²	RE 3 weeks	Hougard et al. 1985
Cx. quinquefasciatus	Ivory Coast: cesspools	SC 2362	10 g/m²	RE 5 weeks	Hougard 1985
Cx. gelidus	Penang, Malaysia: polluted drain	SC 2362	1 liter/ha	82% (1 day); 76% (2 days)	Yap 1985

[a] C = culture, DP = dry powder, LP = lyophilized powder, PP = primary powder, WP = wettable powder, SC = suspension concentrate.
[b] sp/ml = No. spores/ml, kg/ha = kilogram/hectare, g/m^2 = $gram/meter^2$, ppm = parts per million.

TABLE 20.2.
Field Efficacy of Formulations of *B. sphaericus* Strains against *Aedes, Anopheles, Mansonia* and *Psorophora* Species

Species	Locality and Habitat	Strain and Formulation[a]	Application Rate[b]	% Reduction & Residual Effect (RE)[c]	Reference
Ae. nigromaculis, Ae. melanimon	California: pasture plots	WP 1593-4	1.12–3.36 kg/ha	5% (48 hours)	Mulligan, Schaefer, and Miura 1978
Ae. triseriatus	Illinois: tree holes	DP 1593	10 ppm	water remained insecticidal 9 months	Singer 1980
Ae. stimulans	New York: woodland pools	WP 1593	0.2–4.0 ppm	95% (mean 1.2–4 ppm, 72 hours); LC_{50} = 0.26 ppm	Wraight, Molloy, and McCoy 1982
Ae. aegypti	Bangkok: experimental jars	C 1593	2.28 × 10^5 sp/ml	not reported	Silapanuntakul et al. 1983
Ae. melanimon	California: irrigated pasture	PP 2362	0.11–0.56 kg/ha	65% (48 hours, range 4–94%)	Mulla et al. 1985
An. annulipes	Sydney: experimental ponds	DP 1593	0.5, 1, and 2 kg/ha	+95% (3 days at 1–2 kg/ha)	Davidson, Sweeney, and Cooper 1981

An. franciscanus	California: experimental ponds	DP 1593 & 2362	0.12–0.24 kg/ha	no reduction	Davidson et al. 1984
An. subpictus	Pondicherry, India: coir pits	WP VCRC B.6 & B.64	3 g/l/week for 22 weeks	67–100% (24 hours). no RE	Hoti and Balaraman 1984
An. subpictus, An. nigerrimus	Pondicherry, India: rice fields	WP VCRC B.64	0.5–2.0 kg/ha	62–99%; no RE	Kramer 1984
Ma. indiana, Ma. uniformis	Northern Malaysia: rice field ditches;	SC & TP 2362	1 liter or kg/ha	SC, 81% (3 days to 14 days); TP, 94% (3 days to 14 days)	Yap 1985
Ma. uniformis	Penang, Malaysia: experimental plots	SC & TP 2362	0.1–10.0 liter or kg/ha	SC (LC_{50} = 0.70, LC_{90} = 5.91 liter/ha); TP (LC_{50} = 0.19, LC_{90} = 1.41g/ha)	
Ps. columbiae	Florida: field depression	C 1593-4	3.1–19 × 10^4 cell/ml	100% (48 hours)	Ramoska, Burgess, and Singer 1978
Ps. columbiae	California: irrigated pastures	PP 2362	0.28–0.56 kg/ha	+98% (48 hours)	Mulla et al. 1985

[a]C = culture, DP = dry powder, PP = primary powder, WP = wettable powder, SC = suspension concentrate, TP = technical powder.
[b]sp/ml = no. spores/ml, kg/ha = kilogram/hectare, ppm = parts per million.
[c]LC = lethal concentration.

were also scattered reports on field efficacy of *B. sphaericus* against *Aedes, Anopheles, Mansonia,* and *Psorophora* (table 20.2).

Geographical distribution of the field trial locations indicated that most of the trials were carried out in subtropical regions, especially in California and Florida. A few trials were also conducted in tropical regions in western Africa, southern India, Malaysia, and Thailand.

In addition to simulated trials in which artificial containers such as buckets and glass jars were used, the majority of the field trials were conducted in the natural larval habitats of the mosquitoes. Such habitats included tree holes, coir and earthworm collection pits, drains, ditches, cesspools, catch basins, sewage ponds, pasture plots, woodland pools, and experimental ponds (tables 20.1 and 20.2).

Of the strains and formulations of *B. sphaericus* investigated, early trials were carried out using culture preparations. The predominant type of formulations reported was powder, which includes primary, dry, technical, and wettable powders. Efficacy of newer formulations such as pellets and suspension concentrate was also reported (tables 20.1 and 20.2). To date, six strains of *B. sphaericus* have been tested for field efficacy against mosquitoes, including the most common strain, 1593, and the more recently developed potent strain of 2362. A few trials were also carried out using earlier strains of SSII-1 and 1404-9 as well as two strains (B.6 and B.64) developed by the Vector Control Research Center, Pondicherry, India.

Information accumulated so far has indicated that the *Culex* mosquitoes were the most susceptible to *B. sphaericus* when tested in the field. Relative field efficacy of other genera of mosquitoes in decreasing order of susceptibility was *Psorophora, Mansonia, Anopheles,* and *Aedes* (tables 20.1 and 20.2). Residual effects (persistence and recycling) of *B. sphaericus* were also strongly indicated when tested with the *Culex* species. There were conflicting reports concerning the degree of pollution affecting the residual effects of *B. sphaericus* (table 20.1).

All the field trials reported thus far were of small-scale experiments against introduced or naturally breeding mosquito populations. Large-scale field trials of *B. sphaericus* to determine its effect on the fluctuation of mosquito populations and the consequential epidemiology of mosquito-borne diseases have not been undertaken.

20.3 FIELD TRIALS AGAINST *CULEX*

Simulated and field trials of *B. sphaericus* have been reported against nine species of *Culex* mosquitoes: *Cx. annulirostris, Cx. gelidus, Cx. nigripalpus, Cx. peus, Cx. pipiens, Cx. quinquefasciatus, Cx. sitiens, Cx. tarsalis,* and *Cx. tritaeniorhynchus.* The most common species targeted was that of *Cx. quinquefasciatus* (table 20.1), a species that normally develops in polluted water adjacent to human habitations worldwide.

The geographical distributions and the different habitats where *B. sphaericus* trials against *Culex* were conducted were described earlier. Detailed information on test sites is listed in table 20.1. Including simulated trials, almost all the field trials against *Culex* mosquitoes were carried out in an aquatic environment that was considered polluted.

All six strains of *B. sphaericus* described earlier were used for tests against *Culex* species. Application rates (dosages) of *B. sphaericus* used varied with strains, formulations, target mosquito species, and the specific environmental conditions of the field sites. For strain 1593, dosages (spore counts) from culture preparations in actual field conditions varied between 1×10^3 and 1×10^6 spore/ml (table 20.1). A minimum spore count of 1×10^2 spore/ml in field water was reported to achieve adequate control against *Psorophora columbiae* (Ramoska, Burgess, and Singer 1978) and *Cx. nigripalpus* (Ramoska, Burgess, and Singer 1978; Hertlein, Hornby, and Miller 1980; Des Rocher and Garcia 1983).

Field trials using culture preparations of strain 1593 appeared to provide longer residual effects (persistence and recycling) against *Culex* mosquitoes. Hertlein, Levy, and Miller (1979), using a selective medium, were able to retrieve from soil samples viable and infective spores nine months after application of culture preparations of strain 1593 with initial dosage of 5×10^6 spores/ml. Against *Cx. quinquefasciatus* and *Cx. nigripalpus,* culture preparations of strain 1593 were reported to provide longer residual activities (two to three months) in polluted sewage water than in nonpolluted water (one month) (Hertlein, Hornby, and Miller 1980; Hornby, Hertlein, and Miller 1984). However, contradictory results were reported concerning interactions of water quality and residual effect when simulated experiments were conducted against *Cx. quinquefasciatus* (Silapanuntakul et al. 1983) and in field trials using experimental ponds that were organically enriched (Mulla et al. 1984a).

Various powder formulations, including dry powder (DP), primary powder (PP), technical powder (TP), and wettable powder (WP), of different strains of *B. sphaericus* were assessed against *Culex* mosquitoes. The powder formulations demonstrated good initial control against all *Culex* species investigated (table 20.1). Nevertheless, shorter residual effects of three to five weeks (Mulligan, Schaefer, and Wilder 1980; Des Rocher and Garcia 1984; Mulla et al. 1984a; Hougard et al. 1985) or no residual effects (Davidson, Sweeney, and Cooper 1981; Davidson et al. 1984; Hoti and Balaraman 1984; Kramer 1984; Mulla et al. 1984b) were reported for the powder formulations. The dosages of powder formulations used ranged from 0.06 to 2.24 kg/ha.

In general, the various formulations of *B. sphaericus* strains provided good initial control against all *Culex* species assessed. In terms of residual activities, with the existing formulations reported, applications of culture preparations appeared to provide longer residual effects against *Culex* mosquitoes in their often polluted environment.

20.4 FIELD TRIALS AGAINST *AEDES, ANOPHELES, MANSONIA*, AND *PSOROPHORA*

Information from a literature survey reveals that *B. sphaericus* has been tested against six species of *Aedes*, four species of *Anopheles*, two species of *Mansonia*, and one species of *Psorophora* in their respective habitats (table 20.2).

Results of simulated and field tests of *B. sphaericus* against *Aedes* mosquitoes gave overall poor performance (table 20.2). These results confirmed the generally high tolerance of *Aedes* species to various strains of *B. sphaericus* in laboratory susceptibility tests (Lacey and Singer 1982; Mulla et al. 1984b; Cheong and Yap 1985; Yap, Foo, and Tan 1988). The only successful control was reported against *Aedes stimulans* when wettable powder of *B. sphaericus* strain 1593 was applied in woodland pools. A mean mortality value of 95% was reported at concentrations ranging from 1.2 to 4 ppm (Wraight, Molloy, and McCoy 1982).

Using dry powder formulations of *B. sphaericus*, Davidson, Sweeney, and Cooper (1981) were the first to demonstrate the control of *Anopheles* species by this bacillus. The experiments were carried out in sodline experimental ponds (1 m^2) in New South Wales, Australia, at dosages of 0.5 to 2 kg/ha. More than 95% mortality was reported for *Anopheles annulipes* at dosages of 1 to 2 kg/ha. However, no reduction of *Anopheles franciscanus* larvae was reported when dry powders of *B. sphaericus* strains 1593 and 2362 were applied at dosages of 0.12 to 0.24 kg/ha in experimental ponds (35m^2) in California (Davidson et al. 1984). Two trials conducted in southern India using *B. sphaericus* strains B6 and B64 against *Anopheles nigerrimus* and *Anopheles subpictus* gave variable control with no residual effect (Hoti and Balaraman 1984; Kramer 1984).

Against *Mansonia* species, preliminary field trials were conducted in Malaysia using two newer formulations of strain 2362, suspension concentrate BSP1 and technical powder ABG 6184 (Yap 1985). Dose-response experiments were conducted in small plots filled with hosts plants of *Mansonia* larvae, *Eichhornia crassipes*. Using the floating-cage technique as described for field trials of *B.t.i.* and *Mansonia* mosquitoes (Foo and Yap 1983), mean dosage-response values at the 50% mortality level for suspension concentrate (SC) and technical powder (TP) against *Mansonia uniformis* were 0.70 l/ha and 0.19 kg/ha, respectively (Yap 1985). Other experiments were conducted using the above two formulations at dosages of 11 and 1 kg/ha applied to natural populations of *Mansonia* mosquitoes in impounded rice field ditches. Results showed extended residual effects between day 3 and day 14, with accumulated reductions of 81% and 94% for SC and TP, respectively (Yap 1985).

Two field trials were reported for *B. sphaericus* against *Psorophora columbiae* (Ramoska, Burgess, and Singer 1978; Mulla et al. 1985). The first trials were conducted in flooded field depressions using culture preparations of strain 1593-4 with final dosages of 3.1 to 19 × 10^4 viable cells/ml; 100% mortality of the *Psorophora* larvae was obtained at 48 hours (Ramoska, Burgess, and Singer 1978). A more recent trial conducted in irrigated pasture using primary powder of strain 2362 provided 98–99% reduction of larvae at the rates of 0.28 to 0.56 kg/ha (Mulla et al. 1985).

In summary, efficacy of formulations of *B. sphaericus* strains against *Aedes* and *Anopheles* mosquitoes have yielded variable results. Limited trials of *B. sphaericus* against *Mansonia* and *Psorophora* larvae indicated the high potential of using *B. sphaericus* to control mosquitoes in these two genera.

20.5 DISCUSSION

Since the first isolation of *B. sphaericus* from mosquitoes in the early 1960s (Kellen et al. 1965), there have been many more strains reported to have shown potential larvicidal activities against mosquitoes (de Barjac et al. 1985; Davidson 1985; Singer 1985; WHO 1985). Among the various strains described, most recent work has concentrated on strains 1593, 2297, and 2362 (WHO 1985). Laboratory efficacy of various formulations of the above three strains of *B. sphaericus* has been determined for mosquitoes in the genera of *Aedes, Anopheles, Culex, Mansonia,* and *Psorophora* (Lacey and Singer 1982; Mulla et al. 1984b; Cheong and Yap 1985; Yap, Foo, and Tan 1988).

The majority of reported field trials against mosquitoes were conducted using *B. sphaericus* strain 1593 (table 20.1). Due to their more recent introduction and development, the full potential of strains 2297 and 2362 has not been fully explored. Recent laboratory susceptibility tests indicated that strain 2297 has a narrower host range, with *Aedes aegypti* and *Ma. uniformis* exhibiting tolerance to it (Yap, Foo, and Tan 1988). As such, at a recent informal consultation meeting on *B. sphaericus* conducted under the auspices of the UNDP/WB/WHO Special Programme for Research and Training in Tropical Diseases, it was recommended that special emphasis on development be placed on *B. sphaericus* strain 2362 (WHO 1985).

Relative to the formulations of *B. sphaericus,* culture preparations and powder formulations were the more common types used in field trials reported to date (tables 20.1 and 20.2). With the accelerated development in formulation and production of *B. sphaericus,* anticipated formulations such as liquids, granules, briquettes, and pellets should provide more choice in conducting *B. sphaericus* field trials. Results generated from such trials will provide information on the appropriate formulation to be used for specific mosquito species in their respective habitats. For example, comparative field efficacy tests of various commercial formulations of *B.t.i.* have indicated that granules provide best control for *Mansonia* larvae under dense foliage (Foo 1986).

A review of available literature indicates that *Culex* were the most common target mosquitoes chosen for assessments of field efficacy of *B. sphaericus* (table 20.1). Unlike *Aedes* and *Anopheles*, mosquitoes in the genus of *Culex* generally breed in polluted aquatic environment (Bates 1949). The resistance of many *Culex* species to conventional insecticides has been well-documented (WHO 1980). Moreover, due to the polluted habitats, the list of available and environmentally acceptable organic insecticides for the control of *Culex* mosquitoes has been quite limited (Smith 1982). The high efficacy of formulations of *B. sphaericus* against *Culex* in polluted environments and their exceedingly long residual effects have made *B. sphaericus* the *Bacillus* of choice for the control of *Culex* in polluted aquatic environments. In fact, conventional insecticidal larvicides seldom achieve residual activity of more than two to three weeks under practical usages.

Although one of the primary objectives of microbial insecticide development is the control of malaria vectors, there have been surprisingly few field trials conducted using *B. sphaericus* against *Anopheles* mosquitoes (table 20.2). The scarcity of trials against *Anopheles* could be due to their diverse and less-defined larval habitats.

A few species of *Aedes* have demonstrated high tolerance to *B. sphaericus* under laboratory susceptibility tests (Lacey and Singer 1982; Mulla et al. 1984b; Cheong and Yap 1985; Yap et al. 1988). This may explain the lack of field trials on *Aedes* species using *B. sphaericus* (table 20.2). As there were indications of differential susceptibility among *Aedes* species (Mulligan, Schaefer, and Miura 1978; Lacey and Singer 1982; Singer 1985), the use of *B. sphaericus* for *Aedes* control cannot be fully ruled out.

For the *Mansonia* species, laboratory susceptibility tests indicated that the larvae of *Mansonia* mosquitoes were generally more tolerant to common insecticides (Yap, Cutkomp, and Buzicky 1968) and to *B.t.i.* (Foo and Yap 1982). However, laboratory bioassays have indicated that *Ma. uniformis* was highly susceptible to *B. sphaericus* strains 1593 and 2263 (Yap et al. 1988). The peculiar habit *Mansonia* immatures have of attaching to roots of aquatic plants was thought to contribute to less accessibility of larvicides to the immatures. However, larvicidal field trials using *B.t.i.* at higher dosages (Foo and Yap 1983) and *B. sphaericus* strain 2362 at reasonable dosages (Yap 1985) have shown promising results.

Laboratory bioassays have revealed that, next to *Culex*, *Psorophora* was the most susceptible to *B. sphaericus* strains 2013-4 and 2013-6 (Lacey and Singer 1982). Two field trials conducted thus far (Ramoska, Burgess, and Singer 1978; Mulla et al. 1985) indicated the high potential of using *B. sphaericus* for the control of this mosquito.

In contrast to *B.t.i.*, field applications of many formulations of *B. sphaericus* appear to demonstrate long residual activities (Lacey 1985; tables 20.1 and 20.2). Such residual activities could be due to the persistence of spore/crystal complex in the environment and/or the recycling (replication and sporulation) of *B. sphaericus* in mosquito cadavers and their environment. For those trials that showed residual effects, the residual periods ranged from three to nine months. *Bacillus sphaericus* strain 1593 persisted after applications in tree holes and roadside ditches for seven and nine months, respectively (Hertlein, Levy, and Miller 1979; Singer 1980). The recycling of *B. sphaericus* in mosquito cadavers has been demonstrated more recently (Silapanuntakul et al. 1983; Des Rochers and Garcia 1983, 1984; Davidson et al. 1984). Although the recycling of *B. sphaericus* may not provide enough spores for long-term vector control, the implication that subsequent application dosages in the same locality can be lower should be taken into consideration when cost-effectiveness of using *B. sphaericus* for vector control is deliberated. Under normal dosages, *B. sphaericus* has a minimum impact on nontarget organisms (Mulligan, Schaefer, and Miura 1978; Mulla et al. 1984b) and is nontoxic to mammals (Shadduck, Singer, and Lause 1980; de Barjac et al. 1979, 1987).

Information to date indicates that greater numbers of experimental field trials of *B. sphaericus* have been carried out against *Cx. quinquefasciatus* and other *Culex* species in subtropical regions, where mosquitoes are more of a nuisance than a disease vector. With the anticipated production of newer formulations of *B. sphaericus* strains, continued efforts in carrying out more experimental trials against susceptible mosquito species, with emphasis on mosquitoes that are vectors of major human diseases such as filariasis and malaria, should be encouraged.

At present, malaria is still the most prevalent and dangerous vector-borne disease of humans, with millions of cases throughout the tropical world (Gratz 1985). WHO (1984) estimated there were over 90 million cases of filariasis, or more specifically, human lymphatic filariasis, worldwide. Two major types of filariasis occur: Bancroftian filariasis, with 81.6 million cases, is transmitted by *Cx. quinquefasciatus* globally; and Brugian filariasis, with *Mansonia* mosquitoes as main vectors, has more than 8.6 million cases distributed widely in Southeast Asia (WHO 1984). The efficacy of *B. sphaericus* against *Cx. quinquefasciatus* in small-scale field trials is well established. Large-scale field trails of *B. sphaericus* should be carried out against *Cx. quinquefasciatus* in endemic and nonendemic areas of filariasis so as to further ascertain the overall impact of this bacillus on the control of this mosquito. In addition, more experimental trials should be conducted using newer, more appropriate formulations for *Mansonia* control. At present, there is no effective vector control approach for *Mansonia* mosquitoes (Yap 1986).

Concerning the choice of geographical locations for the field trials, trials carried out in subtropical regions have contributed valuable information for the practical control of mosquitoes in the tropical regions; however, variability of efficacy does occur due to differences in environmental conditions. Moreover, conducting *B. sphaericus* trials in tropical endemic areas of diseases will encompass the broader objective of disease control. As such, encouragement should be given to conduct more field trials in tropical areas.

In conclusion, existing formulations of *B. sphaericus* assessed to date have shown adequate initial control and prolonged residual effects against many mosquitoes, especially those in the genera of *Culex, Mansonia, Psorophora,* and *Anopheles.* The environmental impact of using *B. sphaericus* is minimal. We anticipate that in the near future *B. sphaericus,* in conjunction with *B. t. i.,* will have an important role to play in the overall strategy of mosquito control worldwide.

References

Barjac, H. de; Cosmao Dumanoir, V.; Hamon, S.; and Thiery, I. 1987. Safety tests on mice with *Bacillus sphaericus* serotype II5a5b strain 2362. WHO/VBC/1987.948.

Barjac, H. de; Larget, I.; Cosmao Dumanoir, V.; Benichou, L.; and Viviani, G.; Ripouteau, H.; and Papion, S. 1979. Inocuité de *Bacillus sphaericus* souche 1593 pour les mammifères. WHO/VBC/79.731. Mimeo.

Barjac, H. de; Larget-Thiery, I.; Cosmao Dumanoir, V. C.; and Ripouteau, H. 1985. Serological classification of *Bacillus sphaericus* strains in relation with toxicity to mosquito larvae. *Appl. Microbiol. Biotechnol.* 21: 85–90.

Bates, M. 1949. *The natural history of mosquitoes.* New York: Harper & Row.

Cheong, W. C., and Yap, H. H. 1985. Bioassays of *Bacillus sphaericus* (strain 1593) against mosquitoes of public health importance in Malaysia. *Southeast Asian J. Trop. Med. Pub. Hlth.* 16: 54–58.

Davidson, E. W. 1985. *Bacillus sphaericus* as a microbial control agent for mosquito larvae. In *Integrated mosquito control methodologies,* ed. M. Laird and J. W. Miles, 2: 213–226. London: Academic Press.

Davidson, E. W.; Sweeney, A. W.; and Cooper, R. 1981. Comparative field trials of *Bacillus sphaericus* strain 1593 and *B. thuringiensis* var. *israelensis* commercial powder formulations. *J. Econ. Entomol.* 74: 350–354.

Davidson, E. W.; Urbina, M.; Payne, J.; Mulla, M. S.; Darwazeh, H.; Dulmage, H. T.; and Correa, J. A. 1984. Fate of *Bacillus sphaericus* 1593 and 2362 spores used as larvicides in the aquatic environment. *Appl. Environ. Microbiol.* 47: 125–129.

Des Rochers, B., and Garcia, R. 1983. The efficacy of *Bacillus sphaericus* in controlling mosquitoes. In *Proceedings and Papers of the Fifty-first Annual Conference of the California Mosquito and Vector Control Association,* 35–37.

———. 1984. Evidence for persistence and recycling of *Bacillus sphaericus. Mosq. News.* 44: 160–165.

Foo, A. E. S. 1986. Laboratory and field evaluations on the efficacy of *Bacillus thuringiensis* H-14 for the control of *Mansonia* and other vector mosquitoes (Diptera: Culicidae), including some comparative studies with *Bacillus sphaericus.* Ph.D. diss., Universiti Sains Malaysia, Penang, Malaysia.

Foo, A. E. S., and Yap, H. H. 1982. Comparative bioassays of *Bacillus thuringiensis* H-14 formulations against four species of mosquitoes in Malaysia. *Southeast Asian J. Trop. Med. Pub. Hlth.* 13: 206–210.

———. 1983. Field trials on the use of *Bacillus thuringiensis* serotype H-14 against *Mansonia* mosquitoes in Malaysia. *Mosq. News.* 43: 306–310.

Gratz, N. R. 1985. The future of vector biology and control in the World Health Organization. *J. Amer. Mosq. Control Assoc.* 1: 273–278.

Hertlein, B. C.; Hornby, J.; and Miller, T. W., Jr. 1980. Reduction of the St. Louis encephalitis vector in sewage treatment units with *Bacillus sphaericus.* TDR/BCV/SWG.80/WP.09. Mimeo.

Hertlein, B. C.; Levy, R.; and Miller, T. W., Jr. 1979. Recycling potential and selective retrieval of *Bacillus sphaericus* from soil in a mosquito habitat. *J. Invertbr. Pathol.* 33: 217–221.

Hornby, J. A.; Hertlein, B. C.; and Miller, T. W., Jr. 1984. Persistent spores and mosquito larvicidal activity of *Bacillus sphaericus* 1593 in well water and sewage. *J. Ga. Entomol. Soc.* 19: 158–165.

Hoti, S. L., and Balaraman, K. 1984. Recycling potential of *Bacillus sphaericus* in natural mosquito breeding habitats. *Indian J. Med. Res.* 80 (July): 90–94.

Hougard, J. M. 1985. Field evaluation of various *B. sphaericus* strains on *Culex quinquefasciatus* larvae in tropical Africa. WHO/BCV/CSP/WP.9. Mimeo.

Hougard, J. M.; Kohoun, G.; Guillet, P.; Doannio, J.; Duval, J.; and Escaffre, H. 1985. Evaluation en milieu naturel de l'activité larvicide de *Bacillus sphaericus* Neide, 1904 souche 1593-4 dans des gites larvaires à *Culex quinquefasciatus* Say, 1823 en Afrique de l'Ouest. *Cah. ORSTOM, sér. Ent. méd. et Parasitol.* 23 (1): 35–44.

Kellen, W. R.; Clark, T. B.; Lindegren, J. E.; Ho, B. C.; Rogoff, M. H.; and Singer, S. 1965. *Bacillus sphaericus* Neide as a pathogen of mosquitoes. *J. Invertebr. Pathol.* 7: 442–448.

Kramer, V. 1984. Evaluation of *Bacillus sphaericus* and *B. thuringiensis* H-14 for mosquito control in rice fields. *Indian J. Med. Res.* 80: 642–648.

Lacey, L. A. 1984. Production and formulation of *Bacillus sphaericus. Mosq. News* 44: 153–159.

———. 1985. *Bacillus thuringiensis* serotype H-14. *Amer. Mosq. Control Assoc. Bull.* 6: 132–158.

Lacey, L. A., and Singer, S. 1982. The larvicidal activity of new isolates of *Bacillus sphaericus* and *Bacillus thuringiensis* (H-14) against anopheline and culicine mosquitoes. *Mosq. News* 42: 537–543.

Lacey, L. A.; Urbina, M. J.; and Heitzman, C. M. 1984. Sustained release formulations of *Bacillus sphaericus* and *Bacillus thuringiensis* (H-14) for control of container breeding *Culex quinquefasciatus. Mosq. News* 44: 26–32.

Mulla, M. S. 1985. Field evaluation and efficacy of bacterial agents and their formulations against mosquito larvae. In *Integrated mosquito control methodologies,* ed. M. Laird and J. W. Miles, 2: 227–250. London: Academic Press.

Mulla, M. S.; Darwazeh, H. A.; Davidson, E. W.; and Dulmage, H. T. 1984a. Efficacy and persistence of the microbial agent *Bacillus sphaericus* for the control of mosquito larvae in organically enriched habitats. *Mosq. News* 44: 166–173.

Mulla, M. S.; Darwazeh, H. A.; Davidson, E. W.; Dulmage, H. T.; and Singer, S. 1984b. Larvicidal activity and field efficacy of *Bacillus sphaericus* strains against mosquito larvae and their safety to nontarget organisms. *Mosq. News* 44: 336–342.

Mulla, M. S.; Darwazeh, H. A.; Ede, L.; Kennedy, B.; and Dulmage, H. T. 1985. Efficacy and field evaluation of *Bacillus thuringiensis* (H-14) and *B. sphaericus* against floodwater mosquitoes in California. *J. Amer. Mosq. Control Assoc.* 1: 310–315.

Mulligan, F. S., III; Schaefer, C. H.; and Miura, T. 1978. Laboratory and field evaluation of *Bacillus sphaericus* as a mosquito control agent. *J. Econ. Entomol.* 71: 774–777.

Mulligan, F. S., III; Schaefer, C. H.; and Wilder, W. H. 1980. Efficacy and persistence of *Bacillus sphaericus* and *Bacillus thuringiensis* H-14 against mosquitoes under laboratory and field conditions. *J. Econ. Entomol.* 73: 684–688.

Ramoska, W. A.; Burgess, J.; and Singer, S. 1978. Field application of a bacterial insecticide. *Mosq. News* 38: 57–60.

Shadduck, J. A.; Singer, S.; and Lause, S. 1980. Lack of mammalian pathogenicity of entomocidal isolates of *Bacillus sphaericus. Environ. Entomol.* 9: 403–407.

Silapanuntakul, S.; Pantuwatana, S.; Bhumiratana, A.; and Charoensiri, K. 1983. The comparative persistence of toxicity of *Bacillus sphaericus* strain 1593 and *Bacillus thuringiensis* serotype H-14 against mosquito larvae in different kinds of environments. *J. Invertebr. Pathol.* 42: 387–392.

Singer, S. 1980. *Bacillus sphaericus* for the control of mosquitoes. *Biotechnol. Bioeng.* 22: 1335–1355.

———. 1985. *Bacillus sphaericus* (Bacteria). In *Biological control of mosquitoes,* ed. H. C. Chapman, 123–131. Bulletin no. 6., American Mosquito Control Assoc., Fresno, California.

Smith, A. 1982. Chemical methods for the control of vectors of pests of public health importance. WHO/VBC/82.841. Mimeo.

Sutherland, D. J.; Dreyer, R.; Schmidt, R.; Keyes, R.; Ferchak, L.; Rupp, H.; Safranek, A.; O'Connell, T.; Hansen, J.; and McPherson, W. 1986. Field evaluation of *Bacillus sphaericus* as a larvicide in New Jersey, 1985. *N.J. Mosq. Control Assoc. Proc.* 73: 16–25.

WHO. 1979a. Biological control agent data sheet: *Bacillus sphaericus,* strain 1593-4. WHO/VBC/BCDS/79.09. Mimeo.

———. 1979b. Data sheet on the biological control agent *Bacillus thuringiensis* serotype (de Barjac 1978). WHO/VBC/79.750. Mimeo.

_______. 1980. Resistance of Vectors of Disease to Pesticide. Fifth Report of the WHO Expert Committee on Vector Biology and Control. WHO Tech. Rept. Ser. 655.

_______. 1984. Lymphatic Filariasis. Fourth Report of the WHO Expert Committee on Filariasis. WHO Tech. Rept. Ser. 702.

_______. 1985. Informal consultation on the development of *Bacillus sphaericus* as a microbial larvicide. TDR/BCV/SPHAERICUS/85.3. Mimeo.

Wraight, S. P.; Molloy, D.; and McCoy, P. 1982. A comparison of laboratory and field tests of *Bacillus sphaericus* strain 1593 and *Bacillus thuringiensis* var. *israelensis* against *Aedes stimulans* larvae. (Diptera: Culicidae). *Can. Ent.* 114: 55–61.

Yap, H. H. 1985. Simulated and small scale field trials of *Bacillus sphaericus* (strain 2362) against *Culex* and *Mansonia* mosquitoes in Malaysia. WHO/BCV/CSP/WP.16. Mimeo.

_______. 1986. Review on control of Brugian filariasis vectors, especially *Mansonia* species. Proc. WHO Regional Seminar on Brugian Filariasis, Kuala Lumpur, Malaysia, 1–5 July 1985, 131–135.

Yap, H. H.; Cutkomp, L. K.; and Buzicky, A. W. 1968. Insecticidal tests against *Mansonia perturbans* (Walker). *Mosq. News* 28: 504–506.

Yap, H. H.; Ng, Y. M.; Foo, A. E. S.; and Tan, H. T. 1988. Bioassays of *Bacillus sphaericus* (strain 1593, 2297 and 2362) against *Mansonia* and other mosquitoes of public health importance in Malaysia. *Malays. Appl. Biol.* 17: 9–13.

21

Mammalian Safety of *Bacillus sphaericus*

JOEL P. SIEGEL
JOHN A. SHADDUCK

21.1 INTRODUCTION

There has been recent interest in the commercial development of *Bacillus sphaericus* as an adjunct or alternative to chemicals used for the control of mosquitoes. Like their chemical counterparts, microbial insecticides must be evaluated for safety to both animals and humans, although these tests differ from those used for chemicals. Tests of microbial insecticides concentrate on vertebrate toxicity and infectivity, while tests of chemical insecticides focus on acute toxicity, neurotoxicity, and carcinogenicity. The philosophy of microbial testing is discussed in chapter 12 on the safety of *Bacillus thuringiensis* subsp. *israelensis* (*B.t.i.*).

It is especially important to evaluate the mammalian safety of *B. sphaericus* because the genus *Bacillus* contains the mammalian pathogen *Bacillus anthracis.* Furthermore, *B. sphaericus* was implicated in a fatal case of meningitis and generalized Schwartzman reaction (Allen and Wilkinson 1969), and *B. sphaericus* was isolated from a fatal pseudotumor of a human lung (Isaacson et al. 1976). Organisms of the genus *Bacillus* have been associated with infections of traumatic wounds as well (Pearson 1970). Here, we present the results, from least invasive to most invasive, of subcutaneous, intraperitoneal, intradermal, ocular, intraocular, and intracerebral administration of seven isolates of *B. sphaericus* to mice, rats, and rabbits. The main data on three entomicidal isolates have already been published by one of us (Shadduck, Singer, and Lause 1980).

21.2 SOURCE AND PREPARATION OF CULTURES

Three *B. sphaericus* isolates were originally recovered from dead mosquito larvae: in India, SSII-1; in the Philippines, 1404; and in Indonesia, 1593. Isolate 7054 came from dead *Culex quinquefasciatus* larvae, and isolate NCTC 11025 was obtained from an abscess of a human lung.

The insect isolates were grown on *Sphaericus* Synthetic Medium (SSM) slants (Myers and Yousten 1978) for two days and then used to inoculate roller tubes containing 5 ml of SSM broth, which was incubated 16 hours at 30° C (26 rpm) on a tissue-culture roller drum. Seed flasks containing 25 ml SSM broth were then incubated at 30° C for 6 to 9 hours in a rotary shaker bath at 250 rpm, with 1 to 5% (vol/vol) inoculum level. Production flasks were inoculated from the seed flasks and maintained under the same regime. Inocula consisted of whole broth cultures harvested at 18 to 24 hours, and the titer of each inoculum was determined by tube dilution in Brain Heart Infusion (BHI) broth. Isolate NCTC 11025 was grown on BHI agar 48 hours at 30° C, and the plates were scraped to inoculate BHI broth. There was a final concentration at 9.5×10^8 viable organisms per ml of this human isolate.

A commercial suspension of *B. sphaericus* 2362 was received from Solvay and Company, Brussels, Belgium, and additional *B. sphaericus* 2362 was received from Dr. Elizabeth Davidson (Arizona State University, Tempe, Arizona). Fifteen ml of each suspension were mixed with an equal volume of a stock solution (pH 7.2, 0.15M) of sterile phosphate buffered saline (PBS) for a final volume of 30 ml. The suspension was then centrifuged at $1{,}000 \times g$ for 30 minutes in order to pellet the bacteria, and the supernatant was discarded. The pellet was resuspended in PBS to the original volume of 30 ml, then centrifuged at $1{,}000 \times g$; this washing procedure was repeated three more times for a total of four washes. The number of colony-forming units (cfu) per ml was determined by serial dilution and then plating the dilutions onto BHI agar plates, after which the plates were incubated for 24 hours at 30° C. Individual colonies were then counted and the number of cfu per ml determined.

21.3 RESULTS AND DISCUSSION

21.3.1 Subcutaneous and Intraperitoneal Injection

Euthymic mice (CF-W outbred and BALB/cAnNHsd BR), athymic mice (Hsd: Athymic Nude nu AF) and rats (Hsd: Sprague Dawley [SD] BR) were injected subcutaneously and intraperitoneally with two insect isolates (1404, 1593) and one human isolate (NCTC 11025) of *B. sphaericus* (table 21.1). Subcutaneous injection of both viable and autoclaved 1404 resulted in abscesses at the injection site, most likely arising from the presence of foreign material; injection of 1593 and NCTC 11025 did not cause abscesses. There were no signs of clinical illness and no deaths in the animals receiving viable *B. sphaericus*. The inoculum appeared to be localized, because *B. sphaericus* was not recovered from the spleen. None of the 20 rats injected with viable *B. sphaericus* died.

Recent experiments with CD-1 outbred mice injected intraperitoneally with *B. sphaericus* 2362 indicated that 2362 was toxic at doses greater than

TABLE 21.1.
Subcutaneous and Intraperitoneal Injection of 4 Isolates of *B. sphaericus* into Rats and Mice

Species	Isolate	Dose	Results: Subcutaneous	Results: Intraperitoneal
Mouse				
Euthymic	1404	6.9×10^{9}	1/5[a]	—
	1593	9.3×10^{8}	0/5	—
	NCTC 11025[b]	9.5×10^{8}	0/6	—
	NCTC 11025	9.5×10^{7}	0/6	—
	2362	8.0×10^{8}		42/49
Athymic	NCTC 11025	9.5×10^{8}[c]	0/6	—
	NCTC 11025	9.5×10^{6}[c]	0/6	—
Rat				
	1404	1.6×10^{8}	—	0/7
	1593	2.4×10^{8}	—	0/7
	NCTC 11025	9.0×10^{6}	—	0/6

NOTE: Data are given as the number affected/number injected.
[a]Abscess at injection site.
[b]NCTC 11025; recovered from human lung abscess.
[c]Spleens cultured; no *B. sphaericus* recovered.

10^7 cfu (table 21.1). Mortality occurred within 48 hours, and the toxic factor was heat stable. The toxicity appeared spore associated, as injection of filtered supernatant caused no discernible effects. There was no evidence of infection in outbred mice; mice injected with 1.2×10^7 cfu of washed *B. sphaericus* cleared the inoculum over a 67-day period according to the equation $\mathrm{LN}\,Y = \mathrm{LN}\,12.761 - 0.111X \pm 1.20$ ($r^2 = 0.79, p < 0.001$), where *LN Y* is the natural logarithm of the cfu per gram spleen, *X* is the number of days after injection, and $\pm$ 1.20 is the standard error of estimate.

Since an intact immune system may be necessary for protection against *B. sphaericus*, the mammalian pathogenicity of the bacterium was also tested on immune-deficient mice. Athymic mice lack a thymus, hence have no T-lymphocytes (cellular immunity), and, since T-lymphocytes are required for normal B-cell response (humoral immunity) to most complex antigens, athymic mice have impaired antibody responses as well. Injection of as many as 10^8 cfu from the human isolate NCTC 11025 produced no illness or mortality in athymic mice. Athymic mice injected with 10^7 cfu of 2362 cleared the inoculum over 16 days according to the equation $\mathrm{LN}\,Y = 15.22 - 0.218X \pm 0.911$ ($r^2 = 0.642, p = 0.001$), where *LN Y* is the natural logarithm of the cfu per gram spleen, *X* is the number of days after injection, and $\pm$ 0.911 is the standard error of estimate. Heart blood was positive for *B. sphaericus* up to 11 days after injection, but was negative thereafter. Since *B. sphaericus* infections in humans have been associated with immune system depression following treatment with prednisone (Isaacson et al. 1976), it is reassuring that athymic mice are not adversely affected by high doses of *B. sphaericus.*

21.3.2 Intradermal Injection

This experiment was conducted because a worker evaluating the yield of *B. sphaericus* in *Culex pipiens quinquefasciatus* larvae accidentally injected her finger with a suspension of larval fragments, debris, and *B. sphaericus.* There was immediate pain and swelling, which persisted for several weeks. Several factors possibly contributed to these symptoms, and the effects of two isolates of *B. sphaericus* as well as mosquito fragments were evaluated.

Four New Zealand White rabbits were intradermally injected with *B. sphaericus* 7054, *B. sphaericus* SSII-1, and ground uninfected second-instar *Cx. pipiens quinquefasciatus* larvae (table 21.2). Injection of the larval suspension caused the most severe and persistent skin reaction, and isolate SSII-1 caused the mildest reaction. Further experiments are necessary to determine if the apparent difference between SSII-1 and 7054 is significant. The most likely cause of the worker's swollen finger was mosquito fragments, since the reaction of rabbits to *B. sphaericus* decreased with time while the rabbits' reaction to the larval suspension remained the same.

21.3.3 Ocular Irritancy

The eyes of New Zealand White rabbits were exposed to *B. sphaericus* by placing 0.05 ml of whole broth suspension of viable and autoclaved bacteria (1404, SSII-1) into the conjunctival cul-de-sac. Untreated eyes served as controls, and all eyes were observed with a slit lamp 1

TABLE 21.2.
Intradermal Injection (0.05 ml) of 2 Isolates of *B. sphaericus* into Rabbits

Isolate	Day 1 After Injection		Day 3 After Injection	
	Number Affected	Mean Reaction (mm)	Number Affected	Mean Reaction (mm)
7054 + debris	4/4	3.1	1/4	0.75
SSII-1 + debris	4/4	2.0	0/4	0
Debris[a]	4/4	3.5	4/4	3.5

NOTE: All *B. sphaericus* was derived from second-instar *Cx. pipiens quinquefasciatus* larvae.
[a] *Cx. pipiens quinquefasciatus* larvae alone.

TABLE 21.3.
Ocular Irritancy of 2 Isolates of *B. sphaericus* Administered to Rabbits

	Days After Treatment											
	1		2		3		4		5		11	
	A	B	A	B	A	B	A	B	A	B	A	B
Conjunctiva												
Congestion	0	0	0	0	1	0	0	0	0	0	0	1
	△0	0	1	0	1	0	0	0	0	0	0	0
Swelling	0	0	0	0	0	0	0	0	0	0	0	0
	△1	0	0	0	0	0	0	0	0	0	0	0
Discharge	1	1	0	0	0	0	0	0	0	0	0	0
	△0	1	0	0	0	0	0	0	0	0	0	0
Iritis	0	1	0	0	3	0	1	0	1	0	1	0
	△0	1	0	3	2	3	1	2	1	1	1	1
Corneal Opacity												
Intensity	0	0	0	0	0	0	0	0	0	0	0	0
	△0	0	0	0	0	0	0	0	0	0	0	0
Area	0	0	0	0	0	0	0	0	0	0	0	0
	△0	0	0	0	0	0	0	0	0	0	0	0

NOTE: The number of eyes differing from normal eyes are given; each treated eye received 0.05 ml of whole broth bacterial culture. A = isolate 1404-9, six eyes treated; B = isolate SSII-1, five eyes treated; △ = autoclaved.

through 5 and 11 days after treatment. The eyes were scored using a standard system (McDonald and Shadduck 1977). Table 21.3 presents the results in terms of the number of treated eyes differing from untreated eyes. There were no significant differences between the effects of autoclaved and viable *B. sphaericus*; both isolates produced minimal conjunctival congestion and discharge. Additionally, minimal iridal hyperemia was seen in less than half the treated eyes.

Our results indicate that *B. sphaericus* is not an ocular irritant, although it does not necessarily follow that commercial formulations of *B. sphaericus* will be similarly innocuous.

21.3.4 Intraocular Injection

New Zealand White rabbits were injected directly into the eye (globe) with three isolates of viable and autoclaved *B. sphaericus* (SSII-1, 1404, 1593; table 21.4). Viable and autoclaved inocula produced lesions, although the lesions were most severe in animals receiving viable organisms. The effects of injection were dose dependent, and although *B. sphaericus*

TABLE 21.4.
Results of Intraocular Injection (0.05 ml) of 3 Isolates of *B. sphaericus* Into Rabbits

Isolate	Dose Per Eye	Positive Eye Cultures	Eye Lesions	Mean Severity
SSII-1	6×10^6	3/3	3/3	4.0
SSII-1[a]	6×10^6	0/3	2/2	1.0
SSII-1	6×10^4	3/3	3/3	2.7
SSII-1[a]	6×10^4	0/3	3/3	1.0
SSII-1	6×10^1	1/3	2/3	0.7
SSII-1[a]	6×10^1	0/2	0/2	0
1404-9	5×10^7	1/3	3/3	3.0
1404-9[a]	5×10^7	0/3	2/3	1.0
1593-4	5×10^6	3/3	3/3	2.7
1593-4[a]	5×10^6	0/2	2/2	1.5
Culture Medium	—	—	2/2	1.0

NOTE: Data are presented as the number of eyes affected/number of eyes injected. 0 = normal eye. 1 = slight endopthalmitis; 2 = moderate endopthalmitis, slight panopthalmitis; 3 = severe endopthalmitis, moderate panopthalmitis; 4 = severe panopthalmitis.
[a]Autoclaved inoculum.

was recovered from all groups receiving live organisms, there was no evidence of replication inside the eye.

Many of the lesions appeared to be the result of the deposition of large quantities of protein, which adsorbed to the anterior surface of the lens and the posterior surface of the cornea, resulting in damage to these tissues with resultant inflammation. There were also lesions in eyes injected with autoclaved organisms, although these lesions were less severe. Viable inocula may have produced more severe lesions due to the presence of bacterial metabolites and/or toxins; autoclaving deactivates both the bacteria and their toxins.

21.3.5 Ocular Persistence

B. sphaericus was instilled into the left conjunctival cul-de-sac of 12 New Zealand White rabbits (4.48×10^8 cfu in 0.1 ml). Six rabbits had their eyes flushed with 50 ml of lukewarm water 20 seconds after exposure; the duration of the flush was 30 seconds. Six treated eyes remained unflushed. The eyes of two rabbits from each group (flushed and unflushed) were sampled for 8 weeks using sterile swabs, and no rabbit eye was sampled more than twice during the course of the experiment. Swabs were subsequently streaked onto BHI agar plates and the plates incubated at 30° C for 24 hours and the colonies counted. At the end of the experiment the rabbits were killed, and the conjunctival cul-de-sacs were removed and preserved in formalin.

There was no histological evidence of infection by *B. sphaericus.* Moderate heterophilic suppurative conjuctivitis was observed in the tissue, and these lesions were equally distributed between treated and control eyes. *B. sphaericus* was recovered as long as 8 weeks after exposure and was present in both flushed and unflushed eyes. These results demonstrate that *B. sphaericus* can remain in situ on the conjunctival cul-de-sac or on the adjacent bulbar or palpebral conjunctiva for weeks, despite a voluminous flush of water. Thus, it seems reasonable to predict that *B. sphaericus* will be recovered from patients with ocular lesions such as corneal ulcers or conjunctivitis if they are exposed to this entomopathogen, but it seems equally certain that this bacterium will not be the principle causative agent of such lesions.

21.3.6 Intracerebral Injection

Three isolates of *B. sphaericus* (SSII-1, 1404, 1593) were intracerebrally injected into young rats (Hsd: Sprague Dawley [SD] BR) and mice (CF-W outbred). The rats were killed 9 days and the mice killed 14 days after injection in order to evaluate brain lesions. No animals died following intracerebral injection of as many as 10^7 bacteria (table 21.5). Rats had mild lesions, which in most cases consisted of mild meningitis and perivascular cuffs. The severity of the lesions was dose dependent, and in all cases,

TABLE 21.5.
Intracerebral Injection (0.05 ml) of 3 Isolates of *B. sphaericus* into Rats

Isolate	Dose Per Rat	Number Affected/ Number Injected	Mean Brain Lesion Score
SSII-1	6.0×10^6	3/5	0.8
SSII-1[a]	6.0×10^6	2/3	0.7
SSII-1	6.0×10^4	1/3	0.3
SSII-1[a]	6.0×10^4	0/3	0
SSII-1	6.0×10^1	0/3	0
SSII-1[a]	6.0×10^1	0/3	0
1404-9	1.6×10^7	1/2	1.0
1404-9*	1.6×10^7	1/3	0.3
1593-4	2.4×10^7	1/2	1.0
1593-4[a]	2.4×10^7	1/3	0.3

NOTE: 0 = normal brain; 1 = slight meningitis, no encephalitis, focal reactive gliosis; 2 = moderate meningitis, slight encephalitis including perivascular cuffs.
[a] Autoclaved inoculum.

TABLE 21.6.
Mice Experiencing Cerebral Hemorrhage following Intracerebral Injection (0.05 ml) of 3 Isolates of *B. sphaericus*

Isolate	Dose	Number Affected/ Number Injected
SSII-1	1.2×10^7	3/5
1404-9	1.6×10^7	5/5
1593-4	2.4×10^7	4/5
Control[a]		1/5

[a]Sterile culture medium.

injection of viable bacteria produced the most severe lesions. Many of the mice injected with *B. sphaericus* had intracerebral hemorrhages, but sterile culture medium produced one intracerebral hemorrhage as well (table 21.6). In a related experiment (table 21.7), the ability of the entomocidal isolate 1593 to replicate in rat brain tissue was evaluated. Rats were injected and then sacrificed at the specified intervals; their brain tissue was then titrated in order to determine the number of viable *B. sphaericus* present. More than 600 bacteria per 100 mg brain tissue were recovered three days after injection; but the number of viable *B. sphaericus* fell rapidly, and none was recovered two weeks after administration.

These studies demonstrate that large doses of all three isolates (SSII-1, 1404, and 1593) produced mild lesions in the brains of rats and intracerebral hemorrhages in mice. Animals receiving autoclaved inocula also had lesions, which indicates that a component of the lesions is produced by heat-stable

TABLE 21.7.
Quantitative Recovery of *B. sphaericus* 1593 from Rat Brains

Days Post Injection	*B. sphaericus* Recovered per 100 mg Brain
3	600
5	0
7	0
10	0
12	100
14	0
17	0

NOTE: Each rat received 5×10^5 organisms intracerebrally, and groups of 3 rats were killed each day indicated.

foreign material. As with intraocular injection, the increased severity of the lesions produced by live material may result from the presence of bacterial metabolites and toxins in high concentration. *B. sphaericus* was unable to multiply in rat brains, since the number of bacteria recovered decreased with time, but could persist for as long as 12 days; persistence was most likely due to spore survival. Our data indicate that these isolates of *B. sphaericus* were not virulent, invasive pathogens when injected directly into the central nervous system.

21.4 SUMMARY OF OTHER STUDIES

Results obtained by de Barjac et al. (1979) confirm the safety of *B. sphaericus* isolate 1593. The authors evaluated many routes of exposure, such as subcutaneous, intraperitoneal, intracerebral, oral, inhalation, and percutaneous administration of *B. sphaericus*, as well as several rodent species (mice, rats, and guinea pigs). In no test did the animals experience acute or chronic toxicity. Successive passages of *B. sphaericus* 1593 in rodents failed to increase virulence, and *B. sphaericus* 1593 cleared from heart blood within one week after injection. Histological examination of the tissue of animals exposed to *B. sphaericus* revealed no abnormalities.

In recent experiments evaluating *B. sphaericus* 2362, de Barjac et al. (1987) found no evidence of toxicity to Swiss mice. Their inoculum was grown on a nutrient broth and stored at $-20°$ C prior to use. Mice were injected subcutaneously with 7×10^8 cfu and intraperitoneally with 2×10^8 cfu or 5.6×10^8 cfu; no mortality was observed. Additional mice were injected intravenously with similar doses, and other mice were force-fed 5×10^8 cfu without experiencing mortality; in all experiments the mice injected with viable *B. sphaericus* appeared normal and had comparable weight gains to control mice given bacteria-free medium or autoclaved *B. sphaericus*.

21.5 CONCLUSION

Exposure to *B. sphaericus* isolates 1404, 1593, and NCTC 11025 did not result in clinical illness or death. An intact immune system was not essential for protection against *B. sphaericus* as intraperitoneal injection of the human isolate (NCTC 11025) into athymic mice failed to produce mortality. This last result agreed with earlier studies, which found that both intraperitoneal and intravenous injection of a human isolate of *B. sphaericus* into young rabbits was innocuous (Allen and Wilkinson 1969). Ocular irritancy of *B. sphaericus* was minimal, and, although this organism persisted for 8 weeks in the conjunctival cul-de-sac of rabbits, there was no evidence of infection. *B. sphaericus* 1404 and 1593 did produce lesions when injected intraocularly

TABLE 21.8.
Recovery of *B. sphaericus* 2362 Colony-forming units (cfu) from the Spleens of Outbred Female Mice

Days After Injection	N[a]	CFU per Gram Spleen (± standard error)
5	4	109,452 ± 37,390
11	4	168,036 ± 37,433
17	4	52,279 ± 27,338
24	4	40,051 ± 2,322
34	3	103,296 ± 43,205
39	4	8,069 ± 1,755
53	4	625 ± 326
60	4	471 ± 134
67	3	165 ± 96

NOTE: Initial inoculum was 1.2×10^7 cfu per mouse.
[a]Number of spleens sampled.

and intracerebrally. The presence of large quantities of heat-stable foreign material played an important role in lesion formation, but the increased severity of the lesions associated with living material indicates that bacterial metabolites and/or toxins may cause tissue damage when present at high concentrations. *B. sphaericus* 1593 did not multiply in the animals tested, although it persisted in the brain for 12 days. We underscore that persistence does not mean that multiplication occurred, and many entomopathogens can remain viable in mammalian tissue for a period of weeks.

The safety data for *B. sphaericus* 2362 are more variable. In our laboratory, doses greater than 10^7 cfu proved toxic to mice, and the toxic factor was heat stable and spore associated. Studies conducted by de Barjac et al. (1987) using comparable doses and routes of infection found no evidence of toxicity. Both studies are in agreement that there was no evidence of infection, but *B. sphaericus* 2362 could be recovered from the spleens of mice as long as 67 days after injection (table 21.8). Clearance studies conducted by both laboratories indicated that *B. sphaericus* disappeared from the blood within two weeks following parenteral injection. The discrepancy concerning the mammalian toxicity of *B. sphaericus* 2362 may be due to the difference in culture conditions for *B. sphaericus* used by our laboratories, the mouse strain used in both studies, or both of these reasons. However, it is important to note that while *B. sphaericus* 2362 was toxic in our experiments when injected at the high dose of 10^8 cfu, its toxicity still fell within acceptable safety limits and was only evaluated on one strain of mouse and one rodent species. There was no evidence of infection, as even athymic mice cleared this bacterium, and it is unlikely that *B. sphaericus* 2362 poses any hazard to humans.

Acknowledgments

This investigation received support from the Vector Biology and Control component of the UNDP/World Bank/WHO Special Programme for Research and Training in Tropical Diseases.

References

Allen, B. T., and Wilkinson, H. A. 1969. A case of meningitis and generalized Schwartzman reaction caused by *Bacillus sphaericus*. *Johns Hopkins Med. J.* 125: 8–13.

Barjac, H. de; Cosmao Dumanoir, V.; Hamon, S.; and Thiery, I. 1987. Safety tests on mice with *Bacillus sphaericus* serotype H5a,5b, strain 2362. WHO/VBC/87.948. Mimeo.

Barjac, H. de; Larget, I.; Cosmao Dumanoir, V.; Benichou, L.; Viviani, G.; Ripouteau, H.; and Papion, S. 1979. Innocuité de *Bacillus sphaericus* souche 1593 pour les mammifères. WHO/VBC/79.731. Mimeo.

Isaacson, P.; Jacobs, P. H.; Mackenzie, A. M. R.; and Mathews, A. W. 1976. Pseudotumor of the lung caused by infection with *Bacillus sphaericus*. *J. Clin. Pathol.* 29: 806–811.

McDonald, T., and Shadduck, J. A. 1977. Ocular irritation testing. In *Advances in modern toxicology,* ed. H. I. Maiback and F. N. Marzulli, 4: 130–191.

Myers, P., and Yousten, A. A. 1978. Toxic activity of *Bacillus sphaericus* SSII-1 for mosquito larvae. *Infect. Immun.* 19: 1047–1053.

Pearson, H. E. 1970. Human infections caused by organisms of the *Bacillus* species. *Amer. J. Clin. Path.* 53: 506–515.

Shadduck, J. A.; Singer, S.; and Lause, S. 1980. Lack of mammalian pathogenicity of entomicidal isolates of *Bacillus sphaericus*. *Environ. Entomol.* 9: 403–407.

PART 3

The Future

22

The Future of Bacterial Control of Mosquito and Black Fly Larvae

DONALD J. SUTHERLAND

22.1 INTRODUCTION

Within the span of approximately a decade, the chemical control of mosquitoes and black flies has shifted markedly from a reliance on synthetic insecticides toward an increasing usage of *Bacillus thuringiensis* subsp. *israelensis* (*B.t.i.*); pending full registration and commercial availability of *Bacillus sphaericus*, this bacteria undoubtedly will join *B.t.i.* as an operational alternative to synthetic materials for mosquito control. The development of these two bacteria as control agents has been extremely rapid and is dependent on a variety of factors, especially the continuing research efforts dealing with subjects in this volume. Indeed, preparation of this volume has, to a degree, been a contest—a contest for the authors to be aware of and report on recent advances in their research as well as that of other scientists. Other factors, such as resistance to synthetic insecticides and associated environmental concerns, are supporting the research focused on these bacteria and their future development. Many professional vector-control personnel view these agents as solutions to such problems. However, this situation is somewhat reminiscent of the time in 1943–1944 when, with the shortage of pyrethrum and derris, DDT was rapidly investigated and manufactured for the control of arthropod disease vectors. At that time, the value of DDT led Sir Ian M. Heilbron, chemical advisor to the Ministry of Production in Great Britain, to state that "the discovery of DDT indubitably heralds a new era in man's ceaseless fight for mastery against disease" (West and Campbell 1952).

B.t.i. and *B. sphaericus* also represent a new era, not only in the control of disease, but also in the control of mosquitoes and black flies, whose populations can be a nuisance and affect human comfort and general health. With the wisdom offered by history, it is reasonable and appropriate to consider the various factors influencing this new era.

22.2 PERSPECTIVE FOR THE NEXT DECADE

22.2.1 Supportive Factors

Multiple factors will influence the use of *B.t.i.* and *B. sphaericus* during the next ten years (table 22.1). Some of these factors will be supportive of the usage and extrinsic to the bacteria themselves and their efficiency. Resistance to synthetic insecticides, identified as a major stimulus for renewed interest in microbial control agents about a decade ago (Davidson and Sweeney 1983), will continue to foster the use of these bacteria. However, other factors are now playing a greater role. Increasingly, the public attitude toward synthetic insecticides is negative. This attitude is characterized by some viewers as toxiphobia or chemophobia, and refers not only to the operational use of such materials to control arthropod vectors and pests, but also to the manufacture of synthetic insecticides, fungicides, and herbicides and possible generated toxic wastes and spills. Real environmental issues further magnify this negative attitude. In the United States, the Endangered Species Act of 1973 will soon have regulations governing the application of such materials to terrestrial and aquatic habitats of such species; often these habitats are shared by the larval/adult stages of mosquitoes and black flies. In addition, the aquatic resource in general is becoming the subject of wetlands legislation in some states to protect the quality and quantity of water. Consequently, even water management, a valuable component of organized professional mosquito control for many years, is now impeded in larval habitats by state and federal regulations involving lengthy planning of projects and permit processing, which often do not successfully yield the necessary permits.

The amounts of synthetic insecticides used in controlling mosquito and black fly larvae in most areas are dwarfed by the amounts used in general agricultural and structural pest control. During the current process of insecticide reregistration, as governed by the 1988 amendments to the Federal Insec-

TABLE 22.1.
Factors Influencing the Future Use of *B.t.i.* and *B. sphaericus*

Supportive Factors	Counterfactors
Resistance to synthetic insecticides	Efficiency and cost
Public attitude toward synthetic insecticide usage and manufacture	Formulations and application equipment
Endangered species protection	Resistance possibilities
Wetlands legislation governing quality and quantity of water	Applicator perception of safety
Insecticide legislation governing reregistration	Public awareness

ticide, Fungicide, and Rodenticide Act, it is not unexpected that some manufacturers will not choose to satisfy EPA's current data requirements relative to the use of their insecticides in aquatic habitats. Such larvicides (e.g., chlorpyrifos) will then be lost from the list of possible agents for larval control. Although in the past few years there has been a tremendous increase in the use of *B.t.i.* in some areas (chapter 9), all of the above supportive factors, especially the deletion of alternative agents and options for vector control, will promote a very dramatic increase in the foreseeable future. With full registration of *B. sphaericus*, the era of these two microbials in vector control will be firmly established.

22.2.2 Counterfactors

In this era, however, there are also a number of counterfactors or areas of concern, which can influence usage. The first is efficiency and cost. Efficiency of *B.t.i.* has been suitably documented (chapters 9 and 10), and field evaluations of *B. sphaericus* (chapters 18 and 19) support government registration. Since its commercial availability in 1981, the cost of *B.t.i.* formulations generally has become more competitive with that of synthetic larvicides; in 1989 the competitive bid prices to mosquito control agencies in New Jersey for Vectobac CG *B.t.i.* granules 12/14 mesh, Abate® (temephos) 2G granules, and Abate® 5G granules, respectively, were $0.64/lb, $0.87/lb, and $1.30/lb (R. Dreyer, pers. comm.). Generally, *B.t.i.* granules are applied at 5 lbs/acre; Abate® 2G is used for ground application at 5 lbs/acre; Abate® 5G is used for aerial application at 2 lbs/acre. Therefore, for ground application, cost/acre treatment with *B.t.i.* and Abate® respectively is $3.20 and $4.35; consequently, for economy, *B.t.i.* is chosen. However, aerial application allows the treatment of large acreage of larval habitat when larvae are rapidly developing. With *B.t.i.* granules at 5 lbs/acre and Abate® 5G at 2 lbs/acre by helicopter, the cost/acre respectively is $3.20/acre and $2.30/acre. This cost/acre difference is further magnified by the aspect of the helicopter load capacity of 800 lbs; for treatment of equivalent acreage, the use of *B.t.i.* would involve 2.5 times the number of loadings, time interval for application, personnel costs, and fuel costs. Perhaps further formulation development of *B.t.i.* will reduce this differential.

Formulation is the key to the efficiency and economy of a toxic agent. In some cases, an insecticide can be formulated with an insect attractant to draw the insect to the insecticide (e.g., Metcalf et al. 1987). However, in most cases, without a suitable attractant, the insecticide is so formulated that personnel and equipment can efficiently deliver the insecticide to the area of the target insect activity. This is true of agricultural pests as well as arthropod vectors such as mosquitoes and black flies (Sutherland and Carey 1983). The larvae of these latter insects are aquatic; therefore, the applicator generally targets the water surface and depends on the formulation to distribute the toxicant on or within the water column. Some mosquito larvicides/pupicides (e.g., Aerosurf® MLO) remain at the surface to interfere with respiration. But most synthetic larvicides depend on formulation additives and/or granular carriers to distribute the toxicant within the water, where by contact, and possibly ingestion, the larvae receive a lethal dose. For methoprene, the ratio of cuticular/ingested uptake by *Culex pipiens* larvae has been estimated to be approximately 3 to 1 (Brown and Brown 1980). For *B.t.i.* and *B. sphaericus*, formulation for delivery and desired efficiency is even more complicated, since they are toxic only by ingestion. Presenting toxic ingestible particles within the aquatic environment is much like using airborne droplets of an adulticide space spray to make contact with active adult mosquitoes and black flies. Fortunately, the larval populations are more concentrated in the aquatic habitat.

Currently, *B.t.i.* formulations include wettable powders, flowable concentrates, granules, briquettes, and pellets designed to present the *B.t.i.* in the area of larval feeding activity. However, mosquito species vary in their larval activity profiles (Mullen and Hinkle 1988), and not all species feed on the same or single food source, at the same place, and at the same time. Some species, such as *Aedes aegypti* and *Aedes sollicitans,* feed relatively constantly in most instars (Moore and Sutherland, unpublished); others with slow development, such as early-season monovoltine species, may feed rhythmically or discontinuously. Increasingly, research on vector ecology and behavior is focusing attention on the larval stages of mosquitoes and black flies, including aspects of feeding related to the ingestion of toxic bacteria (Currie and Craig 1987; Hart 1987; Dahl, Widahl, and Nilsson 1988; Khawaled, Barak, and Zaritsky 1988).

With the obvious variation in feeding behavior of the target mosquitoes and black flies, it is unfortunate that *B.t.i.* and *B. sphaericus* formulations are not being designed according to target feeding behaviors. Understandably, economy at the present time prevents this. As a result, for example, some granular *B.t.i.* on corn cob carrier for mosquito control is designed to provide a portion (about 75%) of granules that initially remain floating for larvae feeding at or close to the surface and a sinking portion (25%) for bottom feeders (J. Lublinkoff, pers. comm.). However, this can present some practical problems: a thin algal mat (pellicle) often covers the surface of the salt marsh larval habitat of *Ae. sollicitans,* and the light corn cob granules do not completely penetrate the pellicle to affect the larvae. In *Cx. pipiens* and *Aedes vexans* habitats, some of those granules initially floating will be displaced leeward by light surface wind, thereby causing the distribution of *B.t.i.* to be uneven.

It should also be noted selecting corn cob as a stable carrier for *B.t.i.* and *B. sphaericus* often necessitates adjustments or redesign of equipment originally designed for applying Abate® celatom granules (Schmidt 1988). Since granular formulations have been a major and valuable component in organized mosquito control, *B.t.i.* and *B. sphaericus* manufacturers should consider two avenues for fomulation development: a water-dispersible granule, as described by Fleming and Hazen (1983), and a granule for preseason treatment. The latter could involve granular coatings, which would dissolve and release the bacteria coincident with larval activity of early-season mosquito species.

Of all counterfactors, formulation is currently the most important; formulations require an economical and reliable level of efficiency. For aquatic organisms, such as mosquito and black fly larvae, the lipophilic nature of a synthetic insecticide itself leads to rapid absorbtion (Hollinsworth 1976) and contributes to formulation efficiency. *Ae. aegypti* larvae, under laboratory conditions, absorb over 99% temephos within 1 hour after exposure (Leesch and Fukuto 1972); under field conditions, larvae absorbed temephos 100 times over bulk pond concentration (Henry et al. 1971). For *B.t.i.* and *B. sphaericus* much is known of the lethal concentrations (LC) for mosquitoes and black flies (chapters 9, 10, and 14) and crystals/proteins (chapter 4), but little is known of the lethal dose (LD) of spore/crystal/protein per larva. Such knowledge would establish a basis and a need for developing better formulations, and possibly justify the use of greater potency, the use of a more toxic strain, and/or incorporation of phagostimulants.

As the era of *B.t.i.* and *B. sphaericus* continues, questions will be raised about resistance and its possible development. The history of resistance of *Culicidae* and *Simuliidae* to synthetic insecticides is well documented (Georghiou and Mellon 1983), and the question is a valid one. As the usage of these microbials dramatically increases, what are the odds? The significant resistance to *B.t.* in a lepidopteran (McGaughey 1985) has not been observed with mosquitoes. Vasquez-Garcia and Georghiou (cited by Lacey 1985) obtained a 7-fold increase in LC_{50} in 32 generations of *Culex pipiens quinquefasciatus;* Gharib and Szalay-Marzso (1986) reported only a 1.9-fold tolerance development in *Ae. aegypti;* and Goldman, Arnold, and Carlton (1986), with an adjusted selection pressure at LC_{50} of this species for 14 generations, obtained a small but significant shift in response of one strain. The latter authors suggest that resistance development to pathogens may take somewhat longer periods of time and involve concerted genetic change at two or more loci. However, given that resistance of *B.t.i.* and *B. sphaericus* may involve behavioral, metabolic, and site of action mechanisms, resistance is a distinct probability at some place and in some species. Possibly this will involve structural changes in the peritrophic membrane, a mechanism noted in 1 strain of DDT-resistant *Ae. aegypti* (Abedi and Brown 1961); unfortunately, this strain is probably not still available for bioassays with *B.t.i.*

While differences in species susceptibilities may not be attributable to differences in rates of ingestion, dissolution, and proteolysis of *B. sphaericus* toxin (Aly, Mulla, and Federici 1989), we should recognize that modes of feeding behavior of some species (*Ae. aegypti*; Khawaled, Barak, and Zaritsky 1988) are not fixed, and that resistance can involve both feeding behavior and a detoxication mechanism in the same species (*Cx. pipiens pipiens* and methoprene; Brown and Brown 1980). Based on this information, studies on resistance possibilities should increase. These should include bioassays of field populations using standard techniques and protocols (chapters 8, 9, and 18) and laboratory selection using laboratory and wild strains. Such selections should include presence or absence of normal food to investigate the possible role of feeding discrimination in the development of behavioral resistance.

While public perception disfavors synthetic insecticides, this perception is also relevant to *B.t.i.* and *B. sphaericus*. The first level of perception is that of the user, the actual personnel applying such materials. Government regulations in many countries ensure that personnel are informed about the presence, effects, and cautions about "insecticides" in the work place; this includes registered commercial products of these bacteria. The safety of these products is reported in chapters 12 and 21, and their specificity is discussed in other chapters in this volume. However, it may require only one aspect (e.g., odor of a formulation additive) to cause a negative perception by the user. The perception of the general public may be negative also. In 1989 in New Jersey, some citizens have challenged the use of *B.t.* in the control of the gypsy moth (Romano 1989). All involved in the use and development of *B.t.i.* and *B. sphaericus* should be prepared to respond to questions about these bacteria and their safety.

22.3 PROSPECTS FOR THE FUTURE

Within and beyond the next decade, the developments will depend on scientific research, much of which is described and projected by authors in other chapters. Predicting the paths of research is difficult, and brevity in prediction affords the safest position to take. Undoubtedly, new and more virulent strains of these and other bacteria will be sought (chapters 2 and 14); some have already been isolated (Brownbridge and Margalit 1986, 1987a, 1987b). For these, the specificity of their toxicity will have to be examined, paralleling current research with *B.t.i.* and nontarget species such as lepidopterans (Mathavan, Sudha, and Pechimuthu 1989).

Similarly, the production of a bacterium combining the best attributes of both *B.t.i.* and *B. sphaericus* (chapter 13) must be accompanied by parallel studies on specificity of toxicity. Governmental regulations will require these and other studies, since genetically altered agents are synthetic. This is also true for the "ultimate insecticide for mosquito and black fly control" (chapter 7), wherein genes encoding specific toxicity are incorporated into larval food sources, such as algae. This "ultimate" approach, a worthy goal, raises additional questions to be answered by research. Can such synthetic algae survive and compete in the natural environment? And with the recent report that a lepidopteran insect can evolve resistance to *Pseudomonas fluorescens* genetically engineered to express the δ-endotoxin of *B.t.* subsp. *kurstaki* (Stone, Sims, and Marrone 1989), how will mosquitoes and black flies respond to modified algae? The poet D. H. Lawrence (1964) obviously regarded the mosquito as a challenge when he asked the mosquito, "Am I not mosquito enough to out-mosquito you?" The poet addressed the mosquito as "monsieur." Given his failure to recognize the appropriate gender, possibly he was considering both mosquitoes and black flies. In order for us to out mosquito and out black fly such important vectors of disease and nuisances to human life, we must continue to recognize that their control cannot, and should not, rely on a single approach. We must seek management strategies involving not only *B.t.i.* and *B. sphaericus* and their modified offsprings, but also other agents (Davidson and Sweeney 1983) and options that are and will become available.

Acknowledgments

I am indebted to A. Hajek and P. Horan for their aid in manuscript preparation, and to R. Dreyer, Middlesex County Mosquito Extermination Commission, and J. Lublinkoff, PBI-Gordon Corporation, for information cited. New Jersey Experiment Station Publication No. F-40100-01-89 supported by Hatch Act Funds.

References

Abedi, Z. H., and Brown, A. W. A. 1961. Peritrophic membrane as a vehicle for DDT excretion in *Aedes aegypti* larvae. *Ann. Entomol. Soc. Am.* 54: 539–542.

Aly, C.; Mulla, M. S.; and Federici, B. A. 1989. Ingestion, dissolution, and proteolysis of the *Bacillus sphaericus* toxin by mosquito larvae. *J. Invertebr. Pathol.* 53: 12–20.

Brown, T. M., and Brown, A. W. A. 1980. Accumulation and distribution of methoprene in resistant *Culex pipiens pipiens* larvae. *Ent. Exp. & Appl.* 27: 11–22.

Brownbridge, M., and Margalit, J. 1986. New *Bacillus thuringiensis* strains isolated in Israel are highly toxic to mosquito larvae. *J. Invertebr. Pathol.* 48: 216–222.

———. 1987a. Identification of *Bacillus thuringiensis* strains toxic to mosquitoes recently isolated in Israel. *J. Invertebr. Pathol.* 50: 322–323.

———. 1987b. Mosquito active strains of *Bacillus sphaericus* isolated from soil and mud samples collected in Israel. *J. Invertebr. Pathol.* 50: 106–122.

Currie, D. C., and Craig, D. A. 1987. Feeding strategies of larval black flies. In *Black flies: Ecology, population management, and annotated world list,* ed. K. C. Kim and R. W. Merritt, 155–170. University Park, Pa.: Pennsylvania State Univ. Press.

Dahl, C.; Widahl, L.; and Nilsson, C. 1988. Functional analysis of the suspension feeding system in mosquitoes. *Ann. Entomol. Soc. Am.* 81: 105–127.

Davidson, E. W., and Sweeney, A. W. 1983. Microbial control of vectors: A decade of progress. *J. Med. Entomol.* 20: 235–247.

Fleming, J. P., and Hazen, J. L. 1983. Development of water-dispersible granule systems. In *Pesticide formulations and application systems: Third symposium,* ed. T. M. Kaneko and N. B. Akesson, 141–146. ASTM Special Technical Publications 828.

Georghiou, G. P., and Mellon, R. B. 1983. Pesticide resistance in time and space. In *Pest resistance to pesticides,* ed. G. P. Georghiou and T. Saito, 1–46. London: Plenum.

Gharib, A. H., and Szalay-Marzso, L. 1986. Selection for resistance to *Bacillus thuringiensis* serotype H-14 in laboratory strains of *Aedes aegypti* L. In *Fundamental and applied aspects of invertebrate pathology,* ed. R. A. Samson, J. M. Vlak, and D. Peters, 37. Proc. 4th Intern. Colloquiium of Invert. Path. Ponsen and Looijen, Neth.

Goldman, I. F.; Arnold, J.; and Carlton, B. C. 1986. Selection for resistance to *Bacillus thuringiensis* subspecies *israelensis* in field and laboratory populations of the mosquito *Aedes aegypti. J. Invertebr. Pathol.* 47: 317–324.

Hart, D. D. 1987. Processes and patterns of competition in larval black flies. In *Black flies: Ecology, population management, and annotated world list,* ed. K. C. Kim and R. W. Merritt, 109–129. University Park, Pa.: Pennsylvania State Univ. Press.

Henry, P. A.; Schmit, J. A.; Dieckman, J. F., and Murphy, F. J. 1971. Combined high speed liquid chromatography and bioassay for the evaluation and analysis of an organophosphorus larvicide. *Anal. Chem.* 43: 1053.

Hollingsworth, R. M. 1976. The biochemical and physiological basis of selective toxicity. In *Insecticide biochemistry and physiology,* ed. C. F. Wilkinson, 431–506. New York: Plenum.

Khawaled, K.; Barak, Z.; and Zaritsky, A. 1988. Feeding behavior of *Aedes aegypti* larvae and toxicity of dispersed and of naturally encapsulated *Bacillus thuringiensis* var. *israelensis. J. Invertebr. Pathol.* 52: 419–426.

Lacey, L. A. 1985. *Bacillus thuringiensis* serotype H-14. *Amer. Mosq. Control Assoc. Bull.* 6: 132–158.

Lawrence, D. H. 1964. The mosquito. In *The complete poems of D. H. Lawrence,* ed. V. de Sola Pinto and W. Roberts, 1: 332–334. New York: Viking.

Leesch, J. G., and Fukuto, T. R. 1972. The metabolism of Abate in mosquito larvae and houseflies. *Pestic. Biochem. Physiol.* 2: 223–235.

McGaughey, W. H. 1985. Insect resistance to the biological insecticide *Bacillus thuringiensis. Science* 229: 193–195.

Mathavan, S.; Sudha, P. M.; and Pechimuthu, S. M. 1989. Effect of *Bacillus thuringiensis israelensis* on the midgut cells of *Bombyx mori* larvae: a histopathological and histochemical study. *J. Invertebr. Pathol.* 53: 217–227.

Metcalf, R. L.; Ferguson, J. E.; Lampman, R.; and Andersen, J. F. 1987. Dry cucurbitacin-containing baits for controlling diabroticite beetles. *J. Econ. Entomol.* 80: 870–875.

Mullen, G. R., and Hinkle, N. C. 1988. Method for determining settling rates of *Bacillus thuringiensis* serotype H-14 formulations. *J. Amer. Mosq. Control Assoc.* 4: 132–137.

Romano, J. 1989. Parents upset over gypsy moth spray that fell on children. *New York Times,* July 2, sec. 12, 2.

Schmidt, R. 1988. Helicopter calibration of *B. t. i.* granules. *N.J. Mosq. Control Assoc. Proc.* 75: 22–25.

Stone, T. B.; Sims, S. R.; and Marrone, P. G. 1989. Selection of tobacco budworm for resistance to a genetically engineered *Pseudomonas fluorescens* containing the δ-endotoxin of *Bacillus thuringiensis* subsp. *kurstaki. J. Invertebr. Pathol.* 53: 228–234.

Sutherland, D. J., and Carey, W. 1983. Pesticide delivery to aquatic pests. In *Pesticide formulations and application systems: Third symposium,* ed. T. M. Kaneko and N. B. Akesson, 110–120. ASTM Special Technical Publications 828.

West, T. F., and Campbell, G. A. 1952. *DDT and newer persistent insecticides.* New York: Chemical Publishing.

Index

SPRINGER NATURE

GPSR Compliance

The European Union's (EU) General Product Safety Regulation (GPSR) is a set of rules that requires consumer products to be safe and our obligations to ensure this.

If you have any concerns about our products, you can contact us on ProductSafety@springernature.com

In case Publisher is established outside the EU, the EU authorized representative is:

Springer Nature Customer Service Center GmbH
Europaplatz 3
69115 Heidelberg, Germany